Advances in Intelligent Systems and Computing

Volume 536

About this Series

The series "Advances in Intelligent Systems and Computing" contains publications on theory, applications, and design methods of Intelligent Systems and Intelligent Computing. Virtually all disciplines such as engineering, natural sciences, computer and information science, ICT, economics, business, e-commerce, environment, healthcare, life science are covered. The list of topics spans all the areas of modern intelligent systems and computing.

The publications within "Advances in Intelligent Systems and Computing" are primarily textbooks and proceedings of important conferences, symposia and congresses. They cover significant recent developments in the field, both of a foundational and applicable character. An important characteristic feature of the series is the short publication time and world-wide distribution. This permits a rapid and broad dissemination of research results.

More information about this series at http://www.springer.com/series/11156

Jeng-Shyang Pan · Jerry Chun-Wei Lin
Chia-Hung Wang · Xin Hua Jiang
Editors

Genetic and Evolutionary Computing

Proceedings of the Tenth International Conference on Genetic and Evolutionary Computing, November 7–9, 2016 Fuzhou City, Fujian Province, China

Editors
Jeng-Shyang Pan
College of Information Science
and Engineering
Fujian University of Technology
Fuzhou, Fujian
China

Chia-Hung Wang
College of Information Science
and Engineering
Fujian University of Technology
Fuzhou, Fujian
China

Jerry Chun-Wei Lin
School of Computer Science
and Technology
Harbin Institute of Technology (Shenzhen)
Shenzhen, Guangdong
China

Xin Hua Jiang
College of Information Science
and Engineering
Fujian University of Technology
Fuzhou, Fujian
China

ISSN 2194-5357 ISSN 2194-5365 (electronic)
Advances in Intelligent Systems and Computing
ISBN 978-3-319-48489-1 ISBN 978-3-319-48490-7 (eBook)
DOI 10.1007/978-3-319-48490-7

Library of Congress Control Number: 2016954623

Printed on acid-free paper

This Springer imprint is published by Springer Nature
The registered company is Springer International Publishing AG
The registered company address is: Gewerbestrasse 11, 6330 Cham, Switzerland

Preface

This volume comprises the proceedings of the Tenth International Conference on Genetic and Evolutionary Computing (ICGEC 2016), which is hosted by Fujian University of Technology and is held in Fuzhou City, China on 7–9, November, 2016. ICGEC 2016 is technically co-sponsored by Springer, University of Computer Studies, Yangon, University of Miyazaki in Japan, Kaohsiung University of Applied Science in Taiwan, VSB-Technical University of Ostrava, and Taiwan Association for Web Intelligence Consortium. It aims to bring together researchers, engineers, and policymakers to discuss the related techniques, to exchange research ideas, and to make friends.

Thirty seven excellent papers were accepted for the final proceedings. One plenary talk is kindly offered by Prof. Han-Chieh Chao (President of National Dong Hwa University, Taiwan).

We would like to thank the authors for their tremendous contributions. We would also express our sincere appreciation to the reviewers, Program Committee members and the Local Committee members for making this conference successful. Finally, we would like to express special thanks for the financial support from Immersion Co., Ltd, China in making ICGEC 2016 possible.

September 2016

Jeng-Shyang Pan
Jerry Chun-Wei Lin
Chia-Hung Wang
Xin Hua Jiang

Organizing Committee

Honorary Chair

XinHua Jiang   Fujian University of Technology, China

Advisory Committee Chairs

XiaoDong Wang	Fujian University of Technology, China
Jun Murai	Keio University, Japan
Ke-Shou Wu	Xiamen University of Technology, China
Ajith Abraham	Machine Intelligence Research Labs, USA

General Chairs

Jeng-Shyang Pan	Fujian University of Technology, China
Ngo Thanh Long	Le Quy Don University, Vietnam
Ponnuthurai Nagaratnam Suganthan	Nanyang Technological University, Singapore

Program Committee Chairs

Mie Mie Thet Thwin	University of Computer Studies, Yangon, Myanmar
Pyke Tin	University of Computer Studies, Myanmar
Chu-Sing Yang	National Cheng Kung University, Taiwan
Hui Sun	Nanchang Institute of Technology, China
Jerry Chun-Wei Lin	Harbin Institute of Technology (Shenzhen), China

Invited Session Chairs

Thi Thi Zin	University of Miyazaki, Japan
Chia-Hung Wang	Fujian University of Technology, China
Chia-Jung Lee	Fujian University of Technology, China

Local Organizing Chairs

Fu-Min Zou	Fujian University of Technology, China
Zhi-Ming Cai	Fujian University of Technology, China

Electronic Media Chair

Tien-Wen Sung	Fujian University of Technology, China

Publication Chairs

Pei-Wei Tsai	Fujian University of Technology, China
Saw Sanda Aye	University of Technology, China
Xiangwen Liao	Fuzhou University, China
Mao-Hsiung Hung	Fujian University of Technology, China

Finance Chairs

Hong Chen	Fujian University of Technology, China
Alina Wu	Fujian University of Technology, China

Program Committee Members

Amir H. Alavi	Michigan State University, USA
Yogesh Bhalerao	Maharashtra Institute of Technology, Pune, India
Chun-Hao Chen	Tamkang University, Taiwan
M.C. Deo	Indian Institute of Technology, Bombay, India
U.S. Dixit	Indian Institute of Technology Guwahati, India
Philippe Fournier-Viger	Harbin Institute of Technology (Shenzhen), China
Amir Gandomi	Michigan State University, USA
Liang Gao	Huazhong University of Science and Technology, China
Ankit Garg	Indian Institute of Technology Guwahati, India

Contents

Optimization Models and Techniques with Engineering Applications

A Comprehensive Evaluation Model for Traffic Rule

Lin Xiao[1,2,3(✉)], Minqian Tang[2], and Jeng-Shyang Pan[1,3]

[1] School of Information Science and Engineering, Fujian University of Technology, No. 3 Xueyuan Road, University Town, Fuzhou 350108, People's Republic of China
xiaolin201@qq.com

[2] School of Information Management, Jiangxi University of Finance and Economics, No. 665, West Yuping Road, Nachang 330013 People's Republic of China

[3] Key Laboratory of Big Data Mining and Applications of Fujian Province, Fuzhou 350108, Fujian, China

Abstract. This article puts forward a model aimed at evaluating the traffic rule. We build the evaluating models to measure the traffic influencing factors, which can be divided into two kinds, the traffic flow factor and the safety factor. Analyze these factors to judge the performance of the keep-right-except-to-pass rule in light and heavy traffic. Draw the curve about time and other factors, let time be the intermediate variable, by using the figure conversion method, we get the curve in order to analyze the changing situation of each factor in light and heavy traffic. Do the comprehensive analysis of the combination figure by putting all the three curves in one coordinate system. We further set the basic lines as standards to be compared with the observing values. The result shows keep-right-except-to-pass rule performs well in the normal traffic yet badly in the extremely light traffic and the extremely heavy traffic.

Keywords: Traffic rule · Space-occupation ratio · Real traffic capacity · Vehicle headway distance

1 Introduction

Transportation is of vital importance in both ancient and modern human civilization as one of the main ways of communication. With the development of technology, approximately, the prime transportation means changes from carriage to automobile leading to the thriving and prosperous growth period of economy. However, in order to satisfy the increasing demand of economy rise and of human population, more and more automobiles are produced, which results in the high probability of the happening

The work is partially supported by Fujian Province Education Planning projects (FJJKCG15-051) and Fujian University of Technology Foundation Project (CY-Z15092) and Jiangxi Province Graduate Student Innovation Foundation Project (YC2015-B052).

J. Pan et al. (eds.), *Genetic and Evolutionary Computing*, Advances in Intelligent Systems and Computing 536, DOI 10.1007/978-3-319-48490-7_1

of the traffic jam, especially in the big cities such as New York, Beijing, and Rio. People have developed many methods which can be divided into two major different emphases (the hardware method and the software method) to solve such disturbing problems. The hardware method concerns itself mostly with the construction and designs of the roads, that is, building more roads in the city and designing some new kinds of road such as flyovers and tunnels, while the other method with the regulations and rules which are made to guide the behaviors of the drivers, such as the keep-right-except-to-pass rule, which is referred in [1, 2].

In countries (except for Great Britain, Australia, and some former British colonies) where people are used to driving automobiles on the right side of the road should adopt this rule, which requires drivers to drive in the right-most lane unless they are passing another vehicle, in which case they move one lane to the left, pass, and return to their former travel lane. No one knows exactly why most people choose to drive on the right, maybe it could satisfy the right-handers which most people are, and as a result, could guarantee the safety. However, to make sure whether this traditional rule could optimize the traffic situation leading to the best traffic flow, we are supposed to evaluate it by mathematic modeling and computing.

2 Problem Statement and Model Construction

This article puts forward a model aimed at evaluating the traffic rule. We assume the certain cross section of the roadway with two lanes as the simulate traffic situation. Two edges of this cross section (entry and exit, the function of two edges can be exchanged) both allow people to drive in and out straightly.

2.1 Quantitative Estimation Method of the Traffic Situation: Space-Occupation Ratio Model

The city's traffic situation reflects the overall operation condition of the automobiles on the road; it can be roughly divided into two ranges, the light traffic and the heavy one, which is referred in [3]. The heavy traffic happens, according to the definition made by Chicago transport agency, as long as the roadway occupancy rate of one certain road is larger than 30 %. On one certain cross section of the road, the roadway occupancy rate can be classified as R (space-occupation ratio), which equals to S_A (the area of the lane occupied by the automobiles) divided by S_R (the area of the total roadway), referred in [4]. Under the assumption of the dual-lane roadway, two lanes have the equal width, so the simplification form of the space-occupation ratio equation is f (the length of the automobiles passing the cross section during the given observing period) divided by l_r (the total length of the roadway).

As for the parameter f, we cannot neglect the factor that there are many kinds of vehicles differs from size and some other factors. To solve this problem, we use the standardization method. In most countries the transport agency stipulates that one certain kind of automobile as the basic unit of the traffic situation. When counting the quantity of the passing vehicles, other kinds of vehicles should be adjusted by their

Table 1. Type of automobile and the conversion coefficients

Typical type of automobile	Conversion coefficients	Introduction
Basic vehicle	1.0	For passenger car, seat quantity ≤ 19 For freight car, weight ≤ 2[ton]
Middle-sized vehicle	1.5	For passenger car, seat quantity>19 For freight car, 2[ton] < weight < 7 [ton]
Heavy duty car	2.0	Freight car, 7[ton] ≤ weight < 14 [ton]
Trailer	3.0	Freight car, weight ≥ 14[ton]

conversion coefficients to become the quantity of the basic unit. By looking up in the references, we summarized the related information as the table below to show the classification of the vehicle types and the conversion coefficients (Table 1):

Let a, b, c, d respectively represent basic vehicle, middle-sized vehicle, heavy duty car and trailer in order. By using the standardization method which combined conversion coefficients with four different types of automobile, we can calculate the length of the automobiles passing the cross section. As a result, the former simplification form of the space-occupation ratio equation can be rewritten as the final form like:

$$R = \frac{f = \sum a + 1.5\sum b + 2\sum c + 3\sum d}{l_r} \tag{1}$$

2.2 Traffic Flow Evaluation Model: The Real Traffic Capacity Model

According to the related references, the traffic capacity refers to the maximize quantity of automobiles passing the assumed cross section of the roadway during the given observing period, in the case of the certain traffic situation. As a result, we can simplify the real traffic capacity equation turning out to be:

$$N = \frac{1000\overline{v}}{l}(\text{vehicles/h}) \tag{2}$$

where we use N as the symbol of the real traffic capacity, and $\overline{v}$ is the average speed in a certain time interval, l is the vehicle headway distance in a certain time interval.

We use four types of vehicles and the conversion coefficient to calculate the standard traffic flow. a_1, a_2, a_3, a_4 respectively represent the four types' conversion coefficient (basic vehicle = 1, middle-sized vehicle = 1.5, heavy duty car = 2, and trailer = 3) in order, and q_i, similarly, represents the quantities of the certain type of vehicle passing the cross section. As for the vehicle headway distance calculating equation, l_1 is the length of automobiles, and Δt refers to the total time of the given observing period. Under the roadway assumption, when computing l_1, to simplify, we further assume that all the vehicles are the type of basic vehicle and determine according to the relative references its value as 5 m. Then we can get the final form of the real traffic capacity calculating equation, that is:

$$N = \frac{1000\bar{v}}{l_1 + \frac{\bar{v}\Delta t}{n-1}} = \frac{1000(n-1)\bar{v}}{5(n-1) + 60\bar{v}} = \frac{1000(-1 + \sum a_i q_i)\bar{v}}{5(-1 + \sum a_i q_i) + 60\bar{v}} \tag{3}$$

2.3 Safety Evaluation Model (1): The Vehicle Headway Distance Model

Under the previous roadway assumption, the two-lane roadway with the straight driving orientation, there are two main factors which can possibly cause the traffic accident, the vehicle headway distance and the passing sight distance. In this part we first consider the vehicle headway distance model.

The vehicle headway time distance D_t is the difference between two adjacent vehicles' time of before and after passing the cross section of the assumed roadway. It is closely related to the traffic environment and vehicle performance, and also by the impact of traffic control. The vehicle headway distance D_h is the distance between the two near vehicles' head. Where T_f and T_l respectively represent the time used to pass the cross section of the front car and of the later car, and l_1 refers to the length of cars, as we have explained its value is 5 m. Thus the final equation form is:

$$D_h = \left(T_f - T_l\right) \times \bar{v} + 5 \tag{4}$$

2.4 Safety Evaluation Model (2): The Overtaking Sight Distance Model

We know the driving characteristics of dual-lane road are only one lane in the same direction, and a variety of different types of the vehicle running on the road in different speed. As a result, when overtaking, drivers often have to occupy the opposite lane. The whole overtaking process is divided into three stages, namely, the lane changing, passing, and lane returning, referred in [5]. We consider the simplest situation, namely, only one overtaking car and one overtaken car, just as Fig. 1 shows:

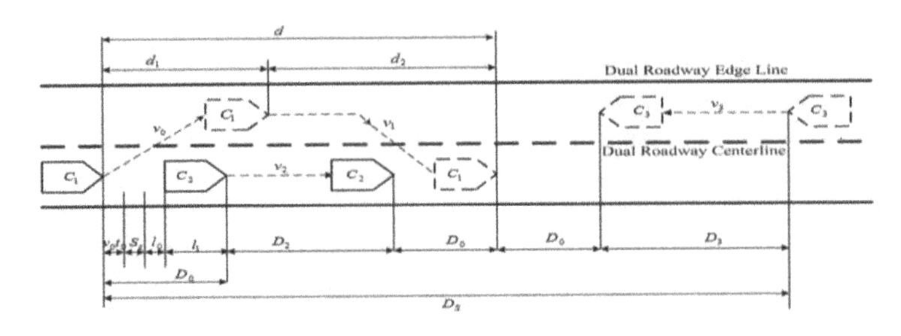

Fig. 1. Overtaking process in details

We call the overtaking sight distance D_S, that is:

$$D_S = 3D_0 + (v_2 + v_3)t = \frac{3v_1 - v_2 + 2v_3}{v_1 - v_2} D_0 + \frac{(v_1 - v_0)^2 (v_2 - v_3)}{2a_a (v_1 - v_2)} \tag{5}$$

This model shows that the overtaking sight distance is determined by the speed of the overtaking car when returning back to its original lane v_1, the speed of the overtaken car v_2, the speed of the coming car in opposite orientation v_3, the average accelerate speed a_a, and the distance between heads of the overtaking and overtaken car D_0. It also indicates that D_S is mostly determined by the value of $(v_1 - v_2)$. When speed of the overtaking and overtaken car is equal, the value of D_S is infinite. This result fits the practical situation, which the overtaking car is unable to overtake.

3 Numerical Computation and Curve Analysis

3.1 Explain the Source of Data

We find a related paper with the similar simulate traffic environment, referred in [6], so we just take the data from it as reference. In the paper the total observing time is 400 s, and the time interval is fixed as 2 s. We summarize the observing values in four large tables, with the all the needed parameters available, namely, the real traffic capacity N, the space-occupation ratio R, the vehicle headway distance D_h, and the average interval velocity V_i, just as Table 2 shows.

Table 2. Observing data of R, N, D_h and V_i

Parameter \ Time(s)	0	2	4	6	8	10	12	…	388	390	392	394	396	398	400
R(%)	0	7	14	14	7	7	7	…	0	0	0	0	0	0	0
N(vehicle/h)	0	14	29	29	14	14	14	…	0	0	0	0	0	0	0
D_h (m/vehicle)	70	70	35	35	70	70	70	…	70	70	70	70	70	70	70
V_i (km/h)	100	90	70	60	55	52	53	…	59	60	62	61	63	65	65

3.2 Figure Conversion

Based on the four separate data tables, we use matlab to find the result respectively in Figs. 2, 3, 4 and 5 which are shown below in order.

We can find that the common independent variable of the four curves is the time. As a result, we can use the method of figure conversion to combine two curves based on the time. By using matlab, $R - N$ curve, $R - V_i$ curve, $R - D_h$ curve are obtained base on the new coordinates, with the ability to analyze the relationship between the two parameters, in other word, from light to heavy traffic, the changes of the real traffic capacity, the average

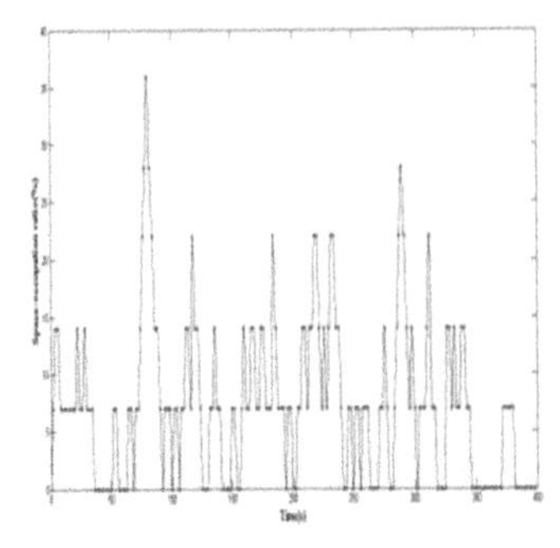

Fig. 2. Time-the space occupation ratio curve

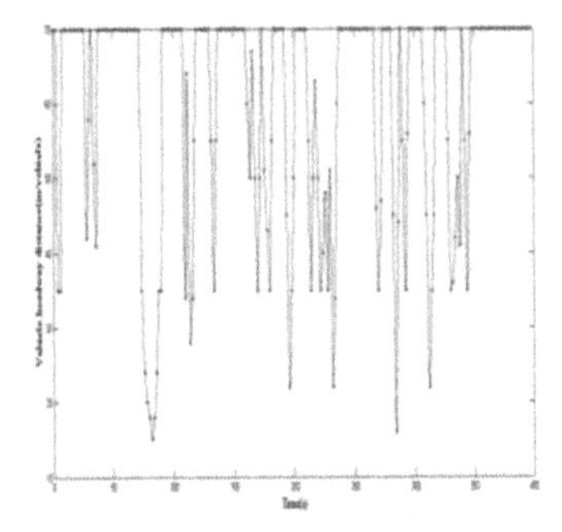

Fig. 3. Time-vehicle headway distance curve

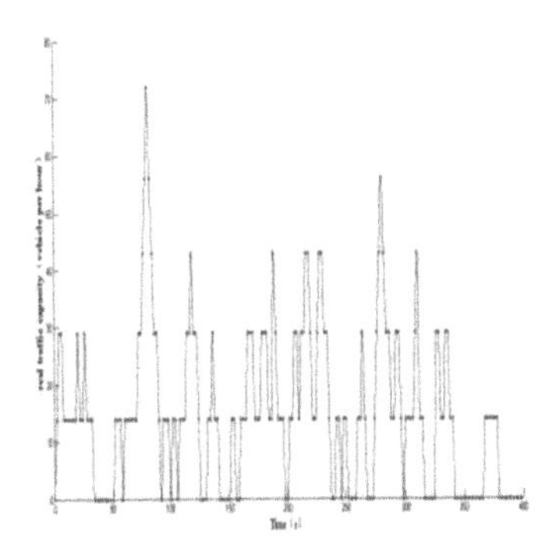

Fig. 4. Time-real traffic capacity curve

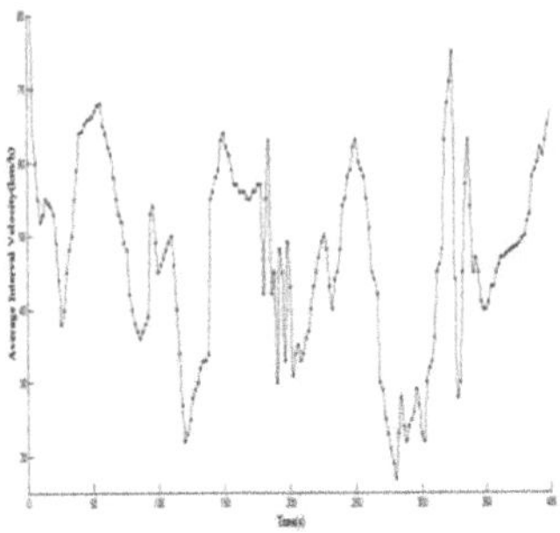

Fig. 5. Time-average interval velocity curve

Table 3. Selected coordinates of the four parameters based on the same time

R(%)	0	7	14	22	28	30	36	38	45	50	67	70	84	90	94	100
V_i(km/h)	100	90	57	37	30	25	23	20	17	14	11	9	7	5	3	2
N(vehicle/h)	0	14	29	43	56	88	72	70	64	60	49	45	32	25	10	0
D_h(m/vehicle)	70	58	42	37	31	24	22	21	14	13	10	9	5	4	3	2

interval and the vehicle headway distance. The required new-form coordinates' data is summarized in Table 3.

By using matlab, we put the $R - N$, $R - V_i$ and $R - D_h$ curve in one figure, in order to evaluate the performance of the keep-right-except-to-pass rule in light and heavy traffic. The combination figure is shown in Fig. 6.

3.3 Curve Analysis

First we analyze the $R - N$ curve by setting the basic line named the general capacity line. According to the related references, the general real traffic capacity is

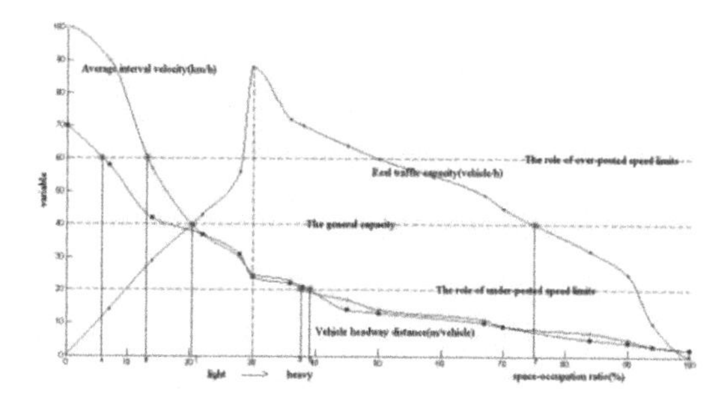

Fig. 6. Combination figure

40 vehicles/h. We can see in the figure that the basic line and the $R-N$ curve have two intersections, the left one and the right one. When $R \in [21\,\%, 76\,\%]$, the $R-N$ curve is above the basic line, which means that the real traffic capacity is better than the general condition. When $R \in [0, 21\,\%)$ and $R \in (76\,\%, 100\,\%]$, however, the $R-N$ curve is below the basic line, which indicates the bad traffic situation. To be more specific, the keep-right-except-to-pass rule performs well in the normal traffic condition and bad in the extremely light traffic and the extremely heavy traffic.

Then we judge the $R-V_i$ curve with the consideration of the role of under- or over-posted speed limits, which are determined as 20 km/h and 60 km/h. These two basic lines intersect the $R-V_i$ curve at two points. When $R \in [12\,\%, 38\,\%]$, the average interval velocity is in the proper speed range, which shows that the keep-right-except-to-pass rule fits well. However, when $R \in [0, 12\,\%)$ and $R \in (38\,\%, 100\,\%]$ the $R-V_i$ curve is out of the proper speed range. Consequently, the rule performs badly in the extremely light traffic and the extremely heavy traffic. In the former condition, the vehicles' speed is too fast to guarantee the safety, while in the latter condition, too slow to guarantee the traffic flow.

Finally we examine the safety by $R-D_h$ curve. From the figure we can find that a large part of the two curves (the $R-D_h$ curve and the $R-V_i$ curve) are almost the same, except for the part of $R \in [0, 23\,\%]$ which we regard as calculation error. So, it's appropriate to assume that the values of two parameters are similar. It's easy to understand that if the speed is big, then the vehicle headway distance is big in consideration of safety. By using the same comparing way as what we use in $R-V_i$, we set two basic line (the upper-limit and the lower-limit), their position are approximately the same as the speed limit line, and the similar conclusion will be shown.

4 Conclusions

To conclude, with the comprehensive analysis of the combination of safety (by D_h), the role of under- or over-posted speed limits and the traffic flow (by N), the keep-right-except-to-pass rule performs well in the normal traffic yet badly in the extremely light traffic and the extremely heavy traffic. So there do exist the optimization opportunity of the rule, we need to discuss further how to design a new rule in order to perform better than the keep-right-except-to-pass rule.

References

1. Khoury, J.E., Hobeika, A.G.: Integrated stochastic approach for risk and service estimation: passing sight distance application. J. Transp. Eng. **138**, 571–579 (2012)
2. Bai, W., Cun-Jun, L.I.: Overtaking model based on different limiting speed. J. Transp. Syst. Eng. Inf. Technol. **13**, 56–63 (2013)
3. Wang, R.Q., Zhou, Y.J., Xiao, C.E.: Calculation method of overtaking sight distance for dual-lane highway. J. Traffic Transp. Eng. **11**, 67–68 (2011)
4. Yue, Y., Luo, S., Luo, T.: Micro-simulation model of two-lane freeway vehicles for obtaining traffic flow characteristics including safety condition. J. Mod. Transp. **11**, 1–9 (2016)
5. Li, S., Zhao, X.J., Gao, Y., et al.: The determination of vehicular trajectories based on video image processing technology. In: 10th International Conference on Electric Technology and Civil Engineering, pp. 3086–3090. IEEE Computer Society, New York (2012)
6. Yao, R.H., Jing, J.: Chain reaction models for vehicle queuing on single-lane road section. Jilin Daxue Xuebao **42**, 892–898 (2012)
7. Bo, Y., Shenghua, X.: The evolutionary game simulation analysis of the multi-agent modeling's virtual enterprise knowledge transfer. Intell. Mag. **29**(5), 20–25 (2010)
8. Jing, Z., Li, X.: Research on topology control system for mesh networks. J. Chin. Comput. Syst. **34**(1), 140–144 (2013)

Composite Probe and Signal Recovery of Compressed Sensing Microarray

Zhenhua Gan[1,2,4], Baoping Xiong[1,3], Fumin Zou[1,4(✉)], Yueming Gao[3], and Min Du[2,3]

[1] College of Information Science and Engineering, Fujian University of Technology, No. 3 Xueyuan Road, University Town, Minhou, Fuzhou, Fujian, China
{ganzh, fmzou}@fjut.edu.cn, xiongbp@qq.com
[2] College of Electrical Engineering and Automation, Fuzhou University, No. 2 Xueyuan Road, University Town, Minhou, Fuzhou, Fujian, China
dm_dj90@163.com
[3] College of Physics and Information Engineering, Fuzhou University, No. 2 Xueyuan Road, University Town, Minhou, Fuzhou, Fujian, China
fzugym@yahoo.com.cn
[4] Key Lab of Automotive Electronics and Electric Drive Technology of Fujian Province, No. 3 Xueyuan Road, University Town, Minhou, Fuzhou, Fujian, China

Abstract. Due to the large number of uncertain factors in hybridization, image capture and processing of the microarray, multiple probes were generally arranged to improve the reliability of the measurement. However, the small area limited the number of probes that were allowed to be added on, so a composite probe would be the better choice. A composite probe contained the linear combination of a variety of gene fragments. It was used so that the microarray could easily realize the repeated gene fragments within a limited region. The number of composite probes would rapidly dwindle when it compared to a traditional microarray. At the same time, since the sparse characteristics of biological gene mutation, the compressed sensing idea is adopted to recovery the gene variation in the composite probes. The 96 fragments can be used with the 48 × 96 sparse random matrix to construct the 48 composite probes when the sparsest level K is no more than 12. Simulation results show that compressed sensing can accurately recover the gene mutation by using the Orthogonal Matching Pursuit (OMP) algorithm.

Keywords: Compressed sensing · Microarray · Composite probe · Sparse random matrix · OMP

1 Introductions

Microarray is a newly technology for high-throughput and quantitative detection in the biology science area. The abilities of microarray to express of thousands of genes simultaneously in a single detection have allowed the application in wide variety of fields, such as molecular biology, genetics, agriculture, disease diagnosis, medical treatment,

J. Pan et al. (eds.), *Genetic and Evolutionary Computing*, Advances in Intelligent Systems and Computing 536, DOI 10.1007/978-3-319-48490-7_2

food safety supervision, and judicial identification [1]. In a traditional microarray, each of the probe represents a complementary gene segment to be used to detect the corresponding gene information [2].

For the measurement noises, multiple probes were usually arranged to improve the reliability of the determination. The same probe was an effective way to avoid the information losses due to the interference of noises, but the repeated arrangement of probes resulted in an increase in the number of probes on the microarray. Thereupon, the weak fluorescence and the small size of probe were producing adverse effects while the density was increased, which also had caused serious irreparable damage for the ability to obtain reliable expression of the probes.

A more efficient method for solving the above problems was to use the composite probes. In this way each composite probe located in a spot was designed to detect the expression of multiple gene fragments simultaneously. The microarray scanner read the intensity of linear combination information from the composite probe, and the message of each gene probe would be obtained via the appropriate recovery algorithm [3].

Traditional cDNA gene sequencing probes produced a large number of mostly useless information, due to the fact that differences in the sequence between the reference sample and test sample were sparse. Because of the sparse characteristics of biological gene mutation, the compressed sensing idea was adopted to recover the gene variation in the composite probes. The compressed sensing theory had provided a strong support for the accurate recovery of the sparse signals, and it had been widely used in biological sensing, radar detection, data compression, image processing, and pattern recognition [4]. The compressed composite probes were constructed based on the compressed sensing ideas. The difference gene sequencing signals could be recovered by observing a small amount of the composite probes [5, 6].

The application of the composite probes on microarray was confirmed by [3]. And a composite probe method for constructing the compressed sensing microarray was proposed in [6]. A sparse low density parity check code (LDPC) as the measurement matrix to construct a compressed sensing microarray, and the recovery algorithm for the gene difference information were also proposed in [6]. For more information, the sparse random matrix in the recovery algorithm had the advantage of being a simple structure, low computational complexity, and easy to update and store in [7].

2 Design of Composite Probe for Compressed Sensing Microarray

2.1 Compressed Sensing

Compressed sensing is a sampling and reconstruction theory for sparse singles. Signal or the signal after a special transformation, with sparse or compressible characteristics, is the premise of compressed sensing [8, 9]. Considering a discrete digital signal $x \in R^{N}$ that has $K << N$ non-zero elements, the signal x is K sparse and N-K elements in the signal x will be 0 or close to 0. Since the signal x is generally not directly measured, we could design an $M \times N$ measurement matrix A to observe M linear combinations of the x, where $K << M << N$.

$$y_{M\times 1} = A_{M\times N}x_{N\times 1} \qquad K << M << N \tag{1}$$

Although the Eq. (1) is a underdetermined system, we also could reconstruct the signal x for K sparse by solving the constrained l_0 minimization,

$$\hat{x} = \arg\min \|x\|_0 \quad s.t. \quad y = Ax \tag{2}$$

where $\|x\|_0$ denotes the l_0-norm.

Unfortunately, solving the l_0 minimization is known as NP-hard. In order to solve this problem, it is usually converted into minimizing the l_1 with the optimization constraints. As long as the measure matrix A satisfies the restricted isometric property (RIP), the Eq. (1) agree with the following constraints,

$$\hat{x} = \arg\min \|x\|_1 \qquad s.t. \quad y = Ax \tag{3}$$

where $\|x\|_1$ denotes the l_1-norm [10, 11].

2.2 Composite Probe for Compressed Sensing Biological Microarray

The biological microarray uses the principle of molecular hybridization, which the gene to bind specific complementary sequences in the microarray probes. Since fluorescent labeling has been achieved already, we can get the fluorescent signal by light excitation. The information of the corresponding gene fragments from the resulting fluorescence signals can also be analyzed.

A typical cDNA microarray is fixed with a large number of probe spots located on the surface, but each probe consists of the single gene fragment, which can only detect specific complementary sequence segments. The detection principle of traditional cDNA is shown in Fig. 1.

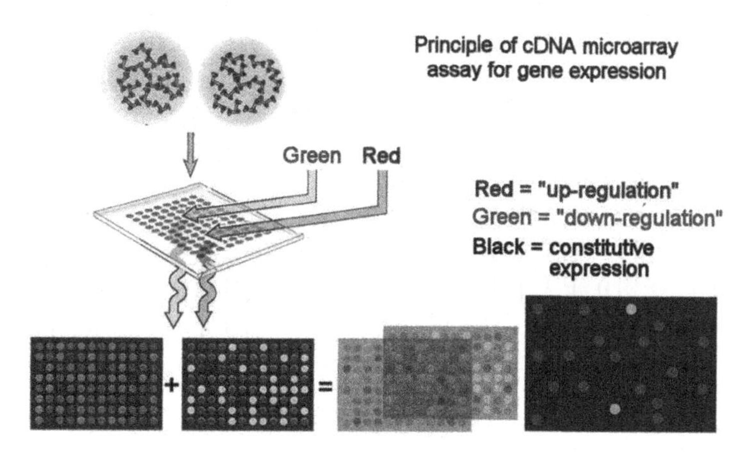

Fig. 1. The principle of traditional cDNA detection

The fluorescence intensity is at its most when the probes are matched normally on the microarray, and the intensity is at its weakest when the probes are mismatched. When the probes are not paired, there is little to no fluorescence intensity. The fluorescence intensity generated by match pairs is 5 times to 35 times more intense than that of a single or two bases mismatch in the probe's sequence. So the accurate determination of the fluorescent intensity is the basis of the specific detection of the biological sequence of microarray probe [12].

The composite probe fluorescence intensity is reflected the cumulative number of fluorescent molecules in various biological fragments fixed in the probe's spot. Literature [7] uses similar techniques as literature [3], which the design of the composite probe is realized by mixing the existing probe molecules according to the linear relationship of the measurement matrix A. This method can be used concurrently with the existing cDNA processing technology.

In particular, there are only a small fraction of the genes to be in a state of mutation. We are considering the difference that the gene expression of test sample is compared with the reference sample. And the difference of the signals which produced by two samples is nature sparse.

In order to construct a compressed sensing microarray with M composite probes, an M × N measurement matrix A with M < < N must be designed for N gene fragments. And we design the measurement matrix A with binary 0/1 elements only to simplify the construction difficulty of the compressed probes.

In two-color microarray of cDNA experiments, the reference sample is labeled by Cy3 while the test sample is labeled by Cy5 [13]. We are comparing two channel's sample by data vectors x_{cy3} and x_{cy5}, and interesting the difference expression of $x = x_{cy3} - x_{cy5}$.

Since there are differences in the small number of gene segments, the distribution of the x is sparse. The compressed sensing idea is relevant to the applications of DNA microarrays in the gene variation. Figure 2 illustrates the structure of the composite probe.

Each row of the matrix A represents a linear combination of the gene fragments. The m-th composite probe is determined by the positions of the gene fragment in the m-th row of matrix A. The combination structure of a composite probe is shown as the following,

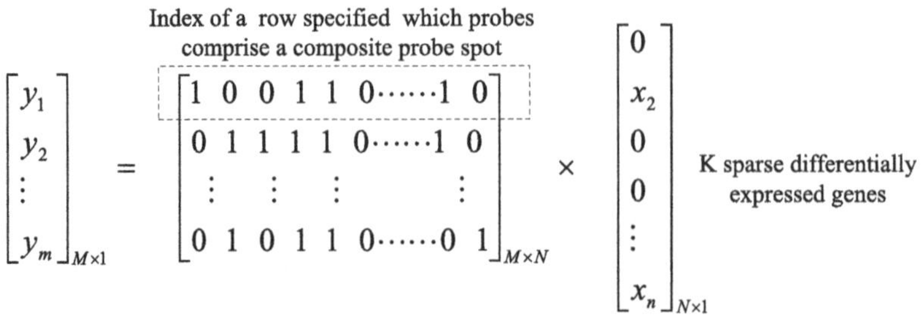

Index of a row specified which probes comprise a composite probe spot

$$\begin{bmatrix} y_1 \\ y_2 \\ \vdots \\ y_m \end{bmatrix}_{M\times 1} = \begin{bmatrix} 1 & 0 & 0 & 1 & 1 & 0\cdots\cdots 1 & 0 \\ 0 & 1 & 1 & 1 & 1 & 0\cdots\cdots 1 & 0 \\ \vdots & & \vdots & & \vdots & & \vdots \\ 0 & 1 & 0 & 1 & 1 & 0\cdots\cdots 0 & 1 \end{bmatrix}_{M\times N} \times \begin{bmatrix} 0 \\ x_2 \\ 0 \\ 0 \\ \vdots \\ x_n \end{bmatrix}_{N\times 1}$$

K sparse differentially expressed genes

Fig. 2. Illustration of the compressed microarray

$$y_j = \sum_{i=1}^{N} a_{ji} x_i, \qquad j = 1, 2, \cdots M \tag{4}$$

where M < < N. Additionally, if the number of nonzero elements is different in each row, the actual mixed solution of probes should be diluted to the specified volume to ensure the consistency of the dilution.

3 Composite Probe Recovery Using Compressed Sensing

3.1 Sparse Random Measurement Matrices

Each column of the random sparse M × N matrix contains only uM non-zero elements with independent and identical distribution [14]. Literature [15, 16] also have pointed out that the recovery effect of sparse random measurement matrix is consistent with the gauss random measurement matrix. Moreover, the literature [14] have further proved that the sparse random matrix satisfies the RIP.

Due to the each row of the matrix represents a linear combination of a probe spot. We limit the elements of the random sparse matrix into binary 1/0 for the sake of constructing simplicity. The configuration process for sparse random matrix is as follows,

(1) Production M × N matrix of zeros;
(2) The position of each column elements is randomly selected according to the sparse coefficient u of the matrix, and these elements would be set to 1.

3.2 Recovery of Variation Gene from Composite Probe

In two-color microarry of cDNA, we are comparing two channel sample by x_{cy3} and x_{cy5}, and interesting in the difference expression of $x = x_{cy3} - x_{cy5}$. By sparse random matrix, the normalized observation value of the composite probe is defined as $y = y_{cy3} - y_{cy5}$.

If the compressed sensing recovery x is obtained directly by the combination method, which is a NP-hard as well known. Formula (3) is an l_1-norm optimization problem, compared to time-consuming convex optimization, the classical sparse approximation methods, such as the Orthogonal Matching Pursuit (OMP) algorithm, would be very suitable.

In the OMP algorithm, the residual vector r, which is the error of approximation vector y, is smaller and smaller after several iterations [17].

Let $x_k = \arg\min_x \|y - A_k x\|_2$, $r_k = y - A_k x$, $A_k = [A_{k-1}\ a_k]$ be a sub-matrix which selected in step k. Then the OMP algorithm process as follows [18, 19].

Input: compressed sampling matrix A, measured value y, the sparsity level K.

Output: reconstruction of the signal ^x, estimated support I.

Initialization: $x_0 = 0$, $r_0 = y$, $k = 0$, estimated support $I = Ø$.

(1) $k \leftarrow k + 1$;
(2) the index that is the best match with the residual vector r_{k-1}, and $\lambda k \leftarrow$ $\text{argmax}_j\{| < r_{k-1}, a_j >|\}$;
(3) update the index $I_k = [I_{k-1}\ \lambda_k]$, and $A_k = [A_{k-1}\ a_k]$;
(4) reconstruction $\hat{}x \leftarrow [A_k]^{-1}\ y$;
(5) update the residual vector as $r_k \leftarrow y - A_k\ (\hat{}x)$;
(6) If $k \leq K$, then execute step (1), otherwise stop at $k > K$.

4 Simulation Results and Analysis

We have designed N = 96 cDNA microarray simulation probes with the idea of array-based comparative genomic hybridization (aCGH). The difference between the reference probes and the test probes, i.e., the sparsity level is K = 12. In the simulation experiments, the differences between the reference probes and the test probes have subjected to random distribution, and the locations of these different composite probes are also subjected to random distribution.

Figure 3a illustrates the reference probe x_{cy3}, and Fig. 3b demonstrates the probe x_{cy5}. Then, the differences between them, i.e., $x = x_{cy3} - x_{cy5}$ are shown in Fig. 3c.

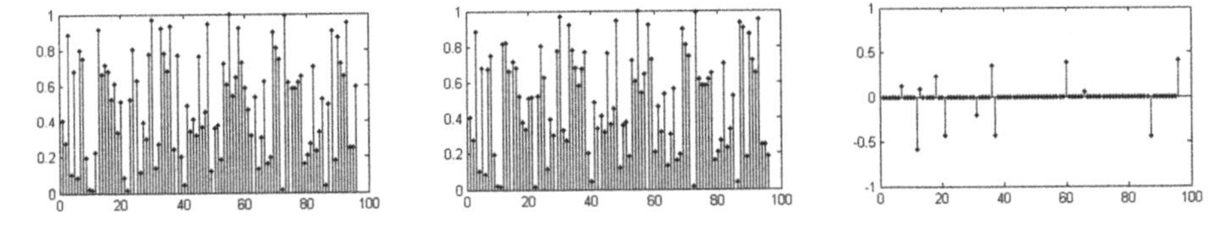

Fig. 3. **a**. The probe *xcy3,* **b**. The probe *xcy5,* **c**. $x = x_{cy3} - x_{cy5}$

We also have designed the sparse random matrix as compressed sensing measurement matrix A and let the elements sparsity coefficient $u = 0.25$. And M = 48 composite probes of compressed sensing microarray are constructed from N = 96 gene fragments by matrix A in the mixed method.

The observations of the composite probes are shown in Fig. 4a and Fig. 4b, while the differences between them, i.e., $y = y_{cy3} - y_{cy5}$ are shown in Fig. 4c.

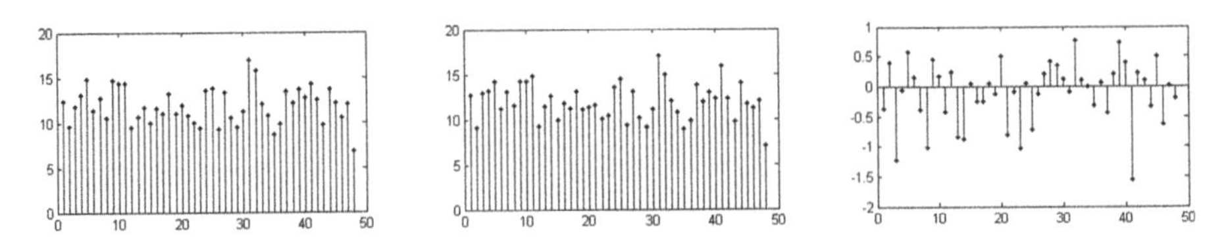

Fig. 4. **a**. The composite probes ycy3, **b**. The composite probes ycy5, **c**. $y = y_{cy3} - y_{cy5}$

We have used the OMP recovery algorithm to successfully reconstruct the gene different vector x, at N = 96, M = 48, K = 24, $u = 0.25$. As shown in Fig. 5, the recovery is so accurate that the relative error is $e = 4.4016 \times 10^{-15}$.

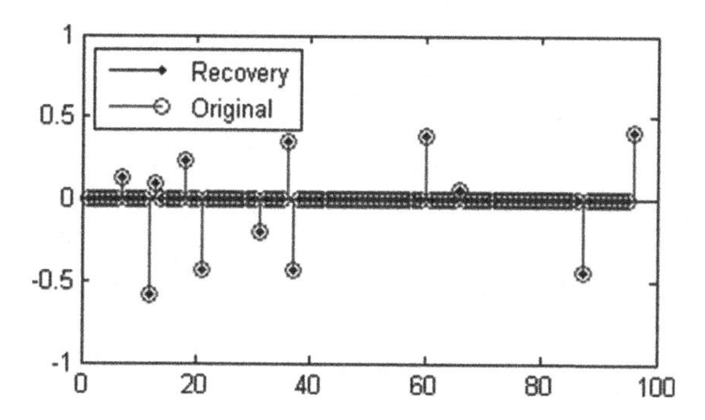

Fig. 5. The recovery of $\hat{}x = x_{cy3} - x_{cy5}$ with $e = 4.4016 \times 10^{-15}$

The structural parameters of cDNA simulation microarray have remained unchanged at N = 96, M = 0.5 N and the sparsity coefficient $u = 0.25$ for matrix A, and sparse K has been changed from zero to M. We still have used the OMP algorithm to recover the vector x. The accurate reconstruction ratios of the simulation signals are shown in Fig. 6.

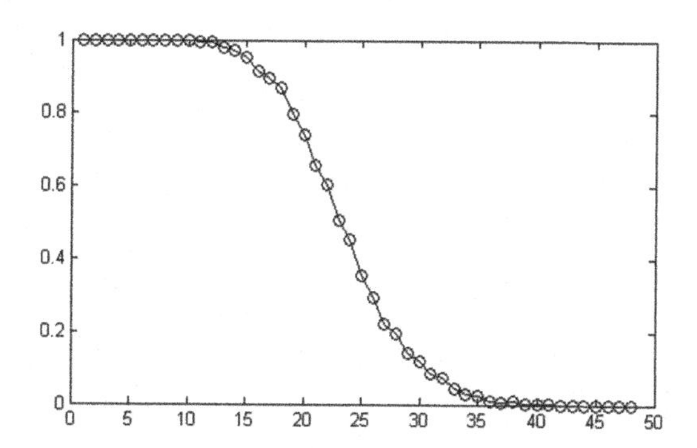

Fig. 6. The accurate reconstruction ratio of $\hat{}x = x_{cy3} - x_{cy5}$ for M = 48

As shown in Fig. 6, the compressed sensing algorithm recovers the probe's difference signals with high accuracy, at N = 96, M = 0.5 N, $u = 0.25$ and $K \leq 12$.

Figure 7 demonstrates the accurate reconstruction ratio of simulation probes under the OMP recovery algorithm, when only the number of composite probe, i.e., M has been changed from zero to N.

It is shown in Fig. 7, compressed sensing algorithm achieves high accurate recovery for difference signals between the reference probes and the sample probes when the sparse random measurement matrix A is used at N = 96, K = 12, $M \geq 48$, $u = 0.25$.

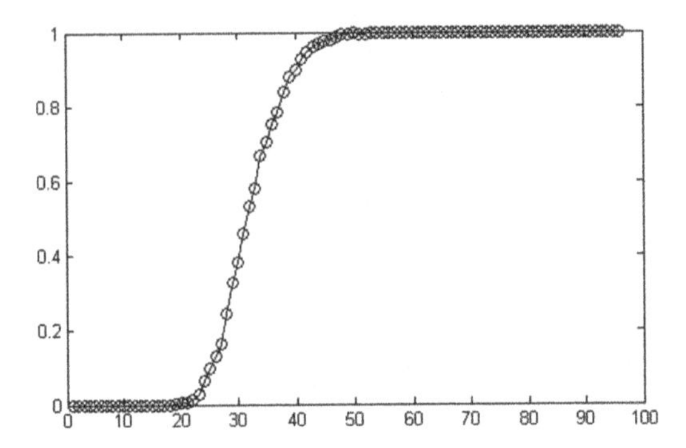

Fig. 7. The accurate reconstruction ratio of $\hat{}x = x_{cy3} - x_{cy5}$ for K = 12

5 Summary and Conclusions

There are a large number of uncertain factors in hybridization, image capture and processing of the microarray. In order to improve the reliability of the measurement, multiple probes are generally arranged to carry out repeated measurements. With a composite probe, a single spot of the compressed sensing microarray can easily and simultaneously measures many gene fragments, so that the repeated measurements of gene fragments can be realized with a limited number of spots. Considering the randomness and sparsity of genetic mutation, the total number of the composite probes installed in the compressed sensing microarray can be sharply reduced compared to that in the traditional microarray. Simulation experiment results show that, by using composite probes with gene fragment at N = 96, M = 0.5 N, and sparse random measurement matrix sparsity coefficient $u = 0.25$, when difference of cDNA probes $K \leq 12$, based on OMP algorithm for compressed sensing, the high accuracy recovery of the difference signal of cDNA can be realized.

Acknowledgments. This work is partially supported by the National Natural Foundation Project (61304199), the Ministry of Science and Technology projects for TaiWan, HongKong and Maco (2012DFM30040), the Major projects in Fujian Province (2013HZ0002-1,2013YZ0002, 2014YZ0001), the Science and Technology project in Fujian Province Education Department (JB13140/GY-Z13088), and the Scientific Fund project in Fujian University of Technology (GY-Z13005,GY-Z13125).

References

1. Ping, L.Y.: Biological sensors and biological chips: the field of biological macromolecules. Chin. J. Lab. Diagn. **9**(4), 645–648 (2005)
2. Guolian, H., Chen, D., Shukuanl, X., et al.: Novel detection system of microbe chip and its application. Acta Optica Sinica **27**(3), 499–504 (2007)
3. Shmulevich, I., Astola, J., Cogdell, D., et al.: Data extraction from composite oligonucleotide microarrays. Nucleic Acids Res. **31**(7), 431–439 (2003)

4. Jiao, L.C., Yang, S.Y., Liu, F., et al.: Development and prospect of compressive sensing. Acta Electronica Sinica **39**(7), 1651–1662 (2011)
5. Sheikh, M.A., Sarvotham, S., Milenkovic, O., et al.: DNA array decoding from nonlinear measurements by belief propagation. In: 2007 IEEE/SP Workshop on Statistical Signal Processing, SSP 2007, pp. 215–219. IEEE (2007)
6. Parvaresh, F., Vikalo, H., Misra, S., et al.: Recovering sparse signals using sparse measurement matrices in compressed DNA microarrays. IEEE J. Sel. Top. Signal Process. **2** (3), 275–285 (2008)
7. Gilbert, A., Indyk, P.: Sparse recovery using sparse matrices. Proc. IEEE **98**(6), 937–947 (2008)
8. Wang, J.-W., Wang, X.: Image reconstruction method based on compressed sensing for magnetic induction tomography. J. Northeast. Univ. (Nat. Sci.) 36(12), 1687–1690 (2015)
9. Dai, Q.-H., Fu, C.-J., Ji, X.-Y.: Research on compressed sensing. Chin. J. Comput. **34**(3), 425–434 (2011)
10. Shi, G.M., Liu, D.H., Gao, D.H., Liu, Z., Lin, J., Wang, L.J.: Advances in theory and application of compressed sensing. Acta Electronica Sinica **37**(5), 1070–1081 (2009)
11. Fei, X., Qingshan, Y.: Neurodynamic optimization method for recovery of compressive sensed signals. Appl. Res. Comput. **32**(8), 2551–2553 (2015)
12. Shen, B., Tu, D.-W., Zeng, A.-H.: DNA chip fluorescence detection by CCD. Opt. Instrum. **27**(5), 16–20 (2005)
13. Xu, Y., Ruan, Q.-F., Li, Y.-P.: Analysis methods of expression genes. J. Food Sci. Biotechnol. **27**(1), 122–126 (2008)
14. Bo, Z., Yu-lin, L., Kai, W.: Restricted isometry property analysis for sparse random matrices. J. Electron. Inf. Technol. **1**, 169–174 (2014)
15. Candes, E.J., Tao, T.: Decoding by linear programming. IEEE Trans. Inf. Theory **51**(12), 4203–4215 (2005)
16. Xiaobo, L.: Research on measurement matrix based on compressed sensing. Beijing Jiaotong University, pp. 16–19 (2010)
17. Tropp, J.A., Gilbert, A.C.: Signal recovery from random measurements via orthogonal matching pursuit. IEEE Trans. Inf. Theory **53**(12), 4655–4666 (2008)
18. Feng, L., Yi, G.: Compressed sensing analysis. Science Press, Beijing, pp. 66–69 (2015)
19. Wang, J.: Support recovery with orthogonal matching pursuit in the presence of noise. IEEE Trans. Signal Process. **63**(21), 5868–5877 (2015)

Curve Fitting Based on Neural Dynamics Optimization

Baoping Xiong[1,2,3,4], Zhenhua Gan[3,4], Fumin Zou[4], Yuemin Gao[2], and Min Du[2(✉)]

[1] Department of Mathematics and Physics, Fujian University of Technology, No. 3 Xueyuan Road, University Town, Minhou, Fuzhou, Fujian, China
xiongbp@qq.com

[2] College of Physics and Information Engineering, Fuzhou University, No. 2 Xueyuan Road, University Town, Minhou, Fuzhou, Fujian, China
fzugym@yahoo.com.cn, dm_dj90@163.com

[3] College of Electrical Engineering and Automation, Fuzhou University, No. 2 Xueyuan Road, University Town, Minhou, Fuzhou, Fujian, China

[4] Key Lab of Automotive Electronics and Electric Drive Technology of Fujian Province, Fujian University of Technology, No. 3 Xueyuan Road, University Town, Minhou, Fuzhou, Fujian, China
{ganzh, fmzou}@fjut.edu.cn

Abstract. Fitting curve is a critical problem in many testing equipment and detection system. But there was larger relative error in fitting curve when the independent variable was relatively small. In this paper, Fitting curve is formulated to a constrained linear programming. a neural dynamics optimization algorithm is obtained by considering the problem in its dual space, and then the dynamic neural network is designed to solve the optimization problem recurrently. The experimental results show that the polynomial coefficients solved by the method is stable, compared with the least square method, the relative error is obviously reduced; The method is simple and requires less samples. It provides a new simple and accurate method of curve fitting for the quantitative detection.

Keywords: Fitting curve · Relative error · Neural dynamics optimization algorithm · The quantitative detection

1 Introduction

Nowadays many testing equipment and detection system require some benchmark data for curve fitting, to establish the approximate function relation between the dependent and independent variables and Convenience for the follow-up testing and application; There are many curve fitting methods, such as the Lagrange interpolation, the newton interpolation, the neural network, the cubic spline, the least square method and some other improved curve fitting method [1], and the least square method has been widely used for its simple and easy to use; But the least squares curve fitting base on minimum

J. Pan et al. (eds.), *Genetic and Evolutionary Computing*, Advances in Intelligent Systems and Computing 536, DOI 10.1007/978-3-319-48490-7_3

sum of error square, it easily leads to the relative error is larger when the detection value is small. This situation will cause the test result is not correct or even a mistake; So bounded constrained optimization that the maximum relative error is the smallest applied to curve fitting, be used to solve this problem;

There are many methods to solve the problem of bounded constrained optimization, such as projection gradient method, spectral projection gradient method, truncated Newton method, limited memory quasi Newton method, and so on, however, all those algorithms suffer from high computational complexity and local optimal solutions [2–5]; So neural dynamics optimization algorithms was used to solve the bounded constrained optimization problem.

The neural dynamics optimization algorithms is first proposed by Hopfield [6], the basic idea of the algorithm is that the solution of the optimization problem is transformed into the solution of corresponding differential equation or differential equations by energy function. The algorithm that has global convergence, low computational complexity, parallel computing, low robustness and is suitable for software and hardware implementation, etc., has been gradually applied in image processing, signal processing, data processing and many other application fields such as machine control [7–12].

The experimental results show that the Polynomial coefficients solved by the proposed method is stable; Compared with the least square, the linear interpolation and the Cubic spline method that the relative error is obviously reduced.

2 Neural Dynamics Optimization Algorithm of Curve Fitting Method

At present, there are many curve fitting methods, and the least square method has been widely used for its simple and easy to use; But the least squares curve fitting base on minimum sum of error square, it easily leads to the relative error is larger when the detection value is small. Even the lager relative error will reach an unacceptable level in the detection process, so we propose a curve fitting method which can reduce the relative error. Its solution process are as follows: first, assume that the experimental data is (x_i, y_i) $(i = 1, 2\ldots n)$, where n is the number of samples. $C_j(j = 0, 1, 2.. m)$ is the polynomial coefficients of curve fitting which to be solved, where m is the order of the polynomial. Then assume: $B = \begin{bmatrix} 1 & x_1 & x_1^2 & \cdots & x_1^m \\ 1 & x_2 & x_2^2 & \cdots & x_2^m \\ \vdots & \vdots & \vdots & \vdots & \vdots \\ 1 & x_n & x_n^2 & \cdots & x_n^m \end{bmatrix}$, $c = \begin{bmatrix} c_0 \\ c_1 \\ \vdots \\ c_m \end{bmatrix}$, $y = \begin{bmatrix} y_0 \\ y_2 \\ \vdots \\ y_n \end{bmatrix}$

So the maximum relative error is smallest expression is as follow:

$$MIN \, \|(Bc - y)/y\|_{\infty} \tag{1}$$

In order to convert 1 into neural dynamics optimization algorithm can be processed form, first assume:

$$d = \|(Bc - y)/y\|_{\infty} \tag{2}$$

Then:

$$-dy < = Bc - y < = dy \tag{3}$$

which is:

$$\begin{cases} Bc - dy \leq y \\ -Bc - dy \leq -y \end{cases} \tag{4}$$

By (2)–(4), (1) can be converted into the following form:

$$\begin{cases} \min f(u) = d = C^T u \\ s.t. \;\; Au \leq b \end{cases} \tag{5}$$

Where:

$u = [c, d]$; $A = \begin{bmatrix} B & -y \\ -B & -y \end{bmatrix}$; $b = \begin{bmatrix} y \\ -y \end{bmatrix}$; $C = [0, 0, \ldots, 0, 1]$; In order to obtain the solution of the vector u, (5) can be converted to equivalent form as follow:

$$\begin{cases} \max \; - C^T u \\ s.t. \;\; Au \leq b \end{cases} \tag{6}$$

Based on the two primal-dual non gradient algorithm optimization of neural dynamics can get the state equation of (6) as follow [7]:

$$\begin{pmatrix} \frac{dq(t)}{dt} \\ \frac{du(t)}{dt} \end{pmatrix} = \lambda \begin{pmatrix} (q(t) + Au(t) - b)^{+} - q(t) - A(A^t q(t) + C) \\ -C - A^T (q(t) + Au(t) - b)^{+} \end{pmatrix} \tag{7}$$

Output equation

$$\mathrm{v(t) = u(t)} \tag{8}$$

Where λ is a constant. **A** block diagram of model (7) is shown in Fig. 1, where the vector *a, b* is the external input, and the vector x, y is the network output.

By the Formula 7 we know that the final result of the global optimal solution u(t) is the polynomial coefficient C and the minimum relative error.

3 Experimental Results and Analysis

In order to verify the feasibility of the proposed method, now suppose that we are given a training set comprising N observations of x, written $x \equiv (x_1, \ldots, x_N)^T$, together with corresponding observations of the values of y, denoted $y \equiv (y_1, \ldots, y_N)^T$. Figure 2

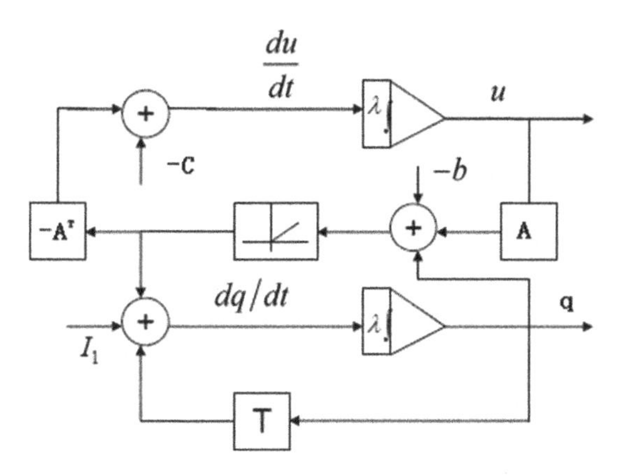

Fig. 1. A block diagram of the algorithm (7), Where $T = -A^TA - I,\ I_1 = -AC$ and I is an unit matrix.

shows a plot of a training set comprising $N = 15$ data points. The input data set $\mathbf{x}$ in Fig. 2 was generated by choosing values of x_n , for $n = 1, \ldots, N$, spaced uniformly in range [0, 4], and the target data set $\mathbf{y}$ was obtained by first computing the corresponding values of the function $10 \ * \ \sin\left(\frac{\pi x}{2}\right) + 15$ and then adding a small level of noise to each such point in order to obtain the corresponding value y_n. Before the experiment, we define the maximum relative error $MRE = \max\left(\left|\frac{y_n - y_n^*}{y_n}\right|\right)\ n = 1, 2\ldots M$. And the mean absolute error $MAE = \frac{1}{M}\sum_{i=1}^{M}\left|y_n - y_n^*\right|$.

Where y_n^* is the predictive value of fitted curve.

For comparison, we perform the linear interpolation (LI), the Cubic spline (CS), the least square (LS), the neural dynamics optimization algorithm (NDOA) to fit the training data shows in Fig. 2. The order of the polynomial which fitted by the least square and the neural dynamics optimization algorithm is four.

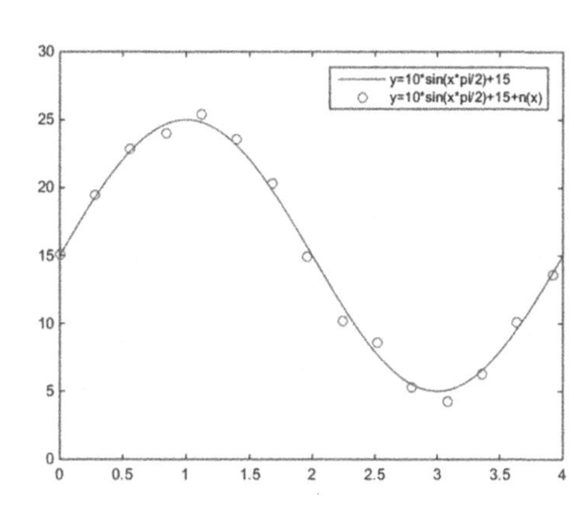

Fig. 2. Training data, plot of a training data set of $N = 15$ points, shown as circles, n(x) is noise.

The computed results were obtained by 1000 tests data spaced uniformly in range [0, 4], the results that the noise added in y is [−1.5, 1.5] random noise are summarized in Table 1 and Fig. 3, the results that the noise added in y is colored noise are summarized in Table 2 and Fig. 4, It can be seen that the MRE of curve which fitted by neural dynamics optimization algorithm is lowest in Tables 1 and 2 and the MAE of curve which fitted by neural dynamics optimization algorithm is highest in Tables 1 and 2. From Figs. 3 and 4, it can be see that the linear interpolation method and the cubic spline method has over-fitting problem.

Table 1. The results that the noise added in y is [−1.5, 1.5] random noise

	LI	CS	LS	NDOA
MRE	48.1973 %	48.0648 %	38.5870 %	28.3020 %
MAE	0.8816	0.8792	0.9048	1.4895

Table 2. The results that the noise added in y is colored noise

	LI	CS	LS	NDOA
MRE	106.3738 %	102.8233 %	59.4661 %	45.5041 %
MAE	1.1463	1.1269	1.0662	2.8822

In order to further verify the feasibility of the proposed method, Experiments using least square method and neural dynamics optimization algorithm to fit the linear and nonlinear two sets of baseline data of testing instruments; The solution trajectory are shown in Figs. 5 and 6, It can be seen that the solution is globally asymptotically stable at the equilibrium point. The maximum relative error results are shown in Tables 3 and 4, It can be seen that the maximum relative error of curve which fitted by neural dynamics optimization algorithm is far lower than the relative error of least squares

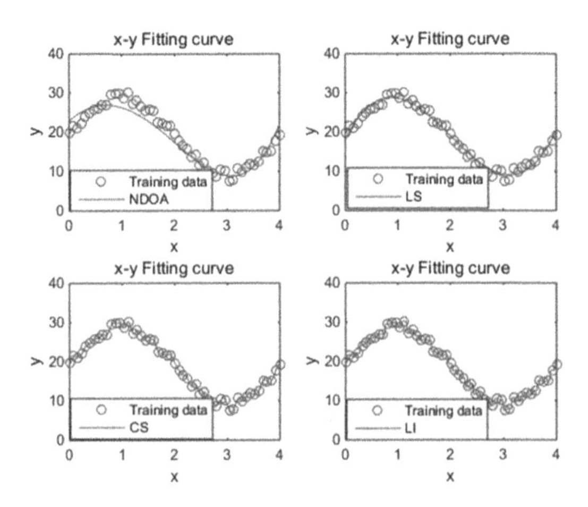

Fig. 3. Fitting curve of training data with [−1.5, 1.5] random noise

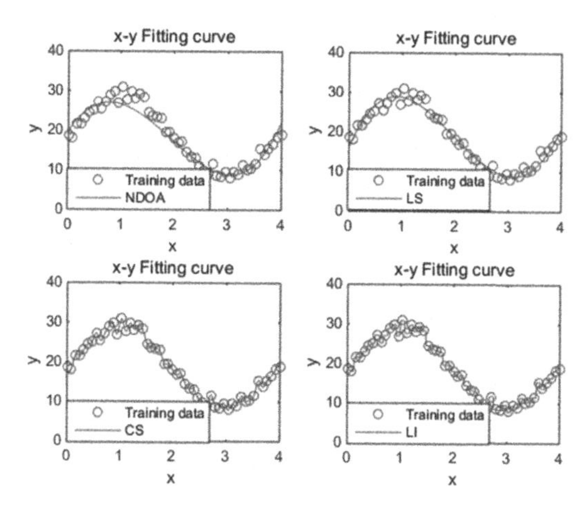

Fig. 4. Fitting curve of training data with colored noise (Color figure online)

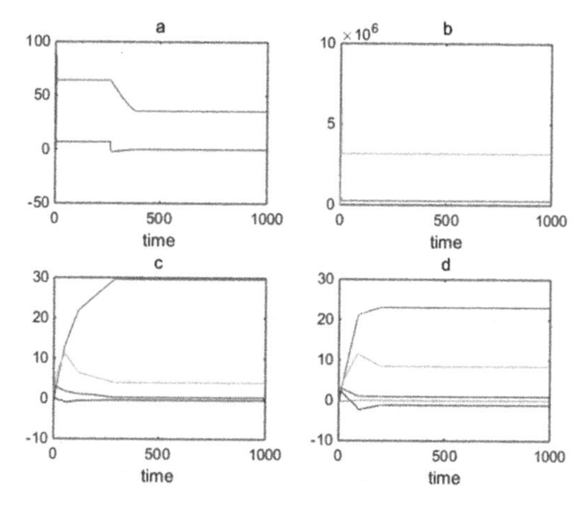

Fig. 5. The solution trajectory of the neural dynamics optimization algorithm for near linear data. a, b, c, d are first order polynomial, quadratic, cubic polynomial, and quartic polynomial

curve fitting. But from Figs. 7 and 8, it can be see that the maximum absolute error of the curve which is fitted by the neural dynamics optimization algorithm is significantly larger than that of the least square method, especially when the y value is larger, the deviation is more obvious. This situation occurs mainly due to that the absolute error with the same relative error, is increases by y value. However, in many detection when the test data exceeds a certain value that the absolute error have little influence on the result of the qualitative, so the maximum absolute error of the curve which is fitted by the neural dynamics optimization algorithm has little influence on test result.

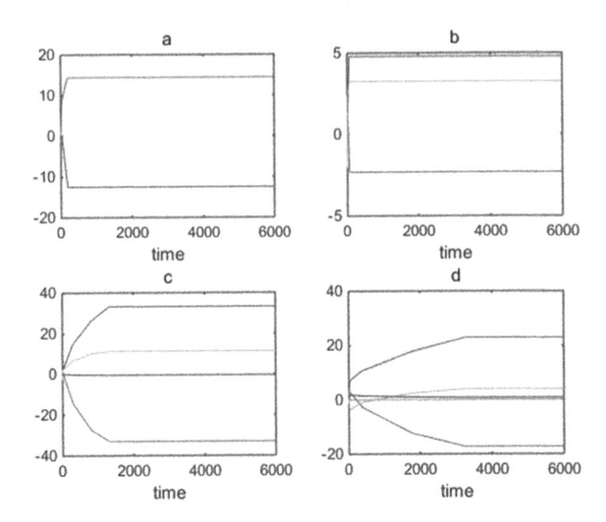

Fig. 6. The solution trajectory of the neural dynamics optimization algorithm for nonlinear data. a, b, c, d are first order polynomial, quadratic, cubic polynomial, and quartic polynomial

Table 3. Maximum relative deviation of nearly linear data curve fitting

	Linear polynomial	Quadratic polynomial	Cubic polynomial	Four times polynomial
Least square method	230.6 %	137.7 %	163.5 %	109.4 %
Neural dynamics method	18.62 %	15.68 %	10.847 %	9.28 %

Table 4. Maximum relative deviation of nonlinear data curve fitting

	Linear polynomial	Quadratic polynomial	Cubic polynomial	Four times polynomial
Least square method	381.42 %	335 %	31.14 %	37.7.4 %
Neural dynamics method	62.35 %	15.68 %	11.9 %	7.1 %

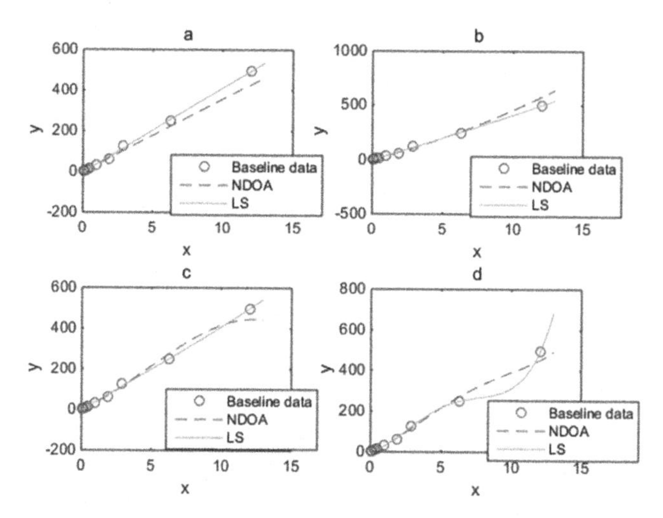

Fig. 7. Fitting curve of near linear data. a, b, c, d are first order polynomial, quadratic, cubic polynomial, and quartic polynomial

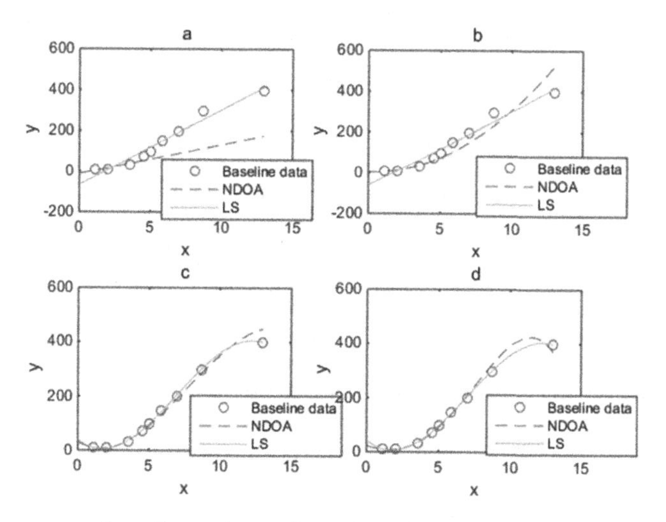

Fig. 8. Fitting curve of nonlinear data. a, b, c, d are first order polynomial, quadratic, cubic polynomial, and quartic polynomial

4 Conclusion

The curve fitting of neural dynamics optimization algorithm method effectively reduces the maximum relative error. But at the same time it makes the absolute error becomes large and is easy to lead to over fitting when the y value is larger. But the absolute error has little effect on the qualitative results when the detection value is large. The method be proposed in this paper provides a new simple and accurate method of curve fitting for the quantitative detection.

Acknowledgments. This work is partially supported by National Natural Foundation Project (61304199), The Ministry of science and technology projects for Hong Kong and Maco (2012DFM30040), Major projects in Fujian Province (2013HZ0002-1, 2014YZ0001).

References

1. Mathews, J.H., Fink, K.D.: Numerical Methods Using MATLAB. Prentice Hall, USA (1998)
2. Xiao, Y.H., Hu, Q.J.: Subspace Barzilai-Borwein gradient method for large-scale bound constrained optimization. Appl. Math. Optim. **58**(2), 275–290 (2008)
3. Andretta, M., Birgin, E.G., Martinez, J.M.: Practical active-set euclidian trust-region method with spectral projected gradients for bound-constrained minimization. Optimization **54**(3), 305–325 (2005)
4. Birgin, E.G., Martinez, J.M., Raydan, M.: Nonmonotone spectral projection gradient methods on convex sets. SIAM J. Optim. **10**, 1196–1211 (2000)
5. Birgin, E.G., Martinez, J.M.: A box-constrained optimization algorithm with negative curvature directions and spectral projected gradients. In: Alefeld, G., Chen, X. (eds.) Topics in Numerical Analysis. Computing Supplementa, vol. 15, pp. 49–60. Springer, Vienna (2001)
6. Hopfield, J.J., Tank, D.W.: Neural computation of decisions in optimization problems. Biol. Cybern. **52**, 141–152 (1985)
7. Xia, Y.: A new neural network for solving linear and quadratic programming problems. IEEE Trans. Neural Netw. **7**(6), 1544–1548 (1996)
8. Malek, A., Alipour, M.: Numerical solution for linear and quadratic programming problems using a recurrent neural network. Appl. Math. Comput. **192**(1), 27–39 (2007)
9. Rahman, S.A., Ansari, M.S.: A neural circuit with transcendental energy function for solving system of linear equations. Analog Integr. Circ. Sig. Process. **66**, 433–440 (2011)
10. Xia, Y., Wang, J., Fork, L.M.: Grasping force optimization for multifingered robotic hands using a recurrent neural network. IEEE Trans. Robot. Autom. **20**, 549–554 (2004)
11. Xia, Y., Chen, T., Shan, J.: A novel iteration method for computing generalized inverse. Neural Comput. **26**, 449–465 (2014)
12. Liu, Q., Wang, J.: Finite-time convergent recurrent neural network with a hard-limiting activation function for constrained optimization with piecewise-linear objective functions. IEEE Trans. Neural Netw. **22**(4), 601–613 (2011)

Certificateless Authentication Protocol for Wireless Body Area Network

Jian Shen[1,2,3(✉)] and Shaohua Chang[1,3]

[1] Jiangsu Engineering Center of Network Monitoring, Nanjing University of Information Science and Technology, Nanjing, China
{s_shenjian, casaha}@126.com

[2] Jiangsu Collaborative Innovation Center on Atmospheric Environment and Equipment Technology, Nanjing University of Information Science and Technology, Nanjing, China

[3] School of Computer and Software, Nanjing University of Information Science and Technology, Nanjing, China

Abstract. In order to address the problems of security communication and vital physiological data for WBANs, in this paper, we propose an effective authentication protocol and secure session key generation method. Based on certificateless cryptography, we present a new certificateless encryption scheme and a new certificateless signature scheme, both of that is provably secure and efficiency. Then, taking the above two scheme as a basis, we propose a novel certificateless authentication protocol which is anonymous and mutual authentication. After successful authenticating, an efficient and secure session key is generated at last for subsequent communication in our protocol.

Keywords: WBANs · CLE · CLS · Authentication · Key generation

1 Introduction

Wireless body sensor networks (WBANs) is a small-sized network but with great practicality. The essence of the WBANs is wireless sensor network, which is mainly used for data acquisition, data processing and data transmission. It can be applied in many fields like smart home, entertainment, military and many other aspects, but mostly applied in medical treatment. In the era of wireless networks, data transmission will encounter a very serious issues, especially for the sensitive data in WBANs, the security of data transmission is of great importance for WBANs. For example, when getting a wrong order from the server, an implantable Blood Glucose Sensor may inject overmuch dose of insulin which may lead to the deteriorate to a patient, or when the medical personnel get a wrong request from a ECG sensor of the patient, they need to respond immediately which waste the medical resources on a great degree [1–4]. The threat to the WBANs applied in medical treatment is very serious for users's life safety. To promote the WBANs better applied in medical treatment, it is vital to protect the integrity, confidentiality and authenticity of the sensitive data of users.

J. Pan et al. (eds.), *Genetic and Evolutionary Computing*, Advances in Intelligent Systems and Computing 536, DOI 10.1007/978-3-319-48490-7_4

To implement the security data transmission for WBANs, many researcher have designed various security authentication between PDA and AP [5–9]. In this paper, we will propose a certificateless encryption authentication method.

1.1 Related Work

In the history of WBANs, public key cryptography is usually used for WBANs field [10–12]. Identity-based cryptography is proposed to avoid the difficulties in establishing and managing public key infrastructure for the traditional public key cryptosystem [12–14]. Identity-based cryptography is advanced in withing no digital certificate, but totally depend on KGC where it may be invaded. Then, based on ID-based cryptography, certificateless cryptography has been proposed for solving the problem of ID-based cryptography in WBANs.

Since 2003, Al-Riyami and Perterson first proposed a new public key cryptography named certificateless public key cryptography [15]. Since then, certificateless encryption is used in many cryptography field. In 2014, Jingwei Liu et al. [16] proposed a certificateless cryptography based remote anonymous authentication protocol which efficiently saves the computation resource. In 2014, Debiao He et al. [17] proposed a cloud assistant based certificateless auditing scheme for WBANs that can provide protection of data integrity. In 2015, Hu Xiong et al. [18] presents a certificateless encryption scheme and certificateless signature scheme, then they build a revocable certificateless anonymous and remote authentication scheme based on the basic of CLE and CLS for WBANs. The advantages of certificateless scheme are solving the key escrow problem, but it may automatically cause computation and resource limited issues in WBANs. In this paper, we propose a security and efficiency authentication protocol forWBANs.

1.2 Our Contribution

In this paper, we propose a certificateless authentication protocol between PDA and AP. In a word, our mainly contributions are in three aspects:

(1) We propose a new certificateless encryption scheme that is efficient to avoid from key replacement attack.
(2) We present a new certificateless signature scheme, and the security of this scheme is based on BDH intractability assumptions.
(3) We show up a novel certificateless anonymous authentication protocol with ECC DL problem that make the protocol more stable while suffering from adversary's attack.

The remainder of the paper is organized as follows: In Sect. 2, we list relevant meanings of abbreviations and symbols and show the basic knowledge for better understanding the paper. In Sect. 3, we display the system model of our paper. In Sect. 4, we present a CLE scheme and CLS scheme. In Sect. 5, we propose our authentication protocol. Section 6 are conclusion.

2 Preliminaries

In this section, we show some relevant basic knowledge used in the paper.

2.1 Bilinear Maps and Difficulty Hypothesis

The $\mathbb{G}_1$ and $\mathbb{G}_2$ is two cyclic group of ($\mathbb{G}_1$, +) and ($\mathbb{G}_2$, •) that with the same prime order p. $\mathcal{G}$ is the generator of $\mathbb{G}_1$. Bilinear map is defined as $\hat{e}$: $\mathbb{G}_1 \times \mathbb{G}_1 \rightarrow \mathbb{G}_2$ satisfied with following natures:

(1) Bilinearity: For all $\mathcal{G}_1$, $\mathcal{G}_2 \in \mathbb{G}_1$, and a, $b \in Z_p, \hat{e}(\mathcal{G}_1^a, \mathcal{G}_2^b) = \hat{e}(\mathcal{G}_1, \mathcal{G}_2)^{ab}$.
(2) Non-degeneracy: Existing $\mathcal{G}_1, \mathcal{G}_2 \in \mathbb{G}_1$, $\hat{e}(\mathcal{G}_1 \mathcal{G}_2) \neq 1$.
(3) Computability: For any $\mathcal{G}_1, \mathcal{G}_2 \in \mathbb{G}_1$, $\hat{e}(\mathcal{G}_1, \mathcal{G}_2) \in \mathbb{G}_2$ can be compute in polynomial time.

3 Model

In order to achieve reliable authentication scheme, we propose a novel certificateless protocol. Next, we will introduce the system architecture of this paper.

3.1 System Architecture

In this paper, we mainly talk about the authentication protocol between PDA and AP. The system architecture is shown as Fig. 1.

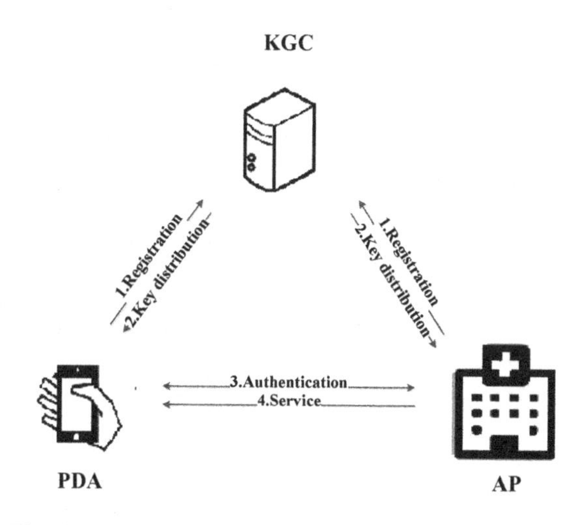

Fig. 1. System architecture in our authentication protocol

4 Design Basis

In this section, we will introduce our new CLE scheme and CLS scheme. This two new schemes are basic design for our new authentication protocol.

4.1 Our Certificateless Encryption Scheme

This scheme includes seven algorithms. Details are as follows:

(1) Setup: Generate two cyclic group $\mathbb{G}_1$, $\mathbb{G}_2$, Define $\mathcal{G}$ as the generator of $\mathbb{G}_1$, and bilinear map $\hat{e}$: $\mathbb{G}_1 \times \mathbb{G}_1 \rightarrow \mathbb{G}_2$, choose elliptic curve $y^2 = x^3 + ax + b$ which is defined on $\mathcal{Z}_p$, choose any nonnegative integer a, b that satisfy $4a^3 + 27b^2 \neq 0$ and both of them less than p, choose hash function defined as follows: $\mathcal{H}_1 : \{0, 1\}^* \rightarrow \mathcal{Z}_p$, $\mathcal{H}_2 : \mathbb{G}_2 \rightarrow \{0, 1\}^n$,
System parameters show as $params = \ <q, \mathcal{G}, \mathbb{G}_1, \mathbb{G}_2, a, b, \mathcal{H}_1, \mathcal{H}_2>$
(2) Partial_Private_Key_Generation: when receiving the identity-$\mathcal{ID}$, KGC record current time $\mathcal{T}$, and choose appropriate a, b to confirm an elliptic curve, and then broadcast the partial private key $(sk_{01}, sk_{02}) = \left(\mathcal{G}^{a \cdot \mathcal{H}1(\mathcal{ID}) + b \cdot \mathcal{H}1(\mathcal{T})}, \mathcal{G}^{a \cdot b}\right)$, to the user.
(3) Set_Secret_Value: After receiving partial private key, user chooses a point (x, y) from elliptic curve as own secret value.
(4) Set_Private_Key: Compute private key pairing $\mathcal{SK} = (sk_1, sk_2) = \left(\mathcal{G}^{a \cdot b \cdot x \cdot y\left(a \cdot \mathcal{H}1(\mathcal{ID}) + b \cdot \mathcal{H}1(\mathcal{T})\right) + a \cdot b}, \mathcal{G}^{a \cdot b}\right)$
(5) Set_Public_Key: User computes $\mathcal{PK} = \mathcal{G}^{x \cdot y}$ as his public key.
(6) Encryption: Choose a random integer $r \in \mathcal{Z}_p$, set $k = \hat{e}(\mathcal{G}, \mathcal{G})^{r \cdot a \cdot b}$, compute ciphertext $\mathcal{C}_0 = \mathcal{H}_2(k) \oplus \mathcal{M}$, $\mathcal{C}_1 = \mathcal{G}^r$, $\mathcal{C}_2 = \mathcal{G}^{r \cdot x \cdot y(a \cdot \mathcal{H}1(\mathcal{ID}) + b \cdot \mathcal{H}1(\mathcal{T}))}$, and compute $t = k \oplus r$. Broadcast ciphertext as $\mathbb{C} = \ <\mathcal{C}_0\, \mathcal{C}_1\, \mathcal{C}_2, t>$.
(7) Decryption: Compute $k = \hat{e}(\mathcal{C}_1, sk_1) / \hat{e}\,(\mathcal{C}_2, sk_2)$, and $r = k \oplus t$. Check whether $\mathcal{C}_1 = \mathcal{G}^r$, compute plaintext $\mathcal{M} = \mathcal{H}_2(k) \oplus \mathcal{C}_0$, if the equality holds, or reject ciphertext if not.

4.2 Our Certificateless Signature Scheme

In this section, we will introduce our certificateless signature scheme. The scheme is displayed as follows:

(1) Setup: This step is the same as described in Sect. 4.1. Here the hash function h_3 is defined as $\mathcal{H}_3$: $\{0, 1\}^* \times \{0, 1\}^* \times \mathbb{G}_1 \times \mathcal{T} \rightarrow \mathbb{G}_1$
(2) Partial_Private_Key_Generation.
(3) Set_Secret_Value
(4) Set_Private_Key.
(5) Set_Public_Key. This algorithm is the same as Sect. 4.1.

(6) Sign: choose a random integer $s \in \mathcal{Z}_p$, compute the signature $\sigma = (\sigma_1, \sigma_2) = (sk_1 \cdot \mathcal{G}^s \cdot \mathcal{H}_3(\mathcal{M}, \mathcal{ID}, \mathcal{PK}, \mathcal{T})^{x \cdot y}, sk_2 \cdot \mathcal{G}^s)$

(7) Verification: Verify whether

$$\hat{e}(\sigma_1, \mathcal{G}) = \hat{e}\left(\mathcal{H}_3(\mathcal{M}, \mathcal{ID}, \mathcal{PK}, \mathcal{T}), \mathcal{PK}\right)\hat{e}\,\mathcal{G}^{a2 \cdot b \cdot \mathcal{H}1(\mathcal{ID})}, \mathcal{PK})\hat{e}\left(\mathcal{G}^{a \cdot b2 \cdot \mathcal{H}1(\mathcal{T})}, \mathcal{PK}\right)\hat{e}(\sigma_2, \mathcal{G}).$$

5 Our Certificatless Authentication Protocol

In this section, we will present our main authentication protocol. Our protocol can provide anonymity, mutual authentication, session key establish, key escrow resilience, non-reputation, forward security and collusion resistance with appropriate computation cost.

5.1 Authentication Protocol

According to Sect. 3.1, we know that our authentication protocol is applied between *PDA* and *AP*, and it take our new CLE and CLS scheme as design basis. The concrete protocol is depicted as following:

(1) Initialization

KGC generates system parameters by executing algorithm like in Sect. 4.1. Three more hash function is defined as $\mathcal{H}_4\colon \{0,1\}^* \times \mathbb{G}_1 \times \mathbb{G}_1 \rightarrow \mathcal{Z}_p$, $\mathcal{H}_5\colon \{0,1\}^* \times \{0,1\}^* \times \{0,1\}^* \rightarrow \mathcal{Z}_p$, $\mathcal{H}_6\colon (x,y) \rightarrow \{0,1\}^*$. And system parameters is $Params = <q, \mathcal{G}\mathbb{G}_1, \mathbb{G}_2, a, b, \mathcal{H}_1, \mathcal{H}_2, \mathcal{H}_3, \mathcal{H}_4, \mathcal{H}_5, \mathcal{H}_6>$.

(2) Registration

In this phase, according to the algorithm **Partial_Private_Key_Generation**, *PDA* generates private key pairing $\mathcal{SK}_{PDA} = (sk^1_{PDA}, sk^2_{PDA}) = \left(\mathcal{G}^{(a \cdot b \cdot x \cdot y(a \cdot \mathcal{H}1(\mathcal{ID}PDA + b \cdot \mathcal{H}1(\mathcal{T})) + a \cdot b}, \mathcal{G}^{a \cdot b}\right)$, *AP* generates private key pairing $\mathcal{SK}_{AP} = (sk^1_{AP}, sk^2_{AP}) = \left(\mathcal{G}^{(a \cdot b \cdot x \cdot y(a \cdot \mathcal{H}1(\mathcal{ID}AP + b \cdot \mathcal{H}1(\mathcal{T})) + a \cdot b}, \mathcal{G}^{a \cdot b}\right)$ According to the algorithm **Set_Private_Key**, **Set_Public_Key**, *PDA* generates public key $\mathcal{PK}_{PDA} = \mathcal{G}^{x1 \cdot y1}$, *AP* generates public key $\mathcal{PK}_{PDA} = \mathcal{G}^{x2 \cdot y2}$.

(3) Authentication phase

Firstly, *PDA* sends authentication request information to corresponding *AP*:

- Choose a random integer $n_{PDA} < n$, and compute $\mathcal{P}_{PDA} = n_{PDA} \times G$.
- Compute a signature pairing $\sigma_{PDA} = (\sigma_1, \sigma_2) = (sk^1_{PDA} \cdot \mathcal{G}^s \cdot \mathcal{H}_3(P_{PDA}, \mathcal{ID}_{PDA}, \mathcal{PK}_{PDA}, \mathcal{T})^{x \cdot y}, sk^2_{PDA} \cdot \mathcal{G}^s)$
- Compute $r = \mathcal{H}_4(\mathcal{ID}_{PDA}, \mathcal{PK}_{PDA}, \sigma_{PDA})$ and an encryption key $k_{PDA} = \hat{e}(\mathcal{G}, \mathcal{G})^{r \cdot a \cdot b}$. Then compute a $\mathcal{MAC}_{PDA} = \mathcal{MAC}_{kPDA}(\mathcal{ID}_{PDA} \| \mathcal{PK}_{PDA} \| \sigma_{PDA})$
- Compute ciphertext $\mathcal{C}_0 = \mathcal{H}_2(k_{PDA}) \oplus \mathcal{MAC}_{PDA}$, $\mathcal{C}_2 = \mathcal{G}^r$, $\mathcal{C}_2 = \mathcal{G}^{r \cdot x1 \cdot y1(a \cdot \mathcal{H}1(\mathcal{ID}AP + b \cdot \mathcal{H}1(\mathcal{T}))}$
- *PDA* sends σ_{PDA}, $\mathcal{MAC}_{PDA}$, $\mathcal{C}_0, \mathcal{C}_1, \mathcal{C}_2$, to *AP*.

Upon receiving the request information, AP executes follow algorithms:

- Compute decryption key $k_{AP} = \hat{e}(C_1, sk_{AP}^1)/\hat{e}(C_2, sk_{AP}^2)$.
- Compute $\mathcal{ID}_{PDA}||\mathcal{PK}_{PDA}||\sigma = \mathcal{H}_2(k_{PDA}) \oplus C_0$, and compare whether the equation holds $\mathcal{MAC}_{AP} = \mathcal{MAC}_{PDA}$.
- Compute $r = \mathcal{H}_4(\mathcal{ID}_{PDA}||\mathcal{PK}_{PDA}||\sigma)$.
- Check whether $C_1 = \mathcal{G}^r$, and verify whether the following equation holds $\hat{e}(\sigma_1, \mathcal{G}) = \hat{e}(\mathcal{H}_3(\mathcal{M}, \mathcal{ID}, \mathcal{PK}, \mathcal{T}), \mathcal{PK}_{PDA}),\ \hat{e}\,\mathcal{G}^{a \cdot b2 \cdot \mathcal{H}1(\mathcal{I}PDA)}, \mathcal{PK}_{PDA}),\ \hat{e}(\sigma_2, \mathcal{G})$. If not equal, interrupt authentication, or continue.
- Choose a random integer $n_{AP} < n$, and compute $\mathcal{P}_{AP} = n_{AP} \times \mathrm{G}$, $\mathcal{Q}_{AP} = n_{AP} \times \mathcal{P}_{PDA}$.
- Compute session key $\mathcal{K}_{AP\_PDA} = \mathcal{G}^{\mathcal{H}_5(\mathcal{ID}AP, \mathcal{ID}PDA, \mathcal{H}_6(\mathcal{Q}AP))}$.
- *AP* sends $\mathcal{MAC}_{\mathcal{K}AP\_PDA}(\mathcal{Q}_{AP}), \mathcal{P}_{AP}$, to *PDA*.

Acquiring the feedback information from *AP*, *PDA* do the following steps to generate session key:

- Compute $\mathcal{Q}_{PDA} = n_{PDA} \times \mathcal{P}_{AP}$.
- Compute session key $\mathcal{K}_{PDA\_AP} = \mathcal{G}^{\mathcal{H}_5(\mathcal{ID}PDA, \mathcal{ID}AP, \mathcal{H}_6(\mathcal{Q}AP))}$.
- Check whether the equation $\mathcal{MAC}_{\mathcal{K}PDA\_AP}(\mathcal{Q}_{PDA}) = \mathcal{MAC}_{\mathcal{K}AP\_PDA}(\mathcal{Q}_{AP})$ holds. If so, the authentication protocol is successful and session key is feasible, if not, reject this session key and output fails.

As mentioned above, if *PDA* and *AP* authenticate each other successful, they will both generate a session key for further information transferring.

6 Conclusion

In this paper, we propose a effective authentication protocol and secure key generation method for WBANs which is based on certificateless cryptography. First, we present a new CLE scheme and a new CLS scheme as a basic, then based on the two schemes, we propose our novel certificateless authentication protocol between PDA and AP in WBANs. Our proposed protocol can provide anonymous and mutual authentication, and generate a secure session key for subsequent communication. The proposed protocol is security and efficiency for protecting wireless communication in WBANs.

Acknowledgments. This work is supported by the National Science Foundation of China under Grant No. 61300237, No. U1536206, No. U1405254, Nos. 61232016 and 61402234, the National Basic Research Program 973 under Grant No. 2011CB311808, the Natural Science Foundation of Jiangsu province under Grant No. BK2012461, the research fund from Jiangsu Technology & Engineering Center of Meteorological Sensor Network in NUIST under Grant No. KDXG1301, the research fund from Jiangsu Engineering Center of Network Monitoring in NUIST under Grant No. KJR1302, the research fund from Nanjing University of Information Science and Technology under Grant No. S8113003001, the 2013 Nanjing Project of Science and Technology Activities for Returning from Overseas, the 2015 Project of six personnel in Jiangsu Province under Grant No. R2015L06, the CICAEET fund, and the PAPD fund.

References

1. Seyedi, M., Kibret, B., Lai, D.T., et al.: A survey on intrabody communications for body area network applications. IEEE Trans. Biomed. Eng. **60**, 2067–2079 (2013)
2. Movassaghi, S., Abolhasan, M., Lipman, J., et al.: Wireless body area networks: a survey. IEEE Commun. Surv. Tutor. **16**, 1658–1686 (2014)
3. Chen, M., Gonzalez, S., Vasilakos, A., et al.: Body area networks: a survey. Mob. Netw. Appl. **16**, 171–193 (2011)
4. Zhang, Z., Wang, H., Vasilakos, A. V., et al.: ECG-cryptography and authentication in body area networks. In: IEEE Transactions on Information Technology in Biomedicine a Publication of the IEEE Engineering in Medicine & Biology Society, vol. 16, pp. 1070–1078 (2012)
5. Fortino, G., Giannantonio, R., Gravina, R., et al.: Enabling effective programming and flexible management of efficient body sensor network applications. IEEE Trans. Hum. Mach. Syst. **43**, 115–133 (2013)
6. Halford, T.R., Courtade, T.A., Chugg, K.M., et al.: Energy-efficient group key agreement for wireless networks. IEEE Trans. Wirel. Commun. **14**, 5552–5564 (2015)
7. Sarvabhatla, M., Vorugunti, C.S.: A secure and robust dynamic ID-based mutual authentication scheme with smart card using elliptic curve cryptography. In: Seventh International Workshop on Signal Design and ITS Applications in Communications. IEEE (2015)
8. Li, M., Yu, S., Lou, W., et al.: Group device pairing based secure sensor association and key management for body area networks. In: Conference on Information Communications, pp. 2651–2659. IEEE Press (2010)
9. Mehmood, Z., Nizamuddin, N., Ch, S.A., et al.: An efficient key agreement with rekeying for secured body sensor networks. In: Second International Conference on Digital Information Processing and Communications, pp. 164–167 (2012)
10. Sangari, A.S., Manickam, J.M.L.: Public key cryptosystem based security in wireless body area network. In: Proceedings of 2014 IEEE International Conference on ICCPCT, pp. 1609–1612 (2014)
11. Li, M., Lou, W., Ren, K.: Data security and privacy in wireless body area networks. IEEE Wirel. Commun. **17**, 51–58 (2010)
12. Ramli, S.N., Ahmad, R.: Surveying the wireless body area network in the realm of wireless communication. In: Proceedings of 2011 IEEE 7th International Conference on IAS, pp. 58–61 (2011)
13. Tan, C.C., Wang, H., Zhong, S., Li, Q.: IBE-Lite: a lightweight identity-based cryptography for body sensor networks. IEEE Trans. Inf. Technol. Biomed. **13**, 926–932 (2009)
14. Hu, K., Xue, J., Hu, C., Ma, R., Li, Z.: An improved ID-based group key agreement protocol. Tsinghua Sci. Technol. **19**(5), 421–428 (2014)
15. Al-Riyami, S.S., Paterson, K.G.: Certificateless public key cryptography. In: Laih, C.-S. (ed.) ASIACRYPT 2003. LNCS, vol. 2894, pp. 452–473. Springer, Heidelberg (2003)
16. Liu, J., Zhang, Z., Chen, X., Kwak, K.S.: Certificateless remote anonymous authentication schemes for wireless body area networks. IEEE Trans. Parallel Distrib. Syst. **25**, 332–342 (2014)
17. He, D., Zeadally, S., Wu, L.: Certificateless public auditing scheme for cloud-assisted wireless body area networks. IEEE Syst. J. **2015** 1–10 (2015)
18. Xiong, H., Qin, Z.: Revocable and scalable certificateless remote authentication protocol with anonymity for wireless body area networks. IEEE Trans. Inf. Forensics Secur. **10**(7), 1442–1455 (2015)

Hierarchical PSO Clustering on MapReduce for Scalable Privacy Preservation in Big Data

Ei Nyein Chan Wai[1,2(✉)], Pei-Wei Tsai[2], and Jeng-Shyang Pan[2]

[1] University of Computer Studies, Yangon, Myanmar
einyeinchanwai@ucsy.edu.mm
[2] Fujian Provincial Key Laboratory of Big Data Mining and Applications, Fujian University of Technology, Fuzhou, Fujian, China
pwtsai@mail.fjut.edu.cn

Abstract. Today organizations are deeply involved in the Big Data era as the amount of data has been exploding with un-predictable rate and coming from various sources. To process and analyze this massive data, privacy is a major concern together with utility of data. Thus, privacy preservation techniques which target at the balance between utility and privacy begin to be one of the recent trends for big data researchers. In this paper, we discuss a technique for big data privacy preservation by means of clustering method. Here, hierarchical particle swarm optimization (HPSO) is used for clustering similar data. To attain scalability for big data, our method is constructed on the novel cloud infrastructure, MapReduce Hadoop. The method is tested by using a novel UCI dataset and the results are compared with an existing approach.

Keywords: Big data · Mapreduce · Privacy preservation · HPSO

1 Introduction

In recent years, Big Data becomes very popular topic to discuss because of the recent advancements in technologies. A huge amount of data consisted of text, images, audio, video and other file types is rapidly increasing and changing. The data is being generated from various sources with un-predictable rate up to trillions of bytes. These data are so called “Big Data”.

Among these massive data, personal privacy sensitive data such as financial transaction records and electronic health records must be taken into account for their security and privacy concerns. In this place, data anonymization plays major role in non-interactive public data sharing process. It refers to hiding identity of sensitive data which makes sure the published data is practically useful for processing (mining) while preserving individuals’ sensitive information.

Normally, there is only one raw data table which includes four types of attributes, namely- identifiers, quasi-identifiers (QID), sensitive attributes (S), and non-sensitive attributes (NS). The identifier attributes are always removed when the data set is published after anonymization. Quasi-identifiers may seem harmless at first glance, but later, the sensitive data can likely be uniquely identified based only on the QIDs.

J. Pan et al. (eds.), *Genetic and Evolutionary Computing*, Advances in Intelligent Systems and Computing 536, DOI 10.1007/978-3-319-48490-7_5

Although anonymization is a popular approach in privacy protection, applying its traditional methods to big data can face with scalability and efficiency challenges. Moreover, while trying to preserve privacy at a certain level, it is also required to hold utility on the other hand. If not, the goal of data distribution will not be achieved.

In this paper, we propose an anonymization approach for big data based on HPSO clustering technique. First, we use HPSO at two phases of clustering to group the data with similar quasi-identifiers. Then, the resulted data groups are anonymized into their general form to achieve privacy. Here, the more similar the data in same group are, the more utility they can retain. To address the scalability issue of big data, the whole approach is built upon MapReduce Hadoop infrastructure.

The remained part of this paper is organized as follows. Section 2 explains some related works that utilize MapReduce for big data privacy preservation. Some preliminaries about big data and HPSO are described at Sect. 3. The detail explanation about our work can be seen at Sect. 4, and its experimental results are at Sect. 5. Section 6 will be conclusion and further implementations.

2 Related Work

Zhang et al. [1] propose a highly scalable MapReduce based median-finding algorithm (MRMondrian) combining the idea of the median of medians and histogram technique. The computation of finding the median of a fixed group can be conducted in a mapper. Then, all the medians are medians are shuffled to one reducer that can find the median of medians.

A privacy model allowing semantic proximity of sensitive values and multiple sensitive attributes is proposed by Zhang et al. [2]. This model combines local recoding and proximity privacy models together to provide an anonymous dataset by means of two-phase clustering approach constructed upon MapReduce framework for scalability. The first phase, t-ancestors clustering, splits an original data set into partitions, so called β clusters that contain similar data records in terms of quasi-identifiers. Then, the proximity-aware agglomerative clustering algorithm locally recodes data partitions in parallel. A proximity-aware distance measure between two data records is defined by combining their distance and proximity index.

The use of K-means clustering for privacy preservation is proposed by Upmanyu et al. [3]. This approach uses Shatter and Merge functions upon which K-means algorithm is run. Each user computes the secret shares of their private data by means of Shatter function, and sends them over to the processing servers. The processing servers then privately collaborate to run the K-means algorithm over the secret shares without reconstructing the actual data. Chinese Remainder Theorem (CRT) is used to reconstruct the secret in Marge function.

Anonymization using Nested Clustering (ANC) for k-anonymity privacy preservation is described at [4]. This approach uses nested clustering and perturbation on each cluster by two phases architecture. During first phase, the original database is clustered efficiently into enough number of sub clusters by grouping and re-clustering repeatedly. The second phase is an anonymization phase in which the numeric values are moved towards the centroid of each of the sub clusters.

Lin et al. [5] proposes a well-known global heuristic search genetic algorithm (GA) based clustering approach for k-anonymization. All chromosomes of the population represent a complete solution to the problem. Each chromosome contains no fewer than k genes, where each gene indicates the index of a record in the data set. A rank-based selection strategy is adopted by sorting all possible pairs of in ascending order, such that a higher-ranked chromosome pair (i.e., two nearby chromosomes) has a higher probability of being selected. Then, the information losses of the two offspring are calculated to determine whether they can replace their parents in the population.

A member of optimization algorithms, bacterial foraging optimization (BFO), is used as clustering approach for l-diversity privacy model is expressed in [6]. This approach modified the chemotaxis step of the BFO algorithm by factorial calculus (FC) to boost the computational performance, and named as FC-BFO.

3 Preliminaries

3.1 Big Data and Its Privacy Models

According to the HACE theorem, big data starts with large-volume, heterogeneous, autonomous sources with distributed and decentralized control, and seeks to explore complex and evolving relationships among data. From IBM website of The Big Data & Analytics Hub, the challenges of big data are discussed in respect of 5 Vs [7] as follows:

1. Volume: huge amount of data; from terabytes to exabytes.
2. Variety: limitless variety of data; text, image, video, audio, social relations, and so on.
3. Veracity: trustworthiness and authenticity of data.
4. Velocity: rapidity of data; batch or streaming.
5. Value: necessity of interdisciplinary cooperation, proportion to veracity.

As big data applications are related to sensitive information, most of the research areas in recent decades tend to emphasize upon the security, especially privacy aspects of this information. The concept of privacy can be expressed as release of information in a controlled way. While maintaining data privacy, privacy preservation techniques must keep in mind the utility for this. The work of [8] said that when the original data is used for measuring utility, we need to measure "utility loss", instead of "utility gain" because privacy should be measured against the trivially-anonymized data whereas utility should be measured using the original data as the baseline.

According to [9], one way of grouping the privacy models is based on the type of attack they are trying to prevent based on two categories: privacy models that counter linkage attacks and probabilistic attacks. The most emphasis models for current research trends are k-anonymity and l-diversity. k-anonymity model is used to prevent from record linkage attacks. A release provides k-anonymity protection if the information for each person contained in the release cannot be distinguished from at least k-1 individuals whose information also appears in the release [10]. Beyond k-anonymity model, l-diversity [11] model is proposed to guarantee privacy against attribute linkage attacks, namely homogeneity attack (positive disclosure) and background knowledge attack (negative disclosure). To attain l-diversity, the values of the sensitive attributes are well-represented in each group.

3.2 MapReduce Model for Big Data

To fulfill the requirements of big data, Google introduced the MapReduce programming model and its open source implementation, Apache Hadoop [12]. MapReduce consists of two different phases; Map phase and Reduce phase. A MapReduce job generally breaks the input data into chunks which are first processed by Map phase in parallel and then by Reduce phase. It works on key-value pairs (key, value) (Fig. 1).

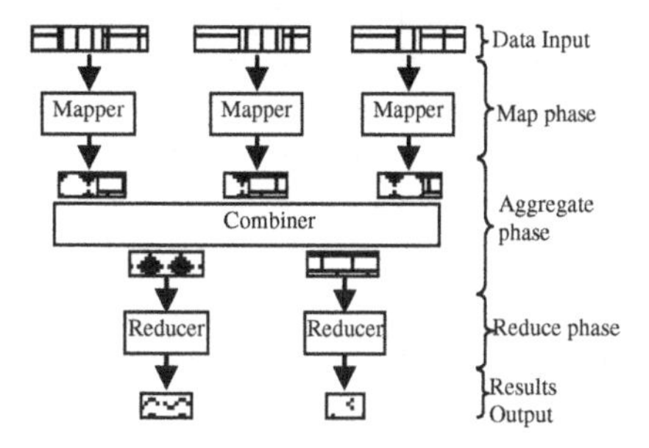

Fig. 1. MapReduce architecture

3.3 Hierarchical Particle Swarm Optimization (HPSO)

PSO is an optimization technique based upon cooperation and coordination among the particles. In PSO, the swarm is initialized to a random solution set. The particles then start moving through the solution space by maintaining a velocity value *V* while keeping track of its best previous position *(pBest)* that achieved so far. Global best *(gBest)* is the best fitness achieved by any of the particles. The fitness of each particle or the whole swarm is evaluated by a fitness function.

HPSO clustering [13] combines both hierarchical clustering and partition clustering techniques and added swarm intelligence to give the novel PSO based hierarchical agglomerative data clustering technique. Initially, the number of particles is kept large for the maximum coverage of the problem space with uniform initialization to the data vector of the data repository using the formula:

$$loc(X(i)) = i * \left[\frac{N}{K} - 1\right] \quad (1)$$

where *loc(X)* represents the location of particle in the repository, *i* is the index of the particle which ranges from 0 to the maximum number of particles *K* and *N* is the total number of data vectors. The Euclidean distance measure is used to find the distance between a particle and a data vector. The velocity of the particle is calculated using the standard PSO velocity update equation:

$$V_i(t+1) = w * V_i(t) + q_1 r_1 (pBest - X_i(t)) + q_2 r_2 (gBest - X_i(t)) + (Y_i(t) - X_i(t)) \quad (2)$$

where $(\mathrm{pBest} - \mathrm{X_i(t)})$ is the cognitive component that controls the movement by keeping track of its best position, $(\mathrm{gBest} - \mathrm{X_i(t)})$ is the social component that indicates the influences of other particles, and $(\mathrm{Y_i(t)} - \mathrm{X_i(t)})$ is self-organizing component takes its inspiration from the other members of that particular cluster. The new position of the particle is based on the previous position of the particle and the velocity of the particle.

The less dense clusters are merged to the nearest well populated cluster. The merging operation takes place once during each generation of the swarm. During a particular generation, a number of iterations are performed to move the particle to the most suitable position, aiming to minimize the intracluster distance. Merging of the particles is based on the average attribute values.

$$X_i = \frac{X_i(nearest) + X_i(loser)}{2} \quad (3)$$

where X_i is the newly formed particle, X_i *(nearest)* is the winner particle and X_i *(loser)* is the less populated particle. This approach starts from a relatively large number of particles and combining down to only one final particle.

When comparing the accuracy, HPSO has improved against hierarchical agglomerative clustering (HAC), and is also better than PSO-clustering and K-means clustering on the experiments done by [14]. Using HPSO to cluster similar data with respect to their quasi-identifiers values can improve the clustering accuracy, an important factor for maintaining utility of that data while preserving privacy. Therefore, we try to construct privacy preservation mechanisms for big data by implementing HPSO clustering on MapReduce infrastructure.

4 HPSO Based Scalable Privacy Preservation for Big Data

The proposed approach consists of two phases. In the first phase, a MapReduce job is done to produce the predefined numbers of intermediate β clusters. Next, a MapReduce job of HPSO lustering is executed on each β cluster (Fig. 2).

Initially, the data are distributed across a number of separated machines (or virtual machines). From these, Particles Initialization step collects the initial data to form the initial particles.

During Phase 1, HPSO clustering is constructed on MapReduce by dividing Map step and Reduce step. In Map step, all data in each partition are assigned to the nearest particle according to their Euclidean distances of quasi-attributes. The Euclidean distance measures in HPSO can only calculate from numerical values. Therefore, all categorical quasi-attributes need to transform into their respective numerical values as in [15]. When the Map step finishes, each particle calculates and updates its fitness,

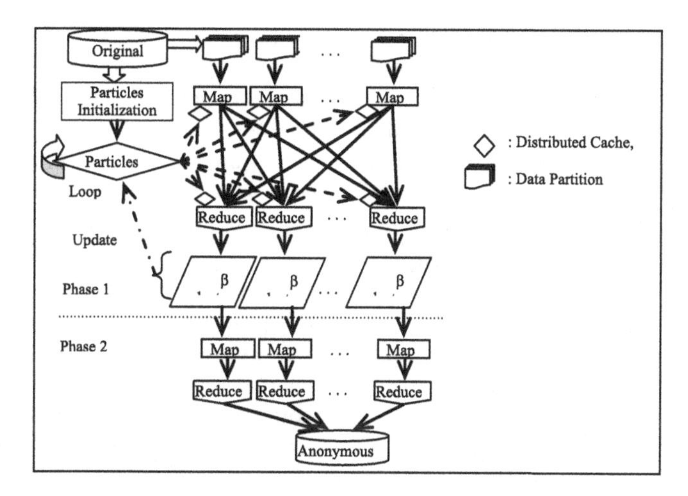

Fig. 2. Process flow of proposed system

velocity and best position in Reduce step. Here, each particle of swarm can be seen as a Reduce of MapReduce structure. After that, a weakest particle, i.e. the particle with minimum number of data members, is searched and consumed by its nearest strong particle. These steps iteratively execute until the number of data members in every particle exceed the predefined k-number to attain k-anonymity. The resulted particles can be seen as the intermediate β-clusters.

Phase 2 is stared with a Map step. This Map step does simply by passing all data members of each β-cluster to its respective Reduce step. The Reduce step in Phase 2 runs the normal HPSO clustering job to produce the small data clusters from the large β-cluster. The results of Phase 2 are data clusters with similar quasi-identifiers values that are then generalized to form their anonymized forms.

5 Experimental Set up and Results

We test out approach on the Hadoop cluster of 4 virtual machines with 1 GB memory and one virtual CPU. A standard UCI Adult dataset is used to test our approach. It consist of 48842 instances with 14 attributes of both categorical, numerical attributes. From these, we uses 9 attributes (6 categorical and 3 numerical) as quasi-identifiers and 3 attributes (1 categorical and 2 numerical) as sensitive data. 10000 data records of Adult dataset are used for testing our approach.

The experiments are done on two purposes. One is to compare the proposed approach with an existing approach of PAC [2], and the other is to analyze the parameters of HPSO and their effects on the utility and privacy of proposed method. To define the information loss, i.e. utility loss, the metric of ILoss [16] is applied on the tested anonymized data.

Figure 3(a) and (b) are the comparisons results of the proposed HPSO based anonymization approach and PAC implemented in Java and tested with data of various sizes, ranging from 300 to 10000 data records. The parameters of PAC are set as k = 10 for anonymity parameter, w_s = 0.5 for weight of proximity, and the number of partition t that varies to makes the sizes of immediate β-clusters in proportion to numbers of data records to 100 for 300, 500, 1000 records, 500 for 3000, 5000, 7000 records, and 1000 for 10000 data records. In the case of HPSO based approach, anonymity parameter k and sizes of immediate β-clusters are the same as PAC, numbers of initial particles p are set p are set from 5 to 30 and reduced to its half in proportion to numbers of data records and sizes of immediate β-clusters.

From Fig. 3(a), we can say that our approach can run on fewer execution times than PAC. This is because our approach does not require taxonomy tree traversal and distances matrix construction as PAC. Instead, we use numerical form of categorical attributes to calculate Euclidean distances between data records and particles positions. Figure 3(b) shows the information loss (iLoss) comparison in normalized form. In this aspect, our approach provides a little higher iLoss than that of PAC. But, it is mostly stable as the size of data increases. From this point, we can observe that our approach can achieve scalability of data, without many changes in iLoss value.

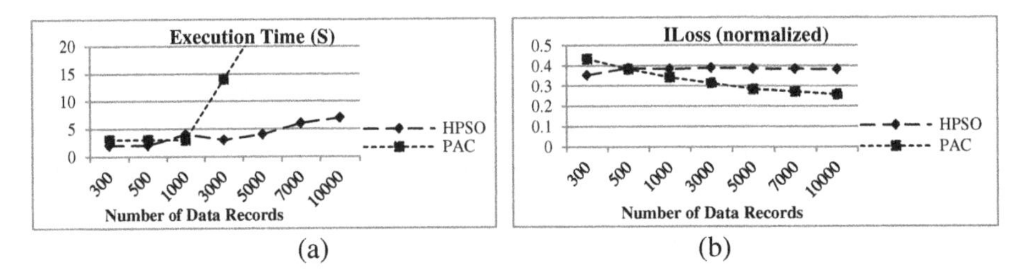

(a) (b)

Fig. 3. Changes of execution time(s) and information loss with respect to number of data records

Figure 4 describes the analysis of the proposed system by means of its execution times, iLoss, and varying anonymity parameter k as 5, 10, and 20. Other parameters are given as the above experiment. By varying anonymity parameter k, the proposed system is analyzed its execution time in seconds and iLoss values. Figure 4(a) shows that the execution of all testing are almost identical because the sizes of β-clusters are the same. According to Fig. 4(b), as anonymity parameter k value increases, iLoss

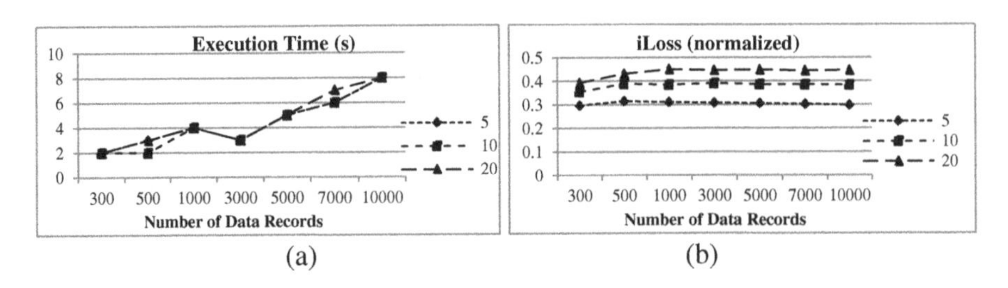

(a) (b)

Fig. 4. Analysis of execution time(s), iLoss and anonymity parameter k

value also increases, and that is also for utility loss. This fact indicates that k value must be as less as it can to attain data utility, without regarding the execution time.

6 Conclusion and Further Extensions

In this paper, we propose an approach for privacy of big data. While maintaining privacy by k-anonymity, utility loss is also kept into account. Another important fact that needed to consider is execution time of anonymization methods. We try to balance all these factors by constructing HPSO clustering based data anonymization on MapReduce Hadoop infrastructure for big data. The execution time weakness of HPSO clustering is reduced by implementing each particle of HPSO as a Reduce of phase 1 in our approach. From the experimental results, we can conclude that our approach can run on fewer execution times than that of PAC with a little higher information loss. The information loss is mostly stable while increasing the data sizes by carefully defining the number of initial particles and anonymity parameter k.

Based on the idea expressed and the experimental results, we plan to construct our approach with large number of virtual machines and test with large data sets to ensure scalability. Next, our trend is to apply role based authentication into our approach to attain both privacy and utility according to the roles of system users.

References

1. Zhang, X., Yang, C., Nepal, S., Liu, C., Dou, W., Chen, J.: A mapreduce based approach of scalable multidimensional anonymization for big data privacy preservation on cloud. In: IEEE Third International Conference on Cloud and Green Computing, pp. 105–112. IEEE Press (2013)
2. Zhang, X., Dou, W., Pei, J., Nepal, S., Yang, C., Liu, C., Chen, J.: Proximity-aware local-recoding anonymization with mapreduce for scalable big data privacy preservation in cloud. IEEE Trans. Computers **64**(8), 2293–2307 (2015)
3. Upmanyu, M., Namboodiri, A.M., Srinathan, K., Jawahar, C.V.: Efficient privacy preserving K-means clustering. In: Chen, H., Chau, M., Li, S.-h., Urs, S., Srinivasa, S., Wang, G. (eds.) PAISI 2010. LNCS, vol. 6122, pp. 154–166. Springer, Heidelberg (2010)
4. Rajalakshmi, V., Mala, G.S.A.: Anonymization based on nested clustering for privacy preservation in data mining. J. Comput. Sci. Eng. (IJCSE) **4**(3), 216–224 (2013)
5. Lin, J.L., Wei, M.C.: Genetic algorithm-based clustering approach for k-anonymization. J. Expert Syst. Appl. **36**, 9784–9792 (2009)
6. Bhaladhare, P.R., Jinwala, D.C.: A clustering approach for the *l*-diversity model in privacy preserving data mining using fractional calculus-bacterial foraging optimization algorithm. J. Adv. Comput. Eng. 2014 (2014)
7. Yin, S., Kaynak, O.: Big data for modern industry: challenges and trends. Proc. the IEEE **103**(2), 143–146 (2015)
8. Li, T., Li, N.: On the tradeoff between privacy and utility in data publishing. In: KDD 2009, Paris, France (2009)

9. Manta, A.: Literature survey on privacy preserving mechanisms for data publishing. M.S. thesis, Department of Intelligence Systems, Delft University of Technology, Delft, Netherland, (2013)
10. Sweeney, L.: k-anonymity: a model for protecting privacy. Int. J. Uncertainty Fuzziness Knowl. Based Syst. **10**(5), 557–570 (2002)
11. Machanavajjhala, A., Gehrke, J., Kifer, D., Venkitasubramaniam, M.: ℓ-diversity: privacy beyond k-anonymity. In: Proceedings of the 22nd International Conference on Data Engineering Workshops (ICDEW 2006) (2006)
12. Ghazi, M.R., Hadoop, D.: MapReduce and HDFS: a developers perspective. J. Procedia Comput. Sci. **48**, 45–50 (2015)
13. Alam, S., Dobbie, G., Riddle, P., Naeem, M.A.: Particle swarm optimization based hierarchical agglomerative clustering. In: IEEE/WIC/ACM International Conference on Web Intelligence and Intelligent Agent Technology, pp. 64–68 (2010)
14. Alam, S., Dobbie, G., Koh, Y.S., Riddle, P., Rehman, S.U.: Research on particle swarm optimization based clustering: a systematic review of literature and techniques. J. Swarm Evol. Comput. **17**, 1–13 (2014)
15. Nouaouria, N., Boukadoum, M.: A particle swarm optimization approach to mixed attribute data-set classification. In: IEEE Symposium on Swarm Intelligence (SIS). IEEE (2011)
16. Xiao, X., Tao, Y.: Personalized privacy preservation. In: ACM SIGMOD International Conference on Management of Data (SIGMOD 2006), pp. 229–240 (2006)

Observation of Unattended or Removed Object in Public Area for Security Monitoring System

Baby Htun(✉) and Myint Myint Sein

University of Computer Studies, Yangon, Myanmar
babyhtun@gmail.com, myintucsy@gmail.com

Abstract. Observation of stationary or unattended objects such as bags, luggage has covered as precaution from some terrorist attacks carrying some explosive things left behind in public areas. One of the important securities monitoring systems is video surveillance system for crowded environmental areas and daily caring and monitoring system. In the proposed system, the unattended object observation is developed for monitoring system. The system input applies the recorded data video files, in order to remove outdoor lighting detection noises controlling and modifying image intensity value before Otsu's method in pre-processing and then convert frame sequences for preprocessing. The system preprocesses to search and detect indoor, outdoor, day lighting in fewer errors by controlling the brightness intensity value of images. The color image processing and morphological operation are performed to observe the object. And then, the system can calculate object statistics using the blob analysis.

Keywords: Observation object · Removed object · Blob analysis · Auto threshold · Outdoor observation

1 Introduction

Several bombing attacks in public areas have happened in recent times. Although we cannot stop these terrorisms, we can avoid the destruction caused by terrorism. Most of the time the destruction is performed by using the unattended objects like a timer bomb, explosives, etc. A smart surveillance system can automatically detect unattended objects in public places. It has been used in many places such as bus stops, train stations, crowded places, and popular buildings, warehouses, shopping malls, common sidewalks or airports. Conventional unattended object observation methods can be separated into two approaches [7, 8]. They are tracking-based methodology and observation-based approach. In the tracking-based methods, tracking foreground objects, searching for candidate static objects and observation to obtain the candidate static objects are involved. In the observation-based approach, using background subtraction and foreground analysis are used to detect the left objects. In this system, unattended object observation is proposed and the system can search and detect unattended object in outdoor recorded files. The system will alert to people responsible for the role such as security guards or staff. It can obtain many benefits in monitoring terrorisms for people and prevent many lives from terrorism bombing attacks.

J. Pan et al. (eds.), *Genetic and Evolutionary Computing*, Advances in Intelligent Systems and Computing 536, DOI 10.1007/978-3-319-48490-7_6

The system avoids and removes dangerous or unknown objects in public areas and captures an alarm for the security guards in suspicious time. It can be efficient in video surveillance system using sharp and brightness image. The system is simple and computationally less intensive as it avoids the use of expensive filters while achieving better detection results. In this paper, Sect. 2 is related works of some references, Sect. 3 is described about proposed system and methods, Sect. 4 is results and discussion and sect. 5 is system conclusion.

2 Related Works

Three steps of data processing are object extracting, classifying object and detection object. In object extracting, involving a background subtraction algorithm which dynamically updates two sets of background. The extracted objects are classified as static or dynamic objects. It can be only interested in ROI (region of interest). The detection of abandoned or removed object achieves by contrast the color information among foreground image, current and original background. Color histograms are calculated for each of three regions using luminance channel YUV. It relies on tracking information to detect drop-off events. This system produced larger errors under bright lighting conditions [1].

Gaussian Mixture Models used for background subtraction are employed to detect static foreground regions without extra computation cost. The test and evaluation demonstrates this method is efficient to run in real time while being robust to quick lighting changes and occlusions in complex environments. The history of background objects in the scene is kept to make the matching algorithm robust to lighting changes in complex videos. The proposed algorithm is being used in real time – IBM Smart Surveillance Solutions. The wait time before triggering alarm for different lighting conditions is set as five minutes [5].

It's used contrast enhancement to improve perceptibility of objects in the scene by enhancing the brightness difference between objects and their background. Next step is to use median filter to reduce noise in the images. Each successive image is subtracted from the background image, and the difference image is thresholded to be converted into a binary image. Morphological operations are affecting the form, structure or shape of an object, applied on binary images. Using Blob analysis, object detection and tracking process are occurred. The further extension is that efforts can be put in a direction to remove limitation given by static thresholds; moreover, the system can be made more robust to work under daylight or outdoor conditions [4]. Object tracking and monitoring implement via the use of distributed wireless sensor networks. It proposed image and sound object tracking. The core algorithm presented detection of change algorithm for detecting a change from a given memory less and stationary process. It extended the algorithm to detect multiple repeated changes among a set of processes and proved asymptotic optimality. The algorithm has been used in various networks applications. A core algorithm presented object identification from images, audio object tracking and distributed network traffic monitoring. It is highly effective and robust, while its applications are also numerous [9, 10, 12].

3 Proposed System

The system can search and detect indoor, outdoor, day lighting by controlling brightness intensity of images in system preprocessing. Using Auto Threshold method in Video Processing, any noise removing methods and shadow removal are not required to reduce noise detection for background segmentation. Using Morphological Processing, the system can clearly remove structures smaller than the structuring element in a binary image.

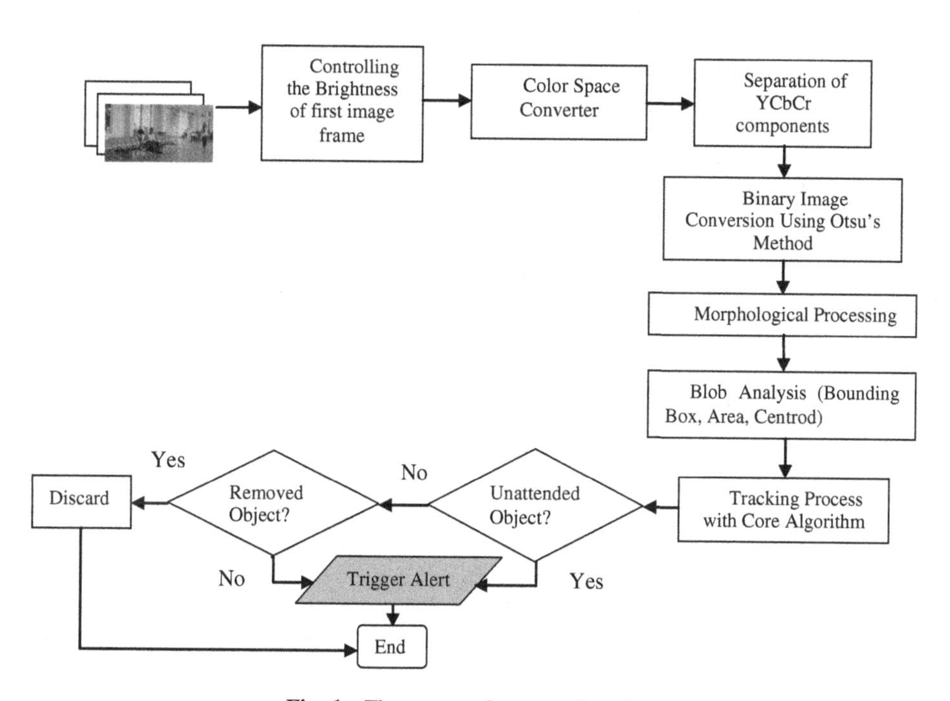

Fig. 1. The proposed system flowchart

Morphological post processing operations such as dilation and erosion are performed to reduce the effects of noise and enhance the detected regions. It provides a comprehensive solution, which can detect the status of an object, whether unattended or removed. The system receives quickly detection using Blob analysis and Core Algorithm on object statistics in 2 s. The system can detect and track several objects in the whole scene of the camera.

In Fig. 1, the recorded movie files are system input and these are converted from 'mov' to 'avi' format files. The first frame of input data file is controlled in preprocessing for recorded video files. If there is for real time files, this modified step is calculated in processing loop step. So, execution time for real time files will take a little long more than recorded files.

The following steps of Sect. 3.1 are controlled and modified in order to initialize threshold for Otsu's method before color space conversion step.

3.1 Calculating Threshold Value

1. Input Video
2. Initialize OtsuThresh (T_o)
3. Read First Frame from Video
4. Compute gray level value by using *graythresh (RGB_firstframe).*
5. Compute mean intensity value of first frame, using *mean2 (graylevel_image).*
   Mean2: Sum of array elements is divided by Number of array elements
6. If mean_intensity_value < T_M (T_M < (Binary White Value/2))
   #For More Darkness Images (Such as Indoor).
   #Darkness value is nearer black value (zero) than white binary value (one).
   Initialize otsu_threshold = T_o ($T_o > 1$)
   Initialize removed_Obj_threshold = T_R ($0 < T_R < 1$)
7. Else
   #For More Brightness Images (Such as Outdoor).
   Initialize otsu_threshold = T_o - (mean_intensity_value) + x
   ($T_o > 1$)
   Where x is constant less than mean intensity value.
   Initialize removed_Obj_threshold = mean_intensity_value + x
   (x< mean_intensity_value)
8. Go To Color Conversion Step.

The system converts the RGB image to Y′CbCr image to reduce color space between images. It supports sharp and brightness image views. The components of Y′ CbCr are Y: Luminance, Cb: Chrominance-Blue and Cr: Chrominance-Red. Luminance is very similar to the grayscale version of the original image. The system gets intensity image from separating and applying three components of YCbCr of an image. For background segmentation, intensity image is converted to binary image by using Otsu's method. In this step, foreground subtraction can get more structure elements by combining morphological operation. Moreover, the system is no need to use any other noise or shadow removal methods. Blob analysis is used for foreground subtraction from complex background. It has passed several parameters to object tracking process.

The object tracking method is calculated the fractions of some parameters and start 'persistent track'. And then if the core algorithm see and catch the motion process it scans all incoming blobs and quantize the values that will be tracking. All the objects and subjects are tracked by bounding boxes with green color. This algorithm returns the bounding box with set of age blobs and then if the unattended object is found in the scene the alarm is set up showing a red filled rectangle. If the subject removed the unattended object, the system will show a message box 'The removed object is found'.

3.2 Color Conversion

To convert the RGB image to Y′CbCr image to reduce color space between images. Y′ CbCr is a family of color image pipeline in video and digital photography system. Y′ is the luma component and C_B and C_R are the blue difference and red difference chroma components. It supports sharp and brightness image views. Y′CbCr signals are created from the corresponding gamma adjusted RGB (red, green, blue) source using two defined constants K_B and K_R as follows [2, 3, 6]:

$$Y' = K_R . R' + (1 - K_R - K_B) . G' + K_B . B'$$

$$P_B = \frac{1}{2} \cdot \frac{B' - Y'}{1 - K_B}$$

$$P_R = \frac{1}{2} \cdot \frac{R' - Y'}{1 - K_R}$$

Where K_B and K_R are ordinarily derived from the definition of the corresponding RGB space. The prime (′) symbols mean gamma correction is being used. The resulting luma (Y) value will have a nominal range from 0 to 1, and the chroma (P_B and P_R) value will have a nominal range from −0.5 to +0.5. Luminance is very similar to the grayscale version of the original image.

3.3 Otsu's Method

The system gets intensity image from separating and applying three components of YCbCr of an image. Using Otsu's method, it determines the threshold by splitting the histogram of the input image to minimize the variance for each of the pixel groups. The object multiplies this scalar value with the threshold value computed by Otsu's method. The result becomes the new threshold value. In this system threshold Scale Factor is 1.3.

1. Compute histogram and probabilities of each intensity level.
2. Set up initial w_i (0) and μ_i (0).
3. Step through all possible thresholds t = 1…maximum intensity
   1. Update w_i and μ_i
   2. Compute $\sigma_b^2(t)$
4. Desired threshold corresponds to the maximum $\sigma_b^2(t)$.
5. Can compute two maxima (and two corresponding thresholds). $\sigma_{b1}^2(t)$ is the greater max and $\sigma_{b2}^2(t)$ is the greater or equal maximum.
6. $Desired\ Threshold = \frac{threshold_1 + threshold_2}{2}$

3.4 Morphological Processing

It is used to fill in small gaps in the detected objects. The two basic operations, dilation and erosion, can be combined into more complex sequences. The most useful of these for morphological filtering are called opening and closing. Closing consists of a dilation followed by erosion and can be used to fill in holes and small gaps. Closing can be used to enhance binary images of objects obtained from thresholding. The closing operation has the effect of filling in holes and closing gaps.

$$A \cdot B = (A \oplus B) \Theta B$$

3.5 Blob Analysis

In computer vision, blob detection methods are aimed at detecting regions in a digital image that differ in properties, such as brightness or color, compared to surrounding regions. Informally, a blob is a region of an image in which some properties are constant or approximately constant; all the points in a blob can be considered in some sense to be similar to each other. The Blob Analysis is blocked to calculate statistics for labeled regions in a binary image. The block returns quantities such as the centroid, bounding box, label matrix, and blob count. One of the first and also most common blob detectors is based on the Laplacian of the Gaussian (LoG). Given an input image $f(x,y)$, this image is convolved by a Gaussian kernel [11, 13].

$$g(x, y, t) = \frac{1}{2\pi t^2} e^{-\frac{x^2+y^2}{2t^2}}$$

At a certain scale t to give a scale space representation $L(x,y;\, t) = g(x,y,t) * f(x,y)$.

3.6 Tracking with Core Algorithm

A lot of parameters from blob analysis are input to the tracking process. It turns some inputs from percentages to fractions. It declares integer valued and persistent elements of a track. The core algorithm is generally used in network sensor, network game, network traffic monitoring and robotic sensor for tracking. In core algorithm, if blobs were found, then process the tracks, scan through all the incoming blobs and quantize the values that it is being tracked. If an existing track, update it and create new track find first unused track, at last fill track information. It determines which objects are stationary and then should trigger alarm [9].

4 Result and Discussion

The following figures Fig. 2 are resulted from using simple fixed threshold for Otsu's Method in outdoor day light. The proposed system can remove these lighting errors in Fig. 2 initializing threshold for Otsu's method using controlled image intensity value. The system cannot be observed when the unattended objects have not very different intensity values with background object. And some similar color objects are not detected by this system because they have nearly same color and intensity values of

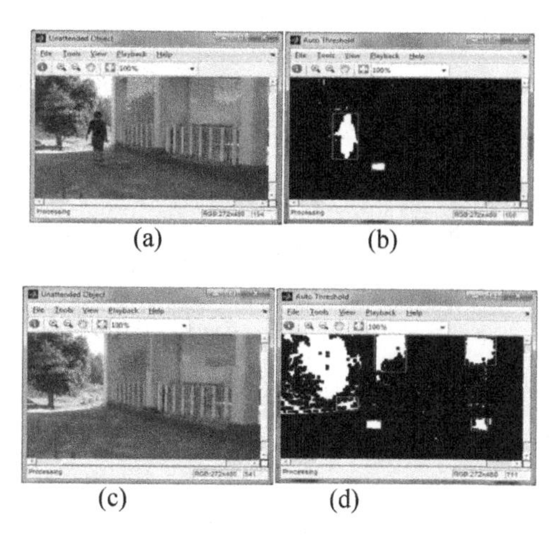

Fig. 2. Results of Using Fixed Simple Otsu's Method (a) Outdoor Image (b) Threshold Image (c) Outdoor Image with errors (d) Threshold Image with Errors

each other. Sometimes mirrors have many reflections in day times. Very tiny objects such as phone, battery cannot be observed by this system.

In Table 1 the number of video clips for outdoor is 8 and the other 42 video files are indoor and number of events is 10 (# of Events: no of events occurred in the test sequence, # of True Positive: number of correct observation, #of False Positive: number of incorrect observation, #False Negative: number of objects remained unobserved). The proposed controlling image intensity value has reduced the number of the incorrect detection.

However, in the some outdoor video files, the background intensity is nearly black (intensity value = zero). So, the foreground object intensity value cannot appear over its background and the foreground value cannot space upon the background value. So, there are some removed object message errors.

The total duration time of this system is following.

- Frame rate per second = 25 frames per second
- Total number of video files = 50
- Total number of duration time in second = 7450 s
- Total number of duration time in minute = 124.1667 min
- Total number of duration time in hours = 2.0694 h
- Total number of frames of input data video files

7450 * 25 = 186,250 image frames

So, the system is tested and observed on 186,250 image frames.

Table 1. Comparing Results of Simple Otsu and Controlling Brightness for Outdoor Light

	Using simple fixed otsu thresholding	Using controlling intensity value
# of Video clips	8	8
# of Events	10	10
# of True positive	8	10
# of False positive	6	0
# of False positive video files	4	0
# of False negative	0	0
# of Removed object error	2	2

5 Conclusion

In security monitoring system, it can save money, lives, destruction and worries for city life. It can prevent from dangerous terrorisms attacks. A computationally efficient and robust method to observe unattended or removed objects in public areas is approved. Background subtraction and foreground analysis are evaluated efficiently. Due to its simplicity the computational cost is kept in low and no training steps are required. Finally, the system can discriminate effectively between unattended or removed by using simple proposed methods. The reliability of proposed system can also be used in public transportation areas indoor or outdoor.

References

1. Antic, B., Ommer, B.: Video parsing for abnormality detection. In: Proceedings of the 13th International Conference on Computer Vision (ICCV 2011). IEEE (2011)
2. Bangare, P.S., Uke, N.J., Bangare, S.L.: Implementation of abandoned object detection in real time environment. Int. J. Comput. Appl. **57**(12), November 2012. (0975 – 8887)
3. Bangare, P.S., Uke, N.J., Bangare, S.L.: An approach for detecting abandoned object from real time video. Int. J. Eng. Res. Appl. (IJERA) **2**(3), 2646–2649 (2012). ISSN: 2248-9622
4. Bayona, A., SanMiguel, J.C., Martínez, J.M.: Comparative evaluation of stationary foreground object detection algorithms based on background subtraction techniques. In: 2009 Advanced Video and Signal Based Surveillance. IEEE (2009). 978-0-7695-3718-4/09. doi:10.1109/AVSS.2009.35
5. Bhargava, M., Chen, C.C., Ryoo, M.S.: Detection of object abandonment using temporal logic. Mach. Vis. Appl. **20**, 271–281 (2009). doi:10.1007/s00138-008-0181-8. Department of Electrical and Computer Engineering, Computer and Vision Research Center
6. Borkar, A., Nagmode, M.S., Pimplaskar, D.: Real time abandoned bag detection using OpenCV. Int. J. Sci. Eng. Res. **4**(11), 660 (2013). ISSN 2229-5518

7. Chitra, M., Geetha, M.K., Menaka, L.: Occlusion and abandoned object detection for surveillance applications. Int. J. Comput. Appl. Technol. Res. **2**(6), 708–713 (2013). ISSN:2319–8656
8. Collazos, A., Fernández-López, D., Montemayor, A.S., Pantrigo, J.J., Delgado, M.L.: Abandoned object detection on controlled scenes using kinect. In: Álvarez Sánchez, J.R., Paz López, F., Toledo Moreo, F., Ferrández Vicente, J.M. (eds.) IWINAC 2013, Part II. LNCS, vol. 7931, pp. 169–178. Springer, Heidelberg (2013)
9. Etellisi, E.A., Burrell, A.T., Papantoni-Kazakos, P.: A core algorithm for object tracking and monitoring via distributed wireless sensor networks. Int. J. Sens. Netw. Data Commun. **1** (2012)
10. Friedman, N., Russell, S.: Image segmentation in video sequences: a probabilistic approach. In: Computer Science Division (1998)
11. Carvajal-González, J., Álvarez-Meza, A., Castellanos-Domínguez, G.: Feature selection by relevance analysis for abandoned object classification. In: Alvarez, L., Mejail, M., Gomez, L., Jacobo, J. (eds.) CIARP 2012. LNCS, vol. 7441, pp. 837–844. Springer, Heidelberg (2012)
12. Hedayati, M., Zaki, W.M.D.W., Hussain, A.: A qualitative and quantitative comparison of real-time background subtraction algorithms for video surveillance applications. J. Comput. Inf. Syst. **8**, 493–505 (2012)
13. Joglekar, U.A., Awari, S.B., Deshmukh, S.B., Kadam, D.M., Awari, R.B.: An abandoned object detection system using background segmentation. Int. J. Eng. Res. Technol. (IJERT), **3**(1), January 2014. ISSN: 2278-0181

Compact Particle Swarm Optimization for Optimal Location of Base Station in Wireless Sensor Network

Jeng-Shyang Pan[1], Thi-Kien Dao[2], Trong-The Nguyen[3(✉)], and Tien-Szu Pan[2]

[1] School of Information Science and Engineering, Fujian University of Technology, Fuzhou, China
[2] Department of Electronics Engineering, National Kaohsiung University of Applied Sciences, Kaohsiung, Taiwan
[3] Department of Information Technology, Hai Phong Private University, Haiphong, Vietnam
vnthe@hpu.edu.vn

Abstract. The computational requirements even in the limited resources of the hardware devices whose small memory size or low price could be addressed by compact optimization methods. In this paper, a compact particle swarm optimization (cPSO) for the base station locations optimization is proposed for wireless sensor networks (WSN). A probabilistic representation random of the collection behavior of swarms is inspired to employ for this proposed algorithm. The real population is replaced with the probability vector updated based on single competition. These lead to the entire algorithm functioning applying a modest memory usage. The experiments to solve the problem of locating the base station in WSN compared with the genetic algorithm (GA) method and the particle swarm optimization (PSO) method show that the proposed method can provide the effective way of using a modest memory.

Keywords: Compact optimization algorithm · Swarm intelligence · Wireless sensor networks

1 Introduction

Wireless sensor networks (WSN) is one of the example types of these devices. WSNs are networks of small, battery-powered, memory-constraint devices named sensor nodes, which have the capability of wireless communication over a restricted area [1]. Due to memory and power constraints, they need to be well arranged to build a fully functional network. The mentioned problem is not enough memory of computational devices to store a population composed of numerous candidate solutions of the computational intelligence algorithms.

The compact algorithm is a promising answer to these problems. An effective compromise used in the compact algorithm is to present some advantages of population-based algorithms but the memory is not required for storing an actual population of solutions. Compact algorithm simulates the behavior of population-based

J. Pan et al. (eds.), *Genetic and Evolutionary Computing*, Advances in Intelligent Systems and Computing 536, DOI 10.1007/978-3-319-48490-7_7

algorithms by employing, instead of a population of solutions with its probabilistic representation. In this way, the space for storing the number of parameters in the memory is smaller. Thus, a run of these algorithms requires less capacious memory devices compared to their correspondent population-based structures.

The location of the base station (BS) is an impacting effective factor to contribute the saving and balancing power consumption of the coordinator nodes (CNs) in heterogeneous WSNs. Figure 1 is an example of WSN model as heterogeneous architecture [2]. The optimal BS location is how to determine the relative position of BS with CNs for distributing balanced energy consumption among CNs while remaining the quality of service network.

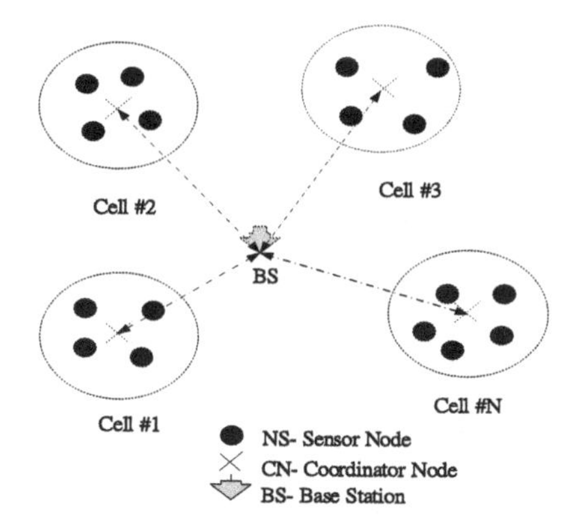

Fig. 1. The wireless sensor network model

In this paper, a compact algorithm based on Particle swarm optimization (PSO) applies to maximize the heterogeneous WSN lifetime through the BS location optimization. A fitness function according to energy consumption per unit time is mathematical modeled based on heterogeneous WSN model.

2 Heterogeneous WSN Model

The heterogeneous sensor networks are typically WSN model that is more practical, and better network performance. The heterogeneous is found in two-tiered WSN that consists of a set of small sensor nodes (SN), a set of coordinator nodes (CN) and at least one base station (BS). The CNs and SNs made up cells or clusters, and in each cluster, there are many SNs and one CN. A small sensor, once triggered by the internal timer or some external signals, starts to capture and encode the environmental phenomena (such as temperature, moisture, motion measure, etc.,) and broadcast the data directly to all CNs within its transmission range and to certain CNs via the relay of some other neighboring sensors. When receiving the raw data from SNs in its cluster, a CN might create an application specific local view for the whole cluster by exploring some correlations among the data sent by different SNs. In the meanwhile, some data fusion can be conducted by CNs to alleviate the redundancy in the raw data from SNs. After a CN creates a local view of the data, it then forwards the information to a BS that generates a comprehensive global view for the entire WSN. Notice that here a CN can communicate directly with a BS, or optionally, CNs can be involved in inter-CN relaying if such activities are needed and applicable. The heterogeneous CNs might have different data transmission rates. If a single SN ran out of energy, its corresponding CN might still have the capability to collect enough information. However, if a CN ran out of energy, the information in its coverage range would be completely lost, which was dangerous to the whole system. Let d be the Euclidean distance from a CN to a BS, and r be the data transmission rate. The energy consumption per unit time can be calculated as following:

$$p(r, d) = r(\alpha_1 + \alpha_2 d^n) \tag{1}$$

where α_1 is a distance-independent parameter and α_2 is a distance-dependent parameter. The energy consumption thus relates to Euclidean distances and data transmission rates. It is assumed each CN has the same α_1, α_2. The lifetime of CN can be calculated as following:

$$l = \frac{e}{p(r, d)} \tag{2}$$

where l is a lifetime of application node, e is the initial energy of coordinator node. For homogenous CNs, the data transmission rate is constant so the center of the minimal circle covering all the CNs is the optimal BS location (with the maximum lifetime). For heterogeneous CNs, the data transmission rates are different. The average rate over a period of time is given as:

$$r = \frac{\int_{T_0}^{T+T_0} r_i(t)}{T} \tag{3}$$

where $r_i(t)$ is a function over the time t, T is a period of time, e.g., one hour or one day or one week, most often it is a constant. The optimal BS location is actually determined by a few critical CNs that run out of energy first. The network lifetime is equivalent to maximize $(\min\{l_i\})$.

$$L = Max(\min\{l_i = e_i/(r_i(\alpha_1 + \alpha_2 d_i^n))\}) \tag{4}$$

where d_i is the distance from CN_i to BS.

3 Compact Algorithm for BS Location Optimization

This section presents the compact algorithm based on the frame of Particle swarm optimization (cPSO) for solving the base station location issue in WSN. We first review briefly PSO and then present the compact method processing for the constrained optimization problem of BS location in WSNs.

3.1 Particle Swarms Optimization

PSO is a powerful evolutionary computational algorithm introduced by Kennedy and Eberhart [3]. The updating policy causes the particle swarm to move toward a region with a higher object value because the particle moves according to an adjusted velocity, which is based on their experience. The updated velocity of particles of the PSO can be expressed as follows.

$$V_i^{t+1} = W^t \times V_i^t + C_1 \times r_1 (x_{pbest_i^t} - X_i^t) + C_2 \times r_2 (x_{gbest^t} - x_i^t) \quad (5)$$

where V_i^t and x_i^t are the velocity and position of the $i-th$ particle at the $t-th$ iteration respectively, $x_{pbest_i^t}$ and x_{gbest^t} are the best previous position of the $i-th$ particle at the $t-th$ iteration and the best position amongst all the particles from the first iteration to the $t-th$ iteration, W^t is the inertia weight at the $t-th$ iteration used to balance the global exploration and local exploitation, C_1 and C_2 are factors of the speed control, and r_1 and r_2 are random variables such that $0 \le r_1, r_2 \le 1$ respectively. The updating position of particles would be presented as following.

$$x_i^{t+1} = x_i^t + V_i^{t+1}, i = 0, 1, ..N - 1 \quad (6)$$

where N is the number of particles or population size, with $Vmax$ is the maximum velocity, $-Vmax \le Vt + 1 \le Vmax$. PSO performs searching via a swarm of particles that update Eqs. (1) and (2) over iterations. The optimal solution is searched by moving each particle in the direction to its previously best ($x_{pbest_i^t}$) position and the global best (x_{gbest^t}) position in the swarm.

3.2 Compact PSO for BS Location

The method of compact algorithm tries to simulate very similarly to searching operators of the population-based methods. A probability distribution is used generating new candidate solutions that being iteratively biased toward an optimal solution. The information-processing objective of the compact algorithm is to simulate the behavior of searching food based on the frame of PSO with a much smaller memory. The actual population base of PSO will be described as a virtual population by encoding within a data structure, namely Perturbation Vector (PV). PV is the probabilistic model of a population of solutions [4]. Candidate solutions are probabilistically generated from the vector, and the competing components toward to the better solutions are used to change the probabilities in the vector.

The virtual population can be configured by considering probability density functions (PDFs) [5]. A probabilistic model of the actual population is the Gaussian distribution that is adapted by truncating PDF. The arrange of the probability density area from -1 to $+1$ is cropped from its domain of $(-\infty, \infty)$, and then normalized to keep its area equal to one [6]. The distribution of the individual in the hypothetical swarms must be based on a PDF. Gaussian PDF with mean μ and standard deviation σ is used to assume a distribution each particle of swarms. The generated trial solutions are allocated in boundary constraints. If the variables of the algorithm are normalized as probability generated trials in interval arrange of −1 to +1, for $j = 1, 2, ..d$, where d is the dimension of the problem, the memory of storing particles will be reduced significantly in the boundaries $[lb_j, ub_j]$. The parameters assume without loss of generality and the design parameters of the PDF for each variable have to be normalized in each search interval arrange of (−1, +1). Therefore, *PV* is a vector of $m \times 2$ of matrix the for specifying the two parameters of the PDF of each design variable as defined.

$$PV^t = [\mu^t, \sigma^t] \tag{7}$$

where μ and σ are mean and standard deviation values of a truncated Gaussian (PDF) within the interval of (−1, +1) respectively. The amplitude of the PDF is normalized by keeping its area equal to 1. The apex t is time steps. The initialization of the virtual population is generated for each design variable i, e.g. μ_i^1 is set to *0* and σ_i^1 is set to k, where k is set as a large positive constant (e.g. referencing k set to *10*). The PDF height normalization is obtained approximately sufficient in well the uniform distribution with a wide shape. A generated candidate solution x_i is produced from $PV(\mu_i, \sigma_i)$. The value of mean μ and standard deviation σ in *PV* are associated equation of a truncated Gaussian PDF.

$$P_i(\mathrm{x}) = \frac{\sqrt{\frac{2}{\pi}} \times \exp(-\frac{(x-\mu_i)^2}{2\sigma_i^2})}{\sigma_i(\mathrm{erf}\left(\frac{\mu_i+1}{\sqrt{2}\sigma_i}\right) - \mathrm{erf}\left(\frac{\mu_i-1}{\sqrt{2}\sigma_i}\right))} \tag{8}$$

where $P_i(\mathrm{x})$ is the value of the PDF corresponding to variable x_i, and *erf* is the error function [7]. The slot memory only needs to store vectors μ_i and σ_i on memory whenever the probability density function is trigged. The PDF in Eq. (8) is then used to compute the corresponding Cumulative Distribution Function (CDF). CDF describes the probability that a real-valued random variable X with a given probability distribution to be found at a value less than or equal to x_i. The newly calculated value of x_i is inversed to from CDF. Two designed variables are used to compete for finding out who is *winner* or *loser*. Toward to the better area in searching space, the winner solution biases the *PV* in the comparison between two design variables for individuals of the swarm. These individuals are also sampled from *PV*. The winner or loser vectors are according to the evaluation of fitness function value that is better or worst. The new solution is then evaluated in Eq. (6) and compared against $x_{pbest_i^t}$ and x_{gbest^t} to determine who is the winning and losing individual. The competing algorithm for winner and loser is based on the elements μ_i^{t+1} and σ^{t+1} of the *PV* are updated to the new solution based on the differential iterative.

$$\mu_i^{t+1} = \mu_i^t + \frac{1}{N_p}(winner_i - loser_i) \tag{9}$$

where *Np* is virtual population size, which is only parameter typical of the compact algorithm and it does not strictly correspond to the population size as in a population-based algorithm. The virtual population size, in real-valued compact optimization, is a parameter which biases the convergence speed of the algorithm. A new candidate solution is obtained based on a comparison between it and the elite. Regarding σ values, the update rule of each element is given as.

$$\sigma_i^{t+1} = \sqrt{(\sigma_i^t)^2 + (\mu_i^t)^2 - (\mu_i^{t+1})^2 + \frac{1}{N_p}(winner_i^2 - loser_i^2)} \tag{10}$$

The processing steps of the compact method are simulated as the behavior of population-based algorithms by sampling probabilistic models.

4 Simulation Results

Wireless sensor model: The heterogeneous WSN of network model could be described as following: S_M a set of SNs, V_N a set of CNs, and at least one BS are randomly distributed in desired areas. The objective function for optimal BS locations in this type of the WSNs is constructed based on the residual energy coordinator nodes, sensor nodes and, contention. Data transmission rate r is considered in the nodes communication. A virtual directed graph is constructed on CNs and iteratively moving SNs belong different cells. If an existing link such as the edge $\overrightarrow{v_i v_k}$ from CN v_i to v_k if there is a sensor s_j that can be moved from the cell of v_i to the cell of v_k. The edge represents a real number as a weight of the contention in the graph. The weight of a CN v_i for assignment x is defined as:

$$w_i(x) = \frac{p_i(\sum_{s_j \varepsilon S_M} r_j \times x_{i,j})}{P_i} \tag{11}$$

where P_i is the initial onboard energy of CN_i, p_i is energy consumption function, r_j is data transmission rate of SN_j and $x_{i,j}$ is energy consumption of the sensor node $i-th$ of the cell $j-th$. In practical in WSNs, several different kinds of sensors cooperate together to fulfill some certain goals. The lifetime of the heterogeneous network is defined as

$$L = \text{Min}(\max(w_i(x))) \tag{12}$$

Subject to constraints

$$x_{i,j} = 0, \forall v_i, \forall s_j \notin N(v_i); \tag{13}$$

$$x_{i,j} \in \{0,1\}, \forall s_j, \forall v_i; \tag{14}$$

$$\sum\nolimits_{v_i} x_{i,j} = 1, \forall s_j; \tag{15}$$

$$\sum\nolimits_{s_j \in S_M} r_j \cdot x_{i,j} \leq k_i, \forall v_i \tag{16}$$

where $k_i = p_i^{-1}(P_i \cdot T)$ let x^{min} be the solution to Eq. (12) and T^{min} be the minimum weight of the CNs. The special case when application nodes are homogeneous; its lifetime is equivalent to minimizing the maximum $\sum_{s_j \in S_M} r_j \cdot x_{i,j}$ subject to constraints (16). If every CN vi satisfies that $N(v_i) = (S_M - v_i)$ then the problem becomes the traditional job scheduling problem [10], which is known to be NP-Hard. Since solving Eq. (12) is NP-hard, the optimal solution is presented by an algorithm approximating by borrowing some ideas from job scheduling [10]. Fitness function is formed by Eq. (12) as following:

$$Fitness(i) = \sum\nolimits_{s_j \in S_M} r_j \cdot x_{i,j} \tag{17}$$

The main steps of the optimization BS positions in WSN are described as follows: (i) A network space is modeled including a solution as S sets to $(s_1, s_2, \ldots, s_N)$, where N is number of sensor nodes. The coordinates of s_i positions is set to (x_i, y_i), where $i = 1, 2, ..N$. Setting the data transmission rate to the network coverage. (ii) The solution is parsing as mapping search agents to a model of the related objective function. (iii) The proposed cPSO run over the fitness function to find optimal deployment WSN of the above model.

Experimental Results: A deployment network is setting in areas to 200 $\times$ 200 m. The number of sensor nodes is set to 200. The remaining energy starts at energy initial for all sensor nodes set to 2.0 J. The energy electronics circuit sets to 50 nJ/bit. The average dissipated energy for each iteration sets to 0.05 pJ/bit; The initial link of the edge $x(i,j)$ randomized from 0 to 1. The parameters setting for PBA are the initial inertia weight $W = (0.9 - 07 * rand)$, coefficients of learning factors c_1, c_2 are set to 2.0. The object functions are valuated fully iterations of 300 repeated by 10 runs in different with random seeds. Simulation results are compared with those similar obtained from the genetic algorithm (GA) [11] and PSO methods for the location optimization of base station in WSN.

Table 1 compares the performance of the proposed cPSo for optimal BS locations in heterogeneous WSNs, with the GA, and PSO methods in terms of quality performance evaluation and speed. It is clearly seen that the average cases of fitness functions in cPSO method are as fast as convergence that original PSO method cases. The average of the obtained best optimization from the proposed method for objective

Table 1. The comparison the proposed cPSO method with the GA method, and the PSO-method in terms of quality performance evaluation and time consumption

Methods	Population size	Avg. obtained values	Compared deviation	Time consumption (m)
GA [11]	40	23.15625	16 %	1.1096
PSO	40	20.269375	2 %	1.0106
cPSO	1	19.806875	0 %	0.8105

functions evaluation in 25 runs are more accuracy optimization than that obtained from the GA, and the PSO methods at 16 % and 2 % respectively.

Moreover, the total time consuming of the proposed cSPO method in 10 runs is quite fast with only 0.8105 m taken. The reason for this fastest speed is the proposed method of cPSO only uses a slot of memory for its solution in the optimal process while the GA and PSO use 40 slots of memory for their solutions. Figure 2 illustrates the comparison of average obtained curves and bars of the proposed cPSO method with PSO-method and GA-method for the objective function.

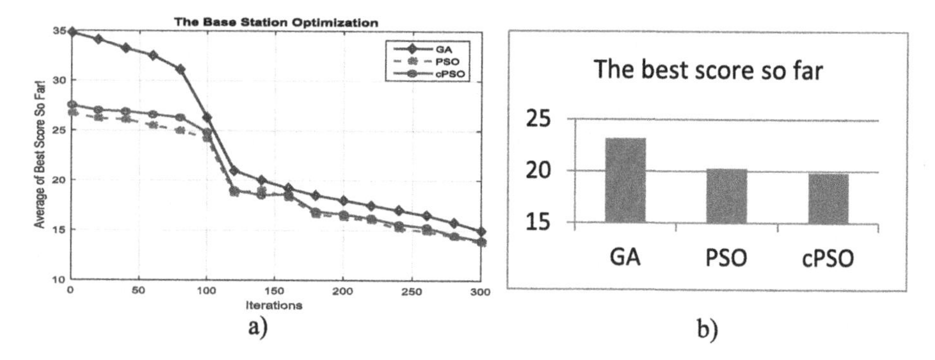

Fig. 2. (a) The comparison curves of the proposed cPSO method with the GA-method and PSO-method for the heterogeneous WSN optimization, (b) The mean of the minimum value of fitness function in of 25 trails in comparisons the proposed cPSO method, with GA-method and PSO- method for optimal heterogeneous WSNs.

5 Conclusion

In this paper, a novel optimal method of using a modest memory for the problem involving base station positioning was proposed for wireless sensor networks (WSN). A compact particle swarm optimization (cPSO) is applied in the proposed method to determine the optimized location for the relocation of the base station in WSN. The model using for the lifetime of the heterogeneous network is described for the objective function. In the proposed approach, a probabilistic representation random of the collection behavior of swarms is inspired to employ for the optimizing process. The real population is replaced with the probability vector updated based on single competition. From the simulation result, the performance of proposed method can increase the network lifetime by optimizing locations of the base station in deploying WSN in term of a normalizing lifetime. The results also are compared with the genetic algorithm (GA) method and the particle swarm optimization (PSO) method show that the proposed method can provide the effective way of using a modest memory.

References

1. Akyildiz, I.F., Su, W., Sankarasubramaniam, Y., Cayirci, E.: A survey on sensor networks. IEEE Commun. Mag. **40**, 102–105 (2002)
2. Antonio, P., Grimaccia, F., Mussetta, M.: Architecture and methods for innovative heterogeneous wireless sensor network applications. Remote Sens. **4**, 1146–1161 (2012)
3. Kennedy, J., Eberhart, R.: Particle swarm optimization. In: Proceedings of ICNN 1995 - International Conference on Neural Networks, p. 16 (1995)
4. Harik, G.R., Lobo, F.G., Goldberg, D.E.: The compact genetic algorithm. IEEE Trans. Evol. Comput. **3**, 287–297 (1999)
5. Billingsley, P.: Probability and Measure, 3rd edn. Wiley, New York (1995)
6. Neri, F., Mininno, E., Iacca, G.: Compact particle swarm optimization. Inf. Sci. (NY) **239**, 96–121 (2013)
7. Abramowitz, M., Stegun, I.A.: Handbook of Mathematical Functions: with Formulas, Graphs, and Mathematical Tables. Courier Corporation, New York (1964)
8. Cody, W.J.: Rational Chebyshev approximations for the error function. Math. Comput. **23**, 631 (1969)
9. Iacca, G., Mallipeddi, R., Mininno, E., Neri, F., Suganthan, P.N.: Super-fit and population size reduction in compact differential evolution. In: IEEE SSCI 2011 - Symposium Series on Computational Intelligence - MC 2011: 2011 IEEE Workshop on Memetic Computing, pp. 21–28. IEEE (2011)
10. Dao, T.-K., Pan, T.-S., Nguyen, T.-T., Pan, J.-S.: Parallel bat algorithm for optimizing makespan in job shop scheduling problems. J. Intell. Manuf. (2015)
11. Mollanejad, A.: DBSR: dynamic base station repositioning using the genetic algorithm in wireless sensor network. Comput. Eng. **7**, 521–525 (2010)

The Reliability and Economic Analysis on System Connection Mode

Chao-Fan Xie[1(✉)], Lin Xu[2], and Lu-Xiong Xu[3]

[1] Network Center, Fuqing Branch of Fujian Normal University, Fuqing, Fuzhou, Fujian, China
119396356@qq.com
[2] Fujian Normal University, Economic Institute, Fuqing, Fuzhou, Fujian, China
xulin@fjnu.edu.com
[3] Electronic Information and Engineering Institute, Fuqing Branch of Fujian Normal University, Fuqing, Fuzhou, Fujian, China
xlxl23456@139.com

Abstract. To improve the reliability of the system, the component can be used as parallel or redundant backup. Among them, the parallel connection mode is simple and the cost is low, but the reliability of increasing rate is small. The other one connection mode is complexity and cost is high, but the reliability is improved greatly. The actual system has many components, through the combination of the original component can produce a variety of ways of connection mode, its composition space is exponential. Every connection mode has its reliability and cost, they are a pair of contradictory goals, how to choose the best connection mode to make the reliability and economy to achieve better, in other words, how much of the number of components are choose for parallel and the number of it for redundant are the best choice? This find the optimal number of for parallel and redundant backups in the sense of Pareto, and analysis the relation between optical connection mode and time.

Keywords: Reliability · Connection mode · Pareto

1 Introduction

Reliability theory is a discipline that can analyze characterize the products specified functions probability of occurrence of random events, which Is the sixties of last century developed new interdisciplinary subject. Thus, the reliability theory is based on probability theory, the first based on field research is machine maintenance problem [1]. At present, the main study of the reliability is the system reliability indices, as well as the reliability of the index on the basis of the optimal detection time to avoid faults, reduce the losses caused by the fault, such as the literature [2–4]. Research which has done basically is qualitative analysis or uses uses numerical analysis to find the approximation value of solution in actual system [5–7]. Ida etc al. proposed a genetic algorithm for solving reliability problems which belongs to intelligent algorithm [8–11]. So all their researches were focusing on engineering method.

J. Pan et al. (eds.), *Genetic and Evolutionary Computing*, Advances in Intelligent Systems and Computing 536, DOI 10.1007/978-3-319-48490-7_8

In N element exponentially distributed parallel system, Xie Chaofan and Xu Luxiong has been analyzed reliability and extreme value, obtained some relevant theoretical guidance. Without considering the economic constraints, the failure rate is equal to all of the components, the system reliability reaches a minimum, in considering the economic constraints, and when the failure rate is unit elastic, the conclusion is the same. And if you choose a good product in parallel with a poor product, the product reliability are more higher than parallel on two moderate of the failure rate. During operation, according to the actual situation, if the failure rate in the envelope line, then the whole system components should be replaced altogether, but if the failure rate is far away from the envelope, then for economic performance considerations, just to update the highest failure rate of several originals out [12]. But in this paper, it only consider parallel connection mode, it doesn't consider redundant backup connection mode.

In actual system, it has many components, through the combination of the original component can produce a variety of ways of connection mode, it's feasible solution space is exponential with the number of components. This article will find the optimal number of for parallel and redundant backups in the sense of Pareto, and analysis the relation between optical connection mode and time.

2 The Definition of the Main Indicators of Reliability

The printing area is 122 mm × 193 mm. The text should be justified to occupy the full line width, so that the right margin is not ragged, with words hyphenated as appropriate. Please fill pages so that the length of the text is no less than 180 mm, if possible.

1. Reliability
   The definition of reliability *R(t)* [13]: it is the probability that product completes the required function under the specified conditions and within the prescribed time. If the life distribution of product is $F(t)$, $t > 0$, the reliability $R(t) = P(T \geq t) = 1 - F(t)$. This is a function of time(t), so it can be called as reliability function. To the components obeying exponential distribution λ, its reliability is $e^{-\lambda t}, t \geq 0$.
2. Failure rate
   $\lambda(t)$: It is the probability of occurring failure in the unit of time after product has worked a period of time(t). According to reliability theory, $\lambda(t) = \frac{f(t)}{1-F(t)}$, when $t > 0$, the failure rate of exponential distribution is constant λ.
3. System parameter specification
   A: represents normal working events of system.
   A_i: represents normal working events of the element i.
   λ_i: represents failure rate of the element i.
   $R_s(t)$: represents system reliability, that is, $P(A) = R_s$.
   $R_i(t)$: represents reliability of the element i, that is, $P(A_i) = R_i$.
   m_s: represents the average lifetime of the system, $m_s = \int_0^{+\infty} R_s(t)dt$.

Parallel system: It is a system consisting of n components. As long as one of these elements works, the system can work; only when all the units fail, the system would fail.

According to the property of probability, the normal working probability of system is as follows:

$$P(A) = P(\bigcup_{i=1}^{n} A_i)$$

$$R_s(t) = 1 - \prod_{i=1}^{n} (1-R_i(t)) = 1 - \prod_{i=1}^{n} (1-e^{-\lambda_i t}) \quad (1)$$

Redundant backup systems, some of the elements work, the other unit does not work, in a waiting or a standby state, When the unit the failure rate of the secondary unit in a standby period is zero, in other words, is a hundred percent reliability during standby. One work, n–1 standby redundant system framework is shown below Fig. 1, where, K is detected and switches.

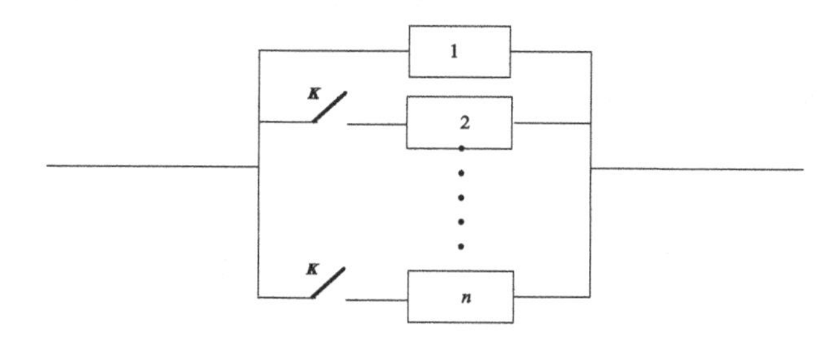

Fig. 1. The logical block diagram of Erlang distribution system.

Use 10-point type for the name(s) of the author(s) and 9-point type for the address(es) and the abstract. For the main text, please use 10-point type and single-line spacing. We recommend the use of Computer Modern Roman or Times. Italic type may be used to emphasize words in running text. Bold type and underlining should be avoided.

The following lemma is assumed that when the switch is absolutely reliable:

Lemma 1.1: $X_1, X_2, \cdots X_n$ is mutually independent random variables, subject to the same parameter λ exponential distribution, then $X = X_1 + X_2 + \cdots + X_n$ obey Erlang distribution of order n, the probability density function is:

$$b_n(u) = \begin{cases} \lambda e^{-\lambda u} \frac{(\lambda u)^{n-1}}{(n-1)!}, & u \geq 0 \\ 0, & u < 0 \end{cases} \quad (2)$$

Proof: Let $p(u)$ is a probability density function of random variables, Since X_i are independent and identically distributed random variables exponentially distributed, then $p(u)$ is exponential distribution probability density function, to make $\varphi(t) = \int_{-\infty}^{+\infty} e^{itu} p(u)du$ is a characteristic function of random variable X_i then $\varphi(t) = (1 - \frac{it}{\lambda})^{-1}$. Because of $X = X_1 + X_2 + \cdots + X_n$, then X shows the probability density function of X_i do the n-fold convolution, Since the Fourier transform can become convolution to multiplication. Therefore, the characteristic function is:

$$\varphi_n(t) = \varphi(t)^n = (1 - \frac{it}{\lambda})^{-n}$$

The probability density function of X is when $u \geq 0$, $b_n(u) = \frac{1}{2\pi} \int_{-\infty}^{+\infty} e^{-itu} \varphi_n(t)dt = \frac{1}{2\pi} \int_{-\infty}^{+\infty} e^{-itu} (1 - \frac{it}{\lambda})^{-n} dt$, points can be obtained from the Division:

$$\frac{1}{2\pi} \int_{-\infty}^{+\infty} e^{-itu} (1 - \frac{it}{\lambda})^{-n} dt = \frac{1}{2\pi} [\frac{\lambda}{(n-1)i} e^{itu} (1 - \frac{it}{\lambda})^{-n+1} \Big|_{-\infty}^{+\infty} + \frac{\lambda u}{n-1} \int_{-\infty}^{+\infty} e^{-itu} (1 - \frac{it}{\lambda})^{-n+1} dt]$$

$$= \frac{1}{2\pi} \frac{\lambda u}{n-1} \int_{-\infty}^{+\infty} e^{-itu} (1 - \frac{it}{\lambda})^{-n+1} dt = \cdots = \frac{1}{2\pi} \frac{(\lambda u)^{n-1}}{(n-1)!} \int_{-\infty}^{+\infty} e^{-itu} (1 - \frac{it}{\lambda})^{-1} dt = \lambda e^{-\lambda u} \frac{(\lambda u)^{n-1}}{(n-1)!}$$

So for the n–1 redundancy elements and system of component life exponentially distributed, its distribution follows n-order Erlang distribution, proof.

Lemma 1.2: When the system life obey the n-order Erlang distribution, the reliability of system is

$$R_s(t) = [1 + \lambda t + \frac{(\lambda t)^2}{2!} + \cdots + \frac{(\lambda t)^{n-1}}{(n-1)!}] e^{-\lambda t} \tag{3}$$

Proof: By Lemma 1.1 shows that probability density function of the system life X is:

$$b_n(u) = \begin{cases} \lambda e^{-\lambda u} \frac{(\lambda u)^{n-1}}{(n-1)!}, & u \geq 0 \\ 0, & u < 0 \end{cases}$$

Since $R_s(t) = P(T \geq t) = 1 - F(t) = 1 - \int_0^t \lambda e^{-\lambda u} \frac{(\lambda u)^{n-1}}{(n-1)!} du$, points can be obtained from the Division:

$$\int_0^t \lambda e^{-\lambda u} \frac{(\lambda u)^{n-1}}{(n-1)!} du = -\int_0^t \frac{(\lambda u)^{n-1}}{(n-1)!} de^{-\lambda u} = -e^{-\lambda u} \frac{(\lambda u)^{n-1}}{(n-1)!} \Big|_0^t + \int_0^t \lambda e^{-\lambda u} \frac{(\lambda u)^{n-2}}{(n-2)!} du$$

$$= -e^{-\lambda t} \frac{(\lambda t)^{n-1}}{(n-1)!} + \int_0^t \lambda e^{-\lambda u} \frac{(\lambda u)^{n-2}}{(n-2)!} du = \cdots = [1 + \lambda t + \frac{(\lambda t)^2}{2!} + \cdots + \frac{(\lambda t)^{n-1}}{(n-1)!}] e^{-\lambda t} + 1$$

we can get the conclusion, proof.

3 Connection Mode Analysis

3.1 Using Multiplication and Division

Now we consider connection mode, suppose there are n_1 element for parallel, n_2 element for Erlang redundancy backup system distribution. According Lemma 1.2 the reliability of it is

$$\left[1+\lambda t+\frac{(\lambda t)^2}{2!}+\cdots+\frac{(\lambda t)^{n_2-1}}{(n_2-1)!}\right]e^{-\lambda t}$$

For the parallel part, the reliability of it is

$$1-(1-e^{-\lambda t})^{n_1}$$

So the whole system reliability is

$$R_s(t)=1-(1-e^{-\lambda t})^{n_1}\left(1-\left[1+\lambda t+\frac{(\lambda t)^2}{2!}+\cdots+\frac{(\lambda t)^{n_2-1}}{(n_2-1)!}\right]e^{-\lambda t}\right) \quad (4)$$

So the whole system cost is

$$C=(n_1+n_2)c+(n_2-1)k \quad (5)$$

c represent the unite cost of element, k represent the unite cost of switch.
So the optimization model is:

$$\begin{cases} \max\ 1-(1-e^{-\lambda t})^{n_1}\left(1-\left[1+\lambda t+\frac{(\lambda t)^2}{2!}+\cdots+\frac{(\lambda t)^{n_2-1}}{(n_2-1)!}\right]e^{-\lambda t}\right) \\ \min\ (n_1+n_2)c+(n_2-1)k \\ s.t.\ \ n_1,\ n_2 \geq 0 \end{cases}$$

We transform the model into another one:

$$\begin{cases} \max\ U(n_1,n_2,t) \\ s.t.\ \ n_1,n_2 \geq 0 \end{cases} = \frac{1-(1-e^{-\lambda t})^{n_1}\left(1-\left[1+\lambda t+\frac{(\lambda t)^2}{2!}+\cdots+\frac{(\lambda t)^{n_2-1}}{(n_2-1)!}\right]e^{-\lambda t}\right)}{(n_1+n_2)c+(n_2-1)k}$$

The function $U(n_1,n_2,t)$ represent the unite cost generate the reliability. Because the variable is discrete, conditions for obtaining extreme value is:

$$\begin{cases} (a)\ U(n_1,n_2,t) \geq U(n_1-1,\ n_2,\ t) \\ (b)\ U(n_1,n_2,t) \geq U(n_1+1,\ n_2,\ t) \\ (c)\ U(n_1,n_2,t) \geq U(n_1,\ n_2-1,\ t) \\ (d)\ U(n_1,n_2,t) \geq U(n_1,\ n_2+1,\ t) \end{cases} \quad (6)$$

From the first Formula (b) we can get that:

$$(1-e^{\lambda t})^{n_1}+n_1 \geq \frac{k-n_2(c+k)}{(1-e^{-\lambda t})c}+\frac{e^{\lambda t}}{\left[1+\lambda t+\frac{(\lambda t)^2}{2!}+\cdots+\frac{(\lambda t)^{n_2-1}}{(n_2-1)!}\right]e^{-\lambda t}}$$

n^* is the Because λt is usually very small, so the inequality approximate

$$n_1 \geq \frac{k-n_2(c+k)}{(1-e^{-\lambda t})c}+\frac{e^{\lambda t}}{(1-e^{-\lambda t})} \tag{7}$$

Similar to the above, from the first Formulas (a), (c), (d), we can get that:

$$n_1-1 \leq \frac{k-n_2(c+k)}{(1-e^{-\lambda t})c}+\frac{e^{\lambda t}}{(1-e^{-\lambda t})} \tag{8}$$

$$n_1 \leq \frac{\frac{(c-1)}{(c+k)}+n_2}{\sum\limits_{k=1}^{n_2-1}\frac{(\lambda t)^k}{k!}-\frac{c}{c+k}} \tag{9}$$

$$n_1-1 \geq \frac{\frac{(c-1)}{(c+k)}+n_2}{\sum\limits_{k=1}^{n_2-1}\frac{(\lambda t)^k}{k!}-\frac{c}{c+k}} \tag{10}$$

When $t \to \infty$ the inequality is approximate:

$$n_1 \leq \frac{\frac{(c-1)}{(c+k)}+n_2}{\frac{k}{c+k}-e^{-\lambda t}} \tag{11}$$

$$n_1-1 \leq \frac{k-n_2(c+k)}{(1-e^{-\lambda t})c}+\frac{e^{\lambda t}}{(1-e^{-\lambda t})} \tag{12}$$

So, the best solution is the intersection of the straight line. And it's the parameter of t, we marks the approximate best solution as $(n_1^*,\ n_2^*)=f(t)$, we use *Crammer Rule* get the best solution $f(t)$ as follows:

$$n_1^* = \frac{\begin{vmatrix} \frac{c-1}{c+k} & -1 \\ k+e^{-\lambda t}c & c+k \end{vmatrix}}{\begin{vmatrix} \frac{k}{c+k} - e^{-\lambda t} & -1 \\ (1-e^{-\lambda t})c & c+k \end{vmatrix}} = \frac{(k+c-1)+ce^{-\lambda t}}{(k+c)-(2c+k)e^{-\lambda t}},$$

$$n_2^* = \frac{\begin{vmatrix} \frac{k}{c+k} - e^{-\lambda t} & \frac{c-1}{c+k} \\ (1-e^{-\lambda t})c & k+e^{-\lambda t}c \end{vmatrix}}{\begin{vmatrix} \frac{k}{c+k} - e^{-\lambda t} & -1 \\ (1-e^{-\lambda t})c & c+k \end{vmatrix}} = \frac{\frac{k^2}{c+k} - ce^{-2\lambda t} - \frac{k^2+c^2-c}{c+k}e^{-\lambda t}}{(k+c)-(2c+k)e^{-\lambda t}} \tag{13}$$

we get all the Pareto solution from t as formula (13), because it's not integer, so we must trans into integer. For example, if unite cost of element $c = 5$, unite cost of switch $k = 3$, we get the best solution as follows:

$n_1^* = \frac{7+5e^{-\lambda t}}{8-13e^{-\lambda t}}$, $n_2^* = \frac{\frac{29}{8}-\frac{29}{8}e^{-\lambda t}-5e^{-2\lambda t}}{8-13e^{-\lambda t}}$, from the formula, it's easy to prove the number for parallel must increase over the time, the number of for Erlang redundancy backup is increase, but it has limit, in this example is 29.

3.2 Using Mathematical Programming

Due to economic constraints are linear constraints, so we choose reliability as the main goal, put economic goal into constraints. We get the model as follows:

$$\begin{cases} \max \; 1-(1-e^{-\lambda t})^{n_1}\{1-[1+\lambda t+\frac{(\lambda t)^2}{2!}+\cdots+\frac{(\lambda t)^{n_2-1}}{(n_2-1)!}]e^{-\lambda t}\} \\ st. \;\; s_1 < (n_1+n_2)c+(n_2-1)k < s_2 \\ n_1 \geq 0, \;\; n_2 \geq 0 \end{cases}$$

and the number part of element n1, n2 is integer, so the feasible solution

Set $f(n_1, n_2) = (1-(1-e^{-\lambda t})^{n_1}\{1-[1+\lambda t+\frac{(\lambda t)^2}{2!}+\cdots+\frac{(\lambda t)^{n_2-1}}{(n_2-1)!}]e^{-\lambda t}\}$

Theorem 3.1: For the function $f(n_1, n_2)$ is increasing by n_1, n_2 and it's limit is 1.

Proof: Set $f_1(n_1) = (1-e^{-\lambda t})^{n_1}$ and $f_2(n_2) = 1-[1+\lambda t+\frac{(\lambda t)^2}{2!}+\cdots+\frac{(\lambda t)^{n_2-1}}{(n_2-1)!}]e^{-\lambda t}$. For the function $f_1(n_1)$, because $1-e^{-\lambda t} < 1$, so $f_1(n_1) \downarrow$, For the function $f_2(n_2)$ part of $1+\lambda t+\frac{(\lambda t)^2}{2!}+\cdots+\frac{(\lambda t)^{n_2-1}}{(n_2-1)!} \uparrow$, so $f_2(n_2) \downarrow$, $f_1(n_1)\cdot f_2(n_2) \downarrow$, now can get $f(n_1, n_2) \uparrow$. From Taylor formula, $\lim_{n_2 \to +\infty} 1+\lambda t+\frac{(\lambda t)^2}{2!}+\cdots+\frac{(\lambda t)^{n_2-1}}{(n_2-1)!} = e^{\lambda t}$, so the limit of $f(n_1, n_2)$ is 1.

From Theorem 3.1, we know the points in the feasible region are all the Pareto optimal solutions. Although the use of multiplication and division are given the Pareto optimal solution, but it only give the only one, and it focus more on thinking from the economic point of view, mathematical programming can give much of the solution, the follow Fig. 2 gives the function of reliability (the parameter $\lambda t = 0.1$):

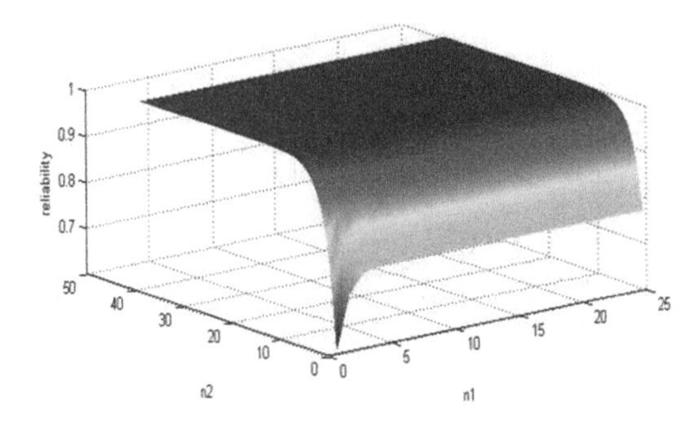

Fig. 2. The Reliability function of n_1, n_2.

4 Conclusion

The reliability of the distribution of Erlang redundancy backup system is always greater than the parallel systems. When you make economic analysis of two systems. How to choose the best connection mode to make the reliability and economy to achieve better, we have three method, one of it is transforming the model into one goal of optimization model, last we get approximate of best solution is the intersection of the straight line, marks the approximate best solution as $(n_1^*,\ n_2^*) = f(t)$, finally we use Crammer Rule get $f(t)$ function form, we can know the number for parallel must increase over the time, the number of for Erlang redundancy backup is increase, but it has limit, and the method of multiplication and division it only get the one best solution of Pareto, in the second method mathematical programming can give much of the solution.

Acknowledgments. This research was partially supported by the School of Mathematics and Computer Science of Fujian Normal University, by the Institute of Innovative Information Industry of Fujian Normal University, by the School of Economic of Fujian Normal University.

References

1. Luss, H.: An inspection policy model for production facilities. J. Manag. Sci. **29**, 101–109 (1983)
2. Bao-he, S.: Reliability and optimal inspection policy of inspected systems. J. OR Trans. **11** (1) (2007)
3. Bao-he, S.: Study on optimal inspection polices based on reliability indices. J. Eng. Math. **25** (6) (2008)
4. Zequeira Romulo, I., Berenguer, C.: On the inspection policy of a two-component parallel system with failure interaction. J. Reliab. Eng. Syst. Saf. **88**(1), 99–107 (2005)
5. Li, F., Zhu, C., Wang, W.: Cost models and optimization of redundant system. J. Syst. Eng. Electr. **22**(9) (2000)
6. Misra, K.B., Ljubojevic, M.D.: Optimal reliability design of a system-a new look. J. IEEE Trans. Reliab. (1973)

7. Yearout, R.D., Reddy, P., Grosh, D.L.: Stand by redundancy in reliability-a review. J. IEEE Trans. Reliab. (1986)
8. Gen, M., Ida, K., Sasaki, M., Lee, J.: Algorithm for solving large-scale 0-1 goal programming and its application to reliability optimization problem. J. Int. J. Comput. Ind. Eng. (1989)
9. Gen, M., Ida, K., Taguchi, T.: Reliability optimization problems: a novel genetic algorithm approach, technical report, ISE93-5. J. Ashikaga Inst. Technol. Ashikaga Jpn. (1993)
10. Gen, M., Ida, K., Kim, J.R.: System reliability optimization with fuzzy goals using genetic algorithm. J. Jpn. Soc. Fuzzy Theory Syst. **10**(2), 356–365 (1998)
11. Yokota, T., Gen, M., Li, Y.: Genetic algorithms for nonlinear mixed integer programming problems and its applications. J. Comput. Ind. Eng. **30**(4), 905–917 (1996)
12. Xu, L., Xie, C.F., Xu, L.X.: Reliability envelope analysis. J. Comput. **25**(4), 26–34 (2014)
13. Zong-shu, W.: Probability and Mathematical Statistics. China Higher Education Press, Beijing (2008)

Teaching Achievement Study of Creative Design Competitions

Rui-Lin Lin(✉)

Department of Commercial Design,
Chienkuo Technology University, Changhua, Taiwan
linrl2002@gmail.com

Abstract. This article presents a study on the teaching and learning process in the Creative Thinking and Methods course and the participation of teacher and students in design competitions and their results. Discussions were made regarding award-winning entries such as slogan, logo design, poster design, visual design, and T-Shirt design. Research results showed that case teaching method helps students become more creative. In addition, the discussion method was used to revise the creation content and encourage students to participate in competitions. Lastly, receiving an award is the greatest affirmation of the student's creation and can improve his motivation and willingness to learn.

Keywords: Slogan · Logo design · Poster design · Visual design · T-Shirt design

1 Introduction

University admission is becoming increasingly diversified. Therefore, during lessons, teachers face a mixed level of student body. The biggest challenge encountered by professors is how to help those without specialized foundation to learn and, at the same time, how to motivate students with strong professional capability.

This article studies the teaching and learning methods in the Creative Thinking and Methods course, where the teacher used case studies to assist students without foundation in developing their creative skills. The discussion method was used to strengthen students with creative capabilities and encourage them to participate in creative design competitions so as to enhance student motivation and willingness to attend classes.

2 Literature Review

The teacher guided the following students to participate in creative design competitions. The slogan designs of Li-Zhu Lin, "Get a glimpse of beauty as you walk into Houli" and Yan-Cheng Lin, "Scent of flowers floats through Houli and spreads its generosity far and wide" were given the excellent work awards during the 2016 Houli Sweet Thoughts Literature and Creative Phrase Category Competition (Fig. 1) [1]. The slogan created by Ya-Yi Su, "Dreams of Taiwanese girls have no sunset" won the first

J. Pan et al. (eds.), *Genetic and Evolutionary Computing*, Advances in Intelligent Systems and Computing 536, DOI 10.1007/978-3-319-48490-7_9

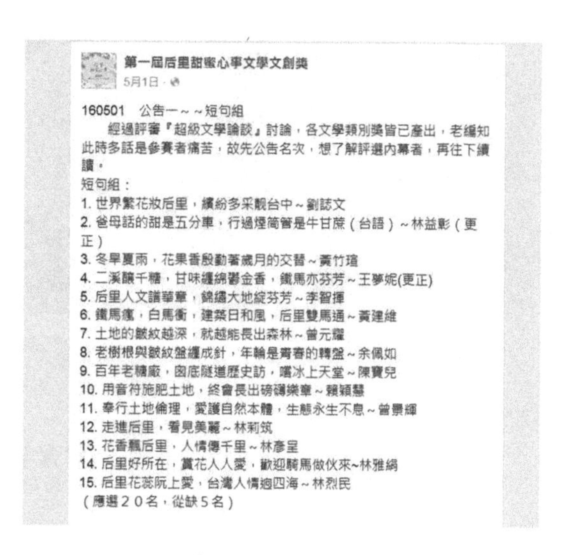
第一屆后里甜蜜心事文學文創獎
5月1日

160501 公告一～～短句組
經過評審『超級文學論談』討論，各文學類別獎皆已產出，老編知此時多話是參賽者痛苦，故先公告名次，想了解評選內幕者，再往下續讀。
短句組：
1. 世界繁花妝后里，繽紛多采靓台中～劉誌文
2. 爸母話的甜是五分車，行過煙茼簹是牛甘蔗（台語）～林益彰（更正）
3. 冬旱夏雨，花果香殷勤著歲月的交替～黃竹瑄
4. 二溪釀千糖，甘味纏綿鬱金香，鐵馬亦芬芳～王夢妮(更正)
5. 后里人文譜華章，錦繡大地綻芬芳～李智揮
6. 鐵馬瘋，白馬衝，建築日和風，后里雙馬通～黃建維
7. 土地的皺紋越深，就越能長出森林～曾元耀
8. 老樹根與皺紋盤纏成針，年輪是青春的轉盤～余佩如
9. 百年老糖廠，囪底隧道歷史訪，嚐冰上天堂～陳寶兒
10. 用音符施肥土地，終會長出磅礴樂章～賴穎慧
11. 奉行土地倫理，愛護自然本體，生態永生不息～曾景耀
12. 走進后里，看見美麗～林莉筑
13. 花香飄后里，人情傳千里～林彥呈
14. 后里好所在，賞花人人愛，歡迎騎馬做伙來~林雅絹
15. 后里花蕊阮上愛，台灣人情遍四海～林烈民
（應選２０名，從缺５名）

Fig. 1. Houli slogan

prize in the Taiwan Girls Day Nationwide Collegiate Slogan Competition held by the Ministry of Education's Taiwan Girls Go (Fig. 2) [2].

The logo design created by Hui-Jun Yang "No to Drag Racing" won the first prize at the Creative Fantasy 2nd Traffic Safety Totem Design Contest, Motorcycle Safety Group (Fig. 3) [3].

For the poster design category, Wei-Lian Chen's "Green Wind" won the second prize at the Changhua City Recycling Contest (Fig. 4) [4]. Yi-Jie Lin's "Breed" was given the third prize and Zi-Wei Pan's "Life is a non-stop journey" the excellent work award (Fig. 6) [5] at the Life Education Poster Competition (Fig. 5).

For visual design, Jia-Juan Zhang's "I like FLOMO" received the first place in the FLOMO 40th Anniversary Creative Packaging Design (Fig. 7) [6]. Zong-Wei Li's "New Born" was awarded the third prize (Fig. 8), Zhi-Yao Xu's "Splash Paint" (Fig. 9) and Jia-Ru Tu's "Ecstasy Faction" (Fig. 10) were honored as Excellent Works at the 2013 Ecstasy Helmet Design Contest [7].

For T-Shirt design, Ya-Wen Yang's "Happy Tumbler" received the Best Design Award at the 2013 Juuway Design Competition (Fig. 11) [8].

3 Creative Design

The teacher selected appropriate competitions according to progress of the course and the students' level of competency. She guided the students to create according to competition methods and led them to brainstorm on their slogan. The winning creations possess features of creative writing and were able to convey the effect of slogan advertising propaganda.

教育部獎狀

建國科技大學
一年級蘇雅貽同學
參加教育部103年度「為臺灣女孩喝采」
臺灣女孩日簡訊標語甄選活動
經評定成績為大專校院組第一名
特頒此狀以資鼓勵

部長 吳思華

中華民國 19 日

Fig. 2. Taiwan Girls Day Nationwide Collegiate Slogan competition

Fig. 3. Logo design- no to drag racing

Figure 3 uses black skull as the background and the color red as a warning to the danger of drag racing. This visual imagery admonishes motorcycle riders to refrain from speeding.

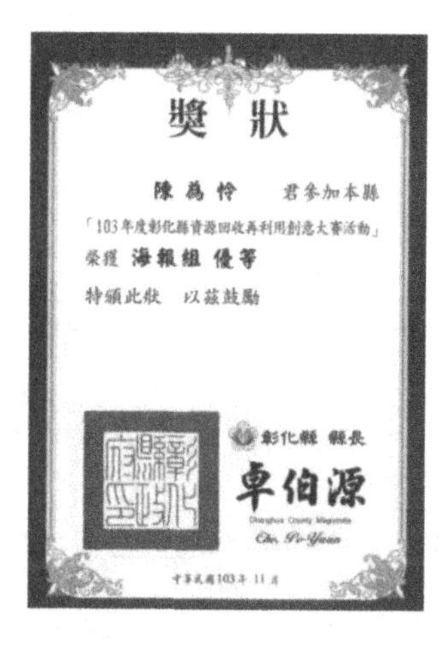
獎 狀

陳為怜 君參加本縣

「103年度彰化縣資源回收再利用創意大賽活動」

榮獲 海報組 優等

特頒此狀 以茲鼓勵

彰化縣 縣長

卓伯源

Changhua County Magistrate

Cho, Po-Yuan

中華民國103年11月

Fig. 4. Poster design- green wind

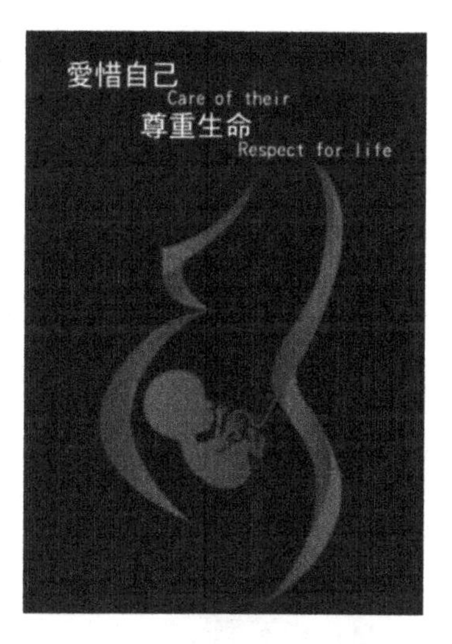

Fig. 5. Poster design- breed

Fig. 6. Poster design- life is a non-stop journey

Figure 4 shows a wind energy storage made of PET bottle with fan blades. This conveys the environmental protection concept of carbon reduction and recycling. In Fig. 5, a sketch of the shape of a pregnant mother was made through lines. The umbilical cord was used to link this image to the birth and beauty of a new life. Figure 6 shows a hand holding a sprout with stout trees as the background to express that human beings should pay attention to the meaning and value of life.

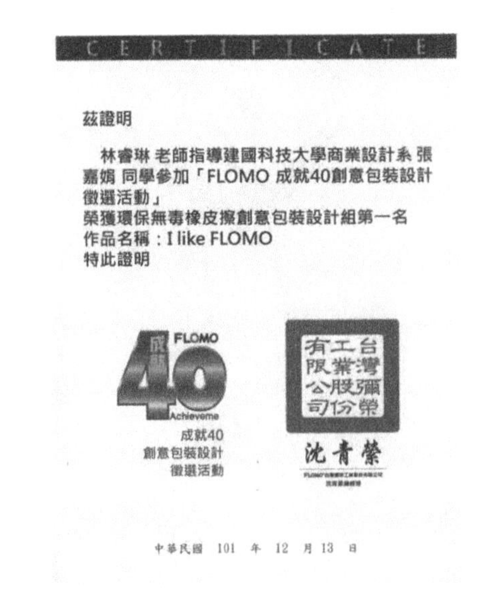
CERTIFICATE

茲證明

林睿琳 老師指導建國科技大學商業設計系 張嘉娟 同學參加「FLOMO 成就40創意包裝設計徵選活動」
榮獲環保無毒橡皮擦創意包裝設計組第一名
作品名稱：I like FLOMO
特此證明

FLOMO 成就 40 Achieveme
成就40
創意包裝設計
徵選活動

台灣彌榮工業股份有限公司
沈青榮

中華民國 101 年 12 月 13 日

Fig. 7. Visual design- I like FLOMO

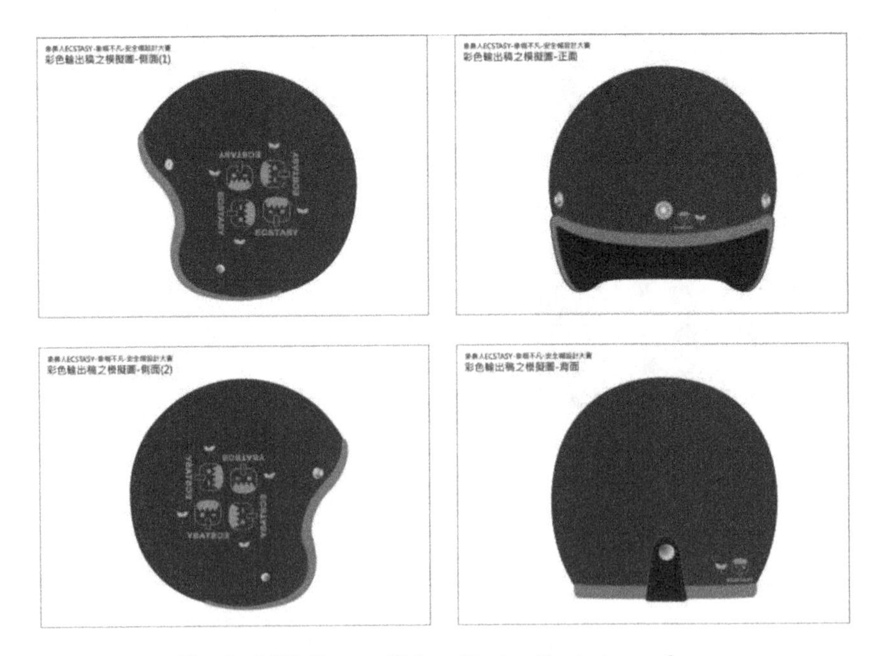

Fig. 8. 2013 Ecstasy Helmet Design Contest- new born

The creative concept used in Fig. 7 is that every day is a FLOMO day for students. This conveys the convenience and importance of daily FLOMO companionship. In Fig. 8, the color brown represents soil, the green color of the sprout symbolizes the

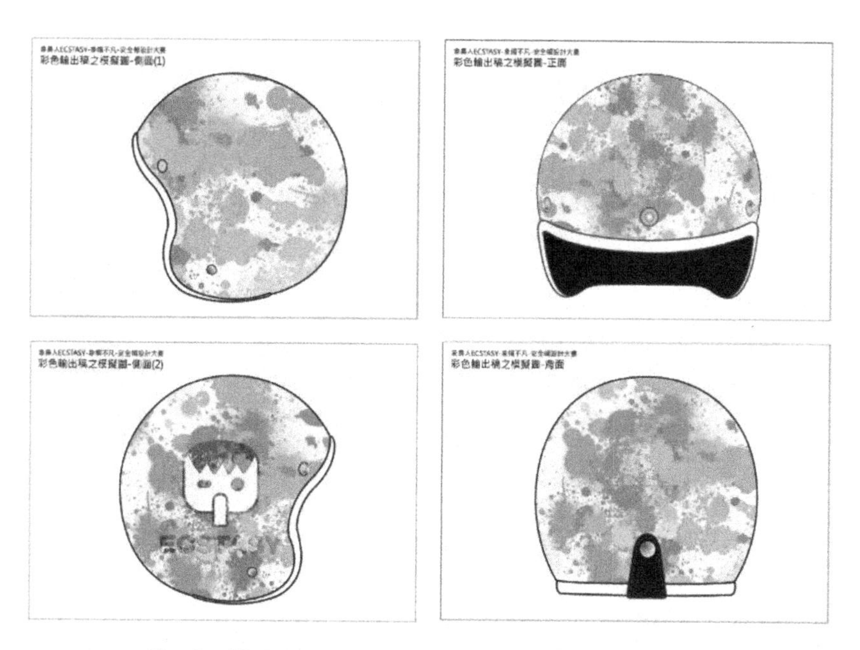

Fig. 9. 2013 Ecstasy Helmet Design Contest- splash paint

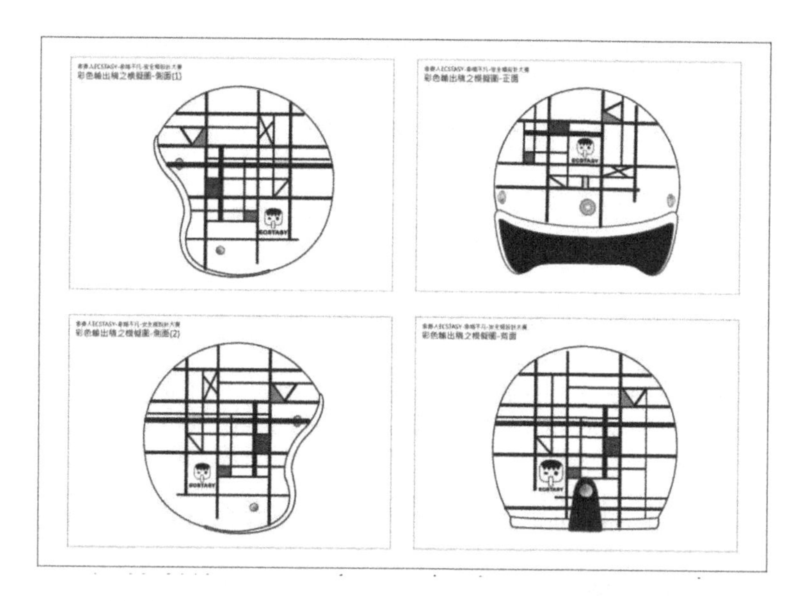

Fig. 10. 2013 Ecstasy Helmet Design Contest- Ecstasy faction

concept of new life. In Fig. 9, vivid colors and casual spraying of paint give a sense of entertainment and gives a joyful and lively imagery. In Fig. 10, Mondrian geometry was the inspiration for the creation. Art Nouveau was used to characterize the creative

Fig. 11. 2013 Juuway Design Competition- happy tumbler

concept behind the new style of Ecstasy. Figure 11 uses an inverted image to communicate that regardless of the side, a tumbler can have a happy expression. It represents the attitude of not being discouraged and getting up despite stumbling.

4 Design Results

The teacher showcased winning entries of students over the years as achievements of teaching research.

5 Conclusions

In general, the results for the innovative research and development of this study are summarized and illustrated below:

(1) The students that participated in competitions and received awards have creative thinking abilities and know how to use creative methods to make their creations simple and powerful, as well as uniquely distinctive and capable of winning.
(2) The students could clearly understand the competition methods and meaning of narration. They were also able to have more discussions with the teacher and make revisions.
(3) Receiving an award in a competition has an important impact on students. It encourages them to have an active learning attitude regarding their professional knowledge and capability.
(4) Winning entries generally received a certificate and cash prize. This can improve student motivation to learn and to participate in competitions in the future.

References

1. 2016 Houli Sweet Thoughts Literature and Creative Phrase Category Competition. https://goo.gl/aSBdHc. 15 August 2016
2. Taiwan Girls Day Nationwide Collegiate Slogan Competition held by the Ministry of Education's Taiwan Girls Go. http://www.edu.tw/. 15 August 2016
3. Creative Fantasy 2nd Traffic Safety Totem Design Contest, Motorcycle Safety Group. http://goo.gl/kP2Qp. 15 August 2016
4. Changhua City Recycling Contest. http://student.tnua.edu.tw/upload/files/%5E]p.pdf. 15 August 2016
5. Life Education Poster Competition. http://s1.ntue.edu.tw/em/. 15 August 2016
6. FLOMO 40th Anniversary Creative Packaging Design. https://bhuntr.com/tw/competitions/73245. 15 August 2016
7. 2013 Ecstasy Helmet Design Contest. https://bhuntr.com/tw/competitions/72892. 15 August 2016
8. 2013 Juuway Design Competition, https://bhuntr.com/tw/competitions/72930. 15 August 2016

A Study of Micro-film Creation: Using the Inspirational Story of a Tomato Farm Micro-film

Jing-Chen Xie(✉)

Department of Digital Media Design and Management, Far East University, No. 49 Zhonghua Road, Xinshi District, Tainan 74448, Taiwan
jing.chen.alen.xie@gmail.com

Abstract. The purpose of the virtual enterprise "i tomato" founded by Far East University is to help students from disadvantaged families. The CEO tomato creative marketing competition held annually makes it possible for consumers to be touched by the stories of love during tomato tasting. The producer won the first prize with this film adaptation of his true life-story. His persistence and determination during film making were very impressive. Nowadays, in the age of decreasing birth rate, micro-film creation using computer technology has become an efficient instrument for schools to enhance their reputation and attract students.

Keywords: Micro-film · Computer technology · Creative marketing competition · Tomato farm

1 Introduction

A good film can touch people deeply. With the world-wide popularity of the Internet, influence of films have become more and more far reaching. However, it takes a lot of time and effort to make a film. Its production cost is very high and cannot be shouldered by a regular organization. In recent years, micro-films have become very popular. The time required to shoot a micro-film is short and the production cost is low. With a good plot, a micro-film can touch the audience in a very short time and leave an unforgettable impression. Moreover, they are highly profitable in terms of advertising [6, 7].

Micro-films are a new type of film. They are short and deep, possessing plot and characters, can be watched while commuting, and capable of leaving a deep impression. Micro-films were developed in 2007. They are marketing works similar to movies in production but much shorter. Micro-films are not short films for artistic purposes only. They differ from movies as they are a method of marketing, a form of storytelling marketing. Micro-films do not have a standard length and can be independent or part of a series [6, 7].

Micro-films are created for public welfare promotion, image advertising, custom-made business purposes, and personal artistic creation. They are quite popular and attractive due to their rather short length and complete plot. In recent years, various

J. Pan et al. (eds.), *Genetic and Evolutionary Computing*, Advances in Intelligent Systems and Computing 536, DOI 10.1007/978-3-319-48490-7_10

forms of micro-films have been used in different fields. In this age of decreasing birth rate, they have particularly become an efficient instrument for schools to recruit students [3].

2 Literature Review

Although micro-films are different from movies, the production process is quite similar. First, in the development stage, a script is written and transformed into a blueprint for filming. The next stage is the pre-production stage, when everything required for filming must be prepared. In this stage, the casts and workforce are hired, locations for filming are decided, and sets are built. Afterwards, shooting begins. The raw footage is then edited and dialogues, scores, sound effects, music, and digital visual effects are added to the film. Finally, the film can be put online for free viewing, usually through a smart phone [5]. Micro-films can also create new business opportunities like movies do [4].

Design and sound effects recording, production and adding computer vision Digital effects. As for the main content of a micro-film production, the producer must pick a story which is probably based on a book, a stage show, a true event, or an original idea. Then he has to determine the subject or the message to be delivered. Afterwards, the content of the script, the style of the plot, and the characters are described and written down. Such films usually do not have too many dialogues or stage directions. Focus is given to the visualization of key illustrations. In the process of filming, the writer may re-write the script several times to make the plot more intense, to make narration clearer, to improve plot structure, to change dialogues, and to enhance the overall style [2].

3 Micro-film Creation

The team involved in the micro-film creation for this study is quite small. The producer had almost no funding and served as both the director and the screenwriter. He was responsible for selecting suitable actors and actresses for particular roles based on the script. Secondly, for every filming location, he had to personally negotiate with related personnel regarding renting the place. He also contacted professionals in related fields to receive guidance and support with regard to equipment, props, sets, costumes, makeup, sound effects, editing, visual arts, and creative presentation [1].

The producer was a student of Far East University. The purpose of the virtual enterprise, “i tomato”, founded by this university, is to spread out love. The financial crisis in 2008 caused parents of many students to lose their job, consequently forcing these students to drop out of school. The university freed up a piece of land and encouraged students to voluntarily plant tomato seedlings and sell ripe tomatoes in order to raise money for the cause of helping students from disadvantaged families. Moreover, part of the income was donated to the Educare Foundation. The process of growing tomatoes could be considered as life education where participating students learn to serve. The course was moving and meaningful. Moreover, during tomato season, the university holds the CEO tomato creative marketing competition annually in order to let consumers have a taste of tomatoes, and at the same time, learn about touching stories of love. The producer won the first prize in the competition with a script based on his personal story. During the two weeks when the wining micro-film was available online, it received more than 10,000 clicks. Many newspapers and magazines reported this story with favorable criticism [8].

4 Result Presentation and Evaluation

The story is about a dispirited young man who became a beggar sleeping in front of a train station. Some passersby despised him, others gave him money out of sympathy. However, there was one who gave him a box of tomatoes everyday (1 & 2) with a note saying "Compared to you, I have been unfortunate since childhood, yet, I pulled myself together. I can do it, so can you (3)." One day, the beggar read the note and was deeply inspired (4). He cleaned himself up (5), shaved his beard (6), then went to the nearby department store (7) in hopes of exchanging some clothes with the tomatoes. As he was rejected (8), a kind customer was filled with compassion and decided to pay for his clothes (9). The beggar was deeply touched by this deed (10).

Then, the beggar went to a hair salon in hopes of exchanging the tomatoes with a haircut. He was refused again. Right before he left with disappointment (11), one of the hairdressers asked the manager to bend the rules and allow him to give the beggar a haircut outside the shop for free. The manager couldn't withstand the hairdresser's pleading and had to agree with this idea (12 & 13). The beggar was very grateful to the hairdresser for his help and shared the tomatoes with everybody. The scene ended at this touching moment. After 6 months, an outstanding and eminent young man stepped into Far East University (14 & 15). He saw the person who used to give him tomatoes working in the tomato farm. The young man asked him, "Can I buy some tomatoes?" (16) They recognized each other and hugged (17). Eventually, this young man continued to spread out this love (18).

The story originated from the producer's childhood. He was an orphan and lived in an orphanage with his younger brother. When he was older, he was taken by his grandparents and raised by them. Being brought up by his grandparents was not easy, coupled with the bad economic situation of the family, hence, life for him was very hard. Later his grandfather passed away and his grandmother remarried. However, their economic situation remained the same. Growing up in this kind of environment made the producer stronger and enabled him to developed the attitude of facing problems with courage. During the production of this micro-film, the producer tried to use all available resources. He asked a male high school classmate to play the beggar, this classmate's girlfriend to act as the department store clerk, his vocational school teacher to be the customer in the department store, his fellow university society member to play the hairdresser, a professional director to be the hair salon manager, and several classmates to play the passersby and other roles.

In addition, the producer was well aware of the strength of teamwork. He found some dirty and smelly clothes for the beggar from the recycle center. He asked a classmate who was good at doing makeup to help create the proper visual representation for the beggar and other roles. He found a shop in the department store willing to close the shop for half a day for shooting and a hair salon willing to lend them the shop

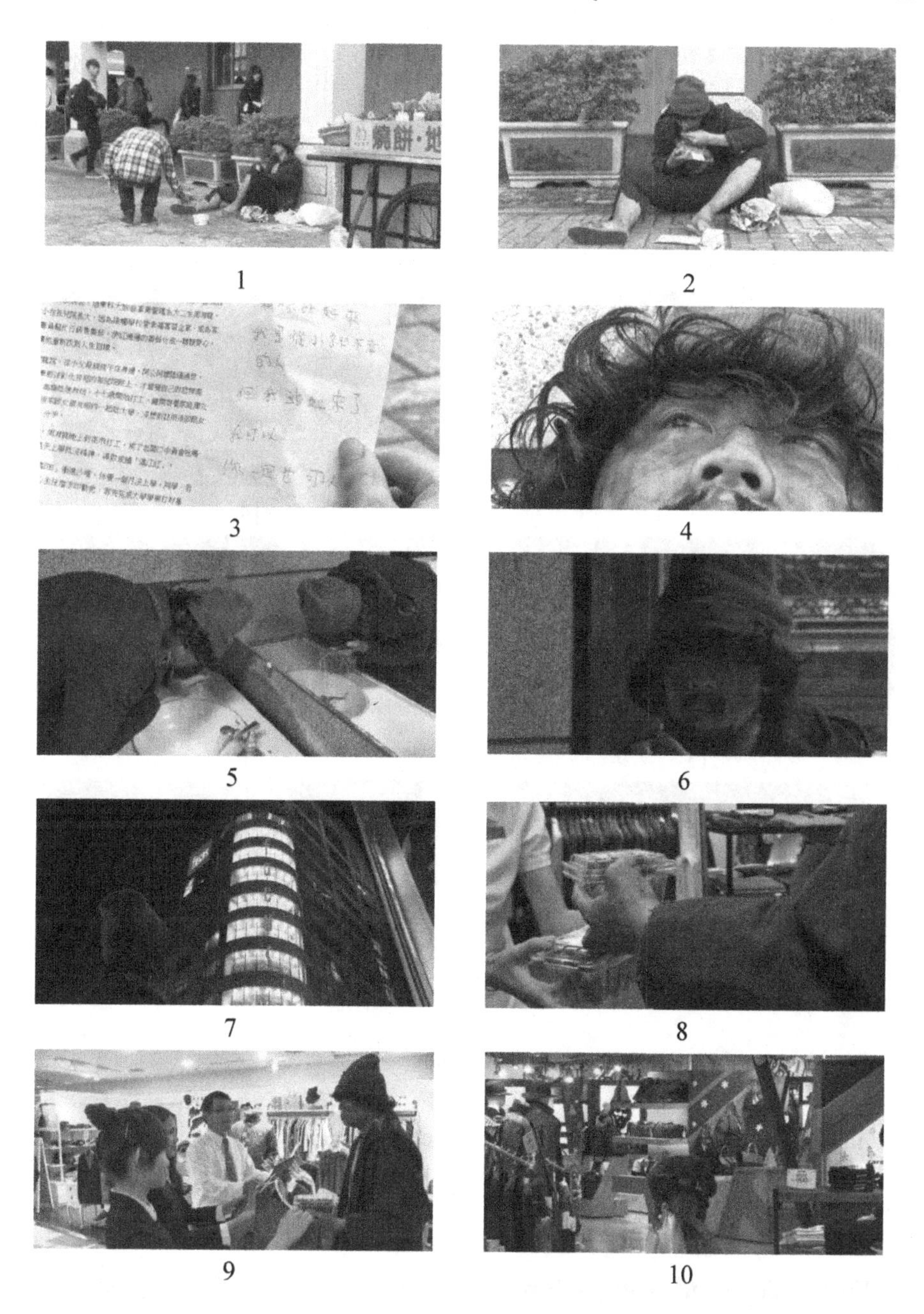

Fig. 1. Drama clip

Fig. 1. (continued)

at night after closing. All these were quite difficult for a sophomore university student. The producer completed an impossible task due to his hardworking attitude. Such character was really very impressive (Fig. 1).

5 Discussion and Suggestions

5.1 Discussion

This study uses "Deliver love through the tomato farm" as the topic for the Far East University micro-film creation. The plot of the film is very touching, and its features are summarized below:

(1) Heart-stirring personal experience: The producer himself had been through some tough times when he was a child. The story based on his personal experience is especially moving.
(2) Realistic costume and makeup with devoted acting: The beggar in the film looked just like a real beggar. While filming, some passersby even gave him money. Furthermore, he was stopped by a guard when he wanted to enter the department store. Apparently he looked like a real beggar.
(3) Genuine feelings and warmth everywhere: In the film, the beggar visited a department store and a hair salon. Because he had no money, he could not pay for clothes or a haircut. Yet, he was helped by others out of love. The tomatoes were able to deliver love and happiness.
(4) Great teamwork with concerted efforts: It was amazing that the producer was able to get so many resources. This fact showed that unity is strength and nothing is impossible if people can work together with one heart.

5.2 Suggestions

However, although the plot is moving, there is still space for improvements. The following suggestions can be used as references for related topics in the future.

(1) The unique experience is touching. However, this kind of life experience is not always necessary to produce a good creation. Reading a lot of books to expand views is also helpful in writing a good script.
(2) The place for shopping is controversial. According to some audience's feedback, the beggar had no money at all and should have gone to a traditional market instead of a department store. This opinion sounded reasonable during the moment. However, after careful consideration, the beggar and the department store gave a prominent contrast and created more conflicts. Yet, care must be given to avoid suspicion of product placement marketing.
(3) The haircut scene is a little bit dragging. It would have been better if the scene ended after the heart-warming act.
(4) Clicks seem to be an important indicator in evaluating an online film. Hopefully the producer will not be satisfied with this success and can create more deeply touching micro-films in the future to deliver more love and warmth.

References

1. CTV News, an orphan, the "prince of par-time jobs", and a second-grade university student changing his life by selling tomatoes. 25 Jan 2015
2. Filmmaking. https://goo.gl/2iGfi2. 8 Sep 2015
3. Hu, G.S., Chen, C.M.: A preliminary study of the application of micro-film in military marketing. Commun. Fu Hsing Kang Acad. J. **103**, 1–23 (2013)

4. Gong, J., Smith, M.D., Telang, R.: Substitution or promotion? the impact of price discounts on cross-channel sales of digital movies. J.Retail. **91**(2), 343–357 (2015)
5. Bakar, M.S.A., Bidin, R.: Technology acceptance and purchase intention towards movie mobile advertising among youth in Malaysia. Procedia-Soc. Behav. Sci. **130**(15), 558–567 (2014)
6. Micro-film. https://goo.gl/n9uYa8. 8 Sep 2015
7. Micro-film, Transfer love and happiness Fans. https://goo.gl/AzghcB. 10 Sep 2015
8. Transfer love and happiness. https://goo.gl/ABA2uL. 8 Sep 2015

Apply Image Technology to River Level Estimation

Ming-Tsung Yeh[1], Yun-Jhong Hu[1], Chien-Wen Lai[2], Chao-Hsing Hsu[3](✉), and Yi-Nung Chung[1]

[1] Department of Electrical Engineering, National Changhua University of Education, Changhua 500, Taiwan
ynchung@cc.ncue.edu.tw

[2] Changhua Christian Hospital, Changhua 500, Taiwan
70672@cch.org.tw

[3] Department of Information and Network Communications, Chienkuo Technology University, Changhua 500, Taiwan
hsu@cc.ctu.edu.tw

Abstract. In this paper, an image based approach of the water flow information measurement is proposed. Applying the image based measurement is safely and efficiently non-contact method. This paper proposes the multiple virtual water level probes (MVWLP) method which can apply in any river environment without ruler where has regular water line on the embankment. This approach mainly applies the color space adjustable technique to reduce noises and uses the adaptive edge detection to extract the water line. Then, it sets some virtual probes on the discovered water line comparing with the preset probes to measure the current water level. We convince that the proposed methods are accurate, robust and adaptable enough to overcome multiple conditions presented in the sites.

Keywords: Water flow information measurement · Multiple virtual water level probes · Smart phone APP system

1 Introduction

In Taiwan, the central mountain range bisects this island that results in most rivers have short watercourses, small basins and rapid flows. These rivers do not have obvious discrimination of upstream, midstream and downstream section due to the distinctive topography feature. The flow volume discharge is fewer during the dry season but the flood often happens during the rainy season that is consequent on the climate change. The water flows of rivers also have vast changes in the typhoon and flood situations, and cause great damage to people lives and properties. Hence, the water regimen of rivers is essential for the water resource management, hazard rescue and mitigation, hydraulic engineering, etc. The information of rivers includes water flow velocity, water level and flow volume discharge. The river infrastructures which have bridges, embankments, dams and reservoirs need these types of information to design and plan the disaster prevention. The water resource agency uses the flow velocity and water level to estimate the early water warning and note the downstream residents to evacuate

J. Pan et al. (eds.), *Genetic and Evolutionary Computing*, Advances in Intelligent Systems and Computing 536, DOI 10.1007/978-3-319-48490-7_11

during flood or typhoon. Therefore, it is important to real time monitor the river flows. All of these stations use the contacting method to measure the water information that requires instruments be placed into water. It can be dangerous in the high flow or flood conditions for handing the instruments. That can endanger the people making the examination and harm instruments for the debris floating on the river. The contacting method also need a lot of manpower for investigation, and furthermore are not real time to update data.

For people and instrument safe, the automatic measuring stations had been constructed and applied the non-contacting water level gauges which have the radar type or acoustic type, and hand-held surface velocity radar gun to obtain the water information since 2007. The non-contacting methods [1, 2] have some advantages such as automatic, convenience and efficiency. They applied the slop-area method to estimate the flow volume discharge. Due to the stations construction cost and the radar mode velocity meter price are very high, there have only two measuring stations to set up multiple linked surface velocity radar to measure the whole flow velocities and estimated the discharge by the flow surface velocity method. These stations use less manpower to operate and achieve long-term measurement that improves the hydraulic investigation efficiency but the cost is higher. Moreover, the computer and digital camera technologies have made great progress recently. For these reasons, it is highly desirable to design an image based measuring method which has non-contacting, safe, automatic, low cost and whole flow field measurement. It is mainly motivation of this study. The water resource agency can obtain this significant flow information real time to monitor and manage the river situation. Especially, the rescue team can use the smart phone to get water flow situation on going and avoid incurring danger. The steps of research algorithm in this paper are described as follows.

2 Image Processing

The image based water level gauge also need clear boundary between the water line and riverbank to determine height. The image enhancement is an efficient and common approach during pre-processing phase [3, 4]. There have many techniques presented to improve the image visibility and be helpful for finding the tracers movement correlation. In this paper, some picked techniques will be implemented. It is useful to apply the intensity adjustment of curve mapping for this case because the image need to weight toward darker. In this paper, we propose to combine the linear and curve transformation that applies the gamma argument to convert again after linear intensity adjustment. It can be flexible to apply the curve gamma argument to reduce the effect of illumination changes. That sets the gamma value lower than 1 to overcome the poor light situation at night time, dusk, dawn or cloudy day. The gamma can be set greater than 1 to mapping the pixel value to darker when the vivid light affects the profile of surface. The experimental results noted the compound method improve the valid velocity vectors about 2 %.

The histogram equalization (HE) is an image processing technique used to adjust the image intensities for enhancing contrast. It assumes the image intensities range is normalized continue numeric between 0 and 1. The $p_r(r)$ is the possibility density function (PDF) of inputted image intensities. The image r is transferred to s via the transformation function T as Eq. (1).

$$s = T(r) = \int_{0}^{r} p_r(x)dx \tag{1}$$

The T assumes differentiable and invertible for simplicity. Then, it can be shown that the PDF of output intensities $p_s(s)$ after transformation is uniformly distributed on range 0 to 1 as Eq. (2) [5, 6].

$$p_s(s) = \begin{Bmatrix} 1 & \text{when } 0 \leq s \leq 1 \\ 0 & \text{others} \end{Bmatrix} \tag{2}$$

After the intensity equalization process, it results in an output image with increasing dynamic distribution range of intensities which tend higher contrast scene. Hence, that can be noted the transformation function T is just a cumulative distributive function (CDF). In general, the histogram of processed image is not exactly flat even though the variables have discrete attribute. The value of normalized histogram is an approximation of possibility each intensity $p_r(r)$. It applies the sum and is equivalent to transform the pixel intensities by the function Eq. (3).

$$s_k = T(r_k) = \sum_{j=1}^{k} p_r(r_j) = \sum_{j=1}^{k} \frac{n_j}{n} \tag{3}$$

Where $p_r(r_j), j = 1, 2, \ldots, L$ denote as the histogram of inputted image intensities, L is the number of possible intensity values, it is often 256, n is the amount of total pixels and n_j is the amount of pixels for intensity r_j. The s_k is the intensity of output image correspond to the intensity r_k of input image and $k = 1, 2, \ldots, L$. The HE is a global method that spread out the histogram components of output image after transformation over all intensity range from 0 to 255 in 8 bit images. It obtains the processed image with approximately flat distribution. Sometimes, that is known to be not uniform for the distribution of the intensities of an image.

Due to the camera electronics and environmental conditions, images are added some noises to the original pixel value. These noises are categorized as salt and pepper noise, impulse noise and Gaussian noise. It is impulse noise for the Gaussian noise but its intensity values are drawn from a Gaussian distribution. The Gaussian noise often presents on the image took in outdoor environment because it has uncontrolled illumination variation especially at night time or vivid lighting. These situations produce the reflections from the river surface and strongly affect the original signals. Hence, that needs to have a filter to filter out noise. Gaussian filter is linear smoothing filter with the weights are computed according to a Gaussian function shown as Eq. (4).

$$g(x) = ce^{-\frac{x^2}{2\sigma^2}} \text{ for one variable.}$$
$$g(x,y) = ce^{-\frac{(x^2+y^2)}{2\sigma^2}} \text{ for two vaiables.} \quad (4)$$

Where c is constant and some scale factor, and σ is user defined. It usually set c a larger number that cause all mask elements are integers to create a mask for filtering. Gaussian smoothing is very effective for removing Gaussian noise drawn from a normal distribution. It is a linear low pass filter. To reduce border blurring, the pixels near border are given higher significance for the weights. There is another adaptive linear filter applied the Weiner filter for smoothing image. This 2D adaptive Wiener method can operate best when the noise is constant power additive noise, a random value is added for each pixel, such as Gaussian noise.

3 Digital Water Level Gauge

In general, the image based measurement of water level use image processing technologies to recognize the numeric of staff ruler and detect the water level. It also needs some image calibrations to convert the pixel distance to real metric unit as meter. Then, using these two measured known parameters estimate the actual water level by linear interpolation method or mapping table. This method has good detection effect on the reservoir environment because that has obvious ruler set on the embankment and steady water line presented. For the outdoor river environment, the staff ruler is not always available in the field or just painted on the pier of bridge with dirty. The water line detection has seriously affected by the dirty gauge. For accurately measuring the water level, it may need excess image processing techniques to reduce noises but have poor efficient. That may need another video camera to capture images from the pier for measuring the water level. Therefore, the paper proposes multiple virtual water level probes (MVWLP) method to estimate water level applied the same images as measuring flow velocity in this study. This method mainly detects the current water line and compare to the pre-measured water level for simply obtaining actual water depth. This approach has main three parts that have image segmentation, image processing and water level measurement. The additional step is to correct the pixel distance to the actual metric as the distance unit calibration.

The images used for the water level measurement are same as the flow velocity detection. It does not need to process entire image because the MVWLP method sets the detecting probes on the particular locations. Therefore, that only needs to segment a portion of image as the ROI where the virtual sensors located for better computing and processing efficiency. Due to the variable outdoor environment, it should set a lot of probes on the probable water lines to prevent single point fault. Hence, it severs two pieces of sub-image from left and right side of original image in this work. The image processing for water level measurement includes color space transformation, image enhancement, noise elimination and morphological process. The image enhancement uses the same image processing techniques. Hence, it focuses on the other three items here. The noise elimination always follows the color space transformation and morphological process. It can be easy to separate the additional noises after transformation. That applies the morphological process to acquire the edge of boundary between the water bodies and riverbank. It is the successful key factor to detect the water line.

In order to remove the shadow of image, the HSV color space is chosen in this paper. The HSV color system uses cylindrical coordinate to present pictures different from the RGB system. It stands for the hue (H), saturation (S) and value (V) components. The HSV color space is more close to the human visual perceptiveness than RGB system. The HSV system uses the lightness value to present the object hue and therefore it can reduce the influence of shadow and ambient light change to overcome the RGB system drawback. The measurement of water level needs to acquire the accurate water line on the riverbank. There have a lot of methods to find the edge of object in the image. The gradient based edge operator is the most frequently used as edge detector. They have two kinds of model for these operators including 1st derivation and 2nd derivation. The 1st derivation method applies 1st differential of image intensity function to compute gradient and mark edge points. In other words, it uses the magnitude of 1st derivation to detect an edge. The 2nd derivation method uses 2nd differential of image intensity function to acquire the variant of gradient and finds edges. It uses the sign of 2nd derivation to exhibit the edge direction. However, the measurement of water level needs simple and obvious water line to estimate. For acquiring accurate water line, it applies the dilation and erosion algorithms of morphological process to clean up unnecessary noises in this study. Applying these structuring elements for morphological processing, it can emphasize the horizontal water line and reduce the noises effects. The additional noises can eliminate by the erosion operation. It has pure water line presented on this picture after filtering out noises but the line has some fractures. Then, that applies the dilation and erosion operation again to smooth the boundary of water line.

4 Set Up Virtual Probes and Estimate Water Level

The segmented image uses the pixels as distance unit that is different from the real metric measurement. Therefore, before measuring the water level on the extracted line, it needs to calibrate and convert the pixel distance to the real metric unit. First, a lot of pre-measuring points and lines need to set on the reference image. As Fig. 1(a) shown, it sets a reference base for the current water line and draw a top watermark for the reference origin that mean the distance is calculated from here. Figure 1(b) shows the real detecting image on the river field which segment from the image of water velocity measurement. The actual water level on the current water base line denotes as W_0 which is used for the further comparing reference value. The unrectified raw image has not mean linear distance distribution and therefore it needs more points on each direction (X and Y) to calibrate the really metric length. In this study, it proposes to use the second order polynomial function with three independent variables to define the variations of actual metric measure related to the vertical pixels distance as (4). The three variables a, b, and c can be obtained by using a least square approach to find the solution as (5). It sets three virtual lines with nine reference points that actually measure the really metric distance m_{ij} of each point from the top watermark as Fig. 2 shown. Then, each virtual line applies three pairs of vertical pixel and metric distance (m_{ij} and y_{ij}) to substitute to (5) and get the solution of three variables.

$$m_i = ay_i^2 + by_i + c \tag{4}$$

$$\begin{bmatrix} m_{i0} \\ m_{i1} \\ m_{i2} \end{bmatrix} = \begin{bmatrix} y_{i0}^2 & y_{i0} & 1 \\ y_{i1}^2 & y_{i1} & 1 \\ y_{i2}^2 & y_{i2} & 1 \end{bmatrix} \begin{bmatrix} a \\ b \\ c \end{bmatrix} \tag{5}$$

Where m_i is the actual metric measure of vertical line i, y_i is the vertical pixels distance of vertical line i calculated from the top watermark. m_{ij} and y_{ij} are the vertical metric and pixel distance of each virtual line from the top watermark.

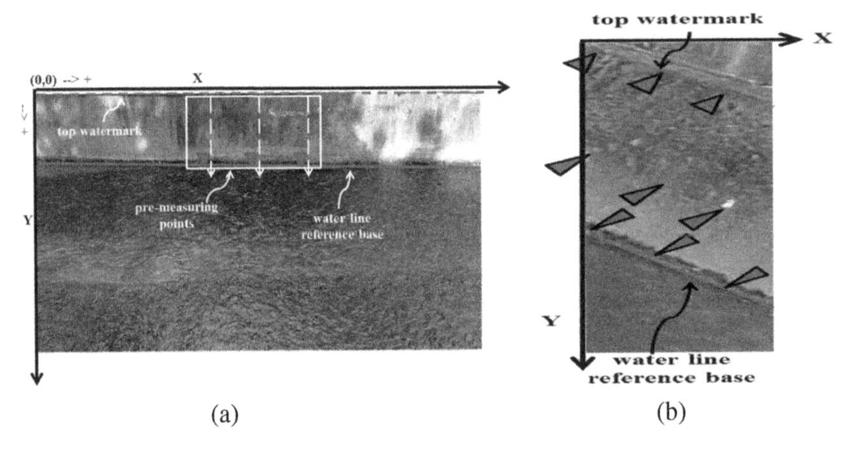

Fig. 1. (a) Pre-defined lines and measuring points for distance calibration (b) The measuring environment set up on the river field

There have a lot of uncertain conditions presented on the outdoor environment. Therefore, it implements three sensor probes on the water line per measuring estimation to prevent single point fault. For simply processing, these probes are put on the same X coordinate with the pre-measuring points on base line. Hence, the distance difference between the current water line and base line can obtain from the subtraction of Y coordinate of probes and corresponding reference points for each virtual line. In Fig. 3, there are three different distance measures for each virtual line respectively. The waterline is not fully flat due to the waves are presented on the surface that causes to have different measuring results for each sensor probe on the same line. To overcome this disadvantage, it proposes to use (6) to calculate the actual water level in this study. It averages all estimating results of sensor probes and then adds the water level of base line. That provides higher precision for the estimation of water level.

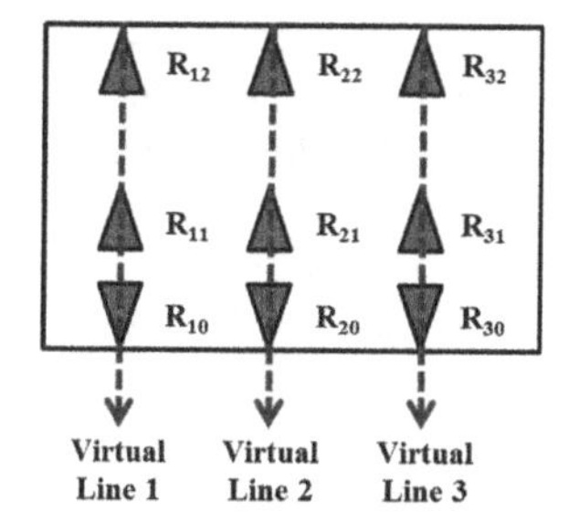

Fig. 2. Virtual line with reference points

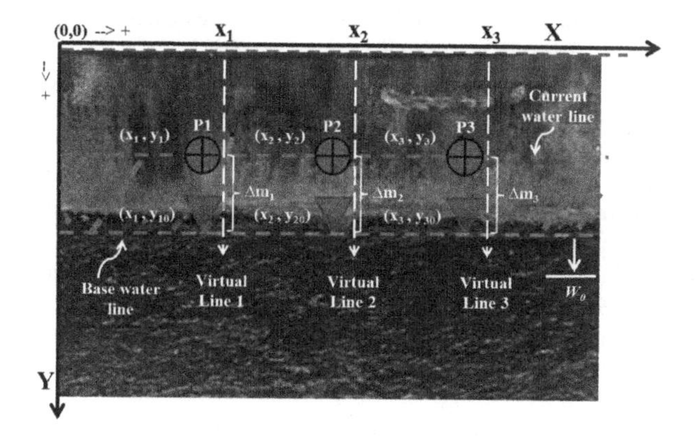

Fig. 3. The principle of the MVWLP

$$\Delta m_i = m_i - m_{i0}$$
$$W = W_0 + \frac{1}{n}\sum_{i=1}^{n} \Delta m_i \tag{6}$$

Where m_i is real metric distance corresponding to y_i, Δm_i is the distance between the position of sensor probe and pre-measuring point on base line, W is the actually current water level, n is the virtual line quantities, and i is the virtual line number that is $1 \sim 3$. Based on the experimental results, this technology can provide fast and accurate information.

5 Conclusion

In this paper, an image based measurement of water level and river flow information are presented. For acquiring the favorable detecting images, the compound image enhancement is used to intensify the tracer images of river surface that is in conjunction with rectified distorted images during the pre-processing steps. The experimental environments include daytime, nighttime, multiple flow and weather conditions that present the most situations for long-term investigation. Based on the experimental results, this technology can provide fast and accurate information. It has proved to be robust and adaptable enough, and have better estimations.

Acknowledgments. This work was supported by the Ministry of Science and Technology under Grant MOST 103-2221-E-018-017- and MOST 105-2221-E-018-023.

References

1. Jodeaua, M., Hauetb, A., Paquiera, A., Le Coza, J., Dramaisa, G.: Application and evaluation of LS-PIV technique for the monitoring of river surface velocities in high flow conditions. Flow Meas. Instrum. **19**(2), 117–127 (2008)
2. Zaragoza, J., Chin, T.-J., Tran, Q.-H., Brown, M.S., Suter, D.: As-Projective-As-Possible image stitching with moving DLT. IEEE Trans. Pattern Anal. Mach. Intell. **36**(7) (2014)
3. Dabiri, D.: Cross-correlation digital particle image velocimetry – a review. Exp. Fluids **46**(2), 191–241 (2009)
4. Gonzalez, R.C., Woods, R.E., Eddins, S.L.: Digital Image Processing Using MATLAB, 2nd edn. Gatesmark Publishing, Upper Saddle River (2009)
5. Keane, R.D., Adrian, R.J.: Theory of cross-correlation analysis of PIV images. Appl. Sci. Res. **49**(3), 191–215 (1992)
6. Keane, R.D., Adrian, R.J., Zhang, Y.: Super-resolution particle image velocimetry. Meas. Sci. Technol. **6**, 754–768 (1995)

Arena Simulation for Aviation Passenger Security-Check Systems

Chia-Hung Wang[1,2](✉)

[1] College of Information Science and Engineering, Fujian University of Technology, Fuzhou 350118, Fujian, China
jhwang728@hotmail.com
[2] Fujian Provincial Key Laboratory of Big Data Mining and Applications, Fuzhou 350118, Fujian, China

Abstract. In this paper, we develop an Arena simulation model to study the optimal design of security-check systems for screening passengers in airports. With the increasing of terrorist attacks, the airport safety has become more and more important, but the enhanced security increases transportation costs and reduces airport efficiency. Through the numerical experiments with our Arena simulation models, it demonstrates the effects of the risk threshold for further inspection on the passengers' average waiting time and the airport's safety level.

Keywords: Simulation · Aviation security · Passenger checkpoint screening · Queueing system · Waiting time

1 Introduction

The rise of terrorist attacks in the last several years has made the security screening of airline passengers an essential feature of air travel [1, 2]. Within the aviation security-check system, the preeminent objective of passenger and baggage screening is to prevent prohibited items from entering the airport terminal and getting put onboard a commercial aircraft. The security screening procedures must have a high degree of reliability, because the possible loss of life and aircraft are enormous.

However, the passenger screening is costly to implement and operate [3]. The congestion and delay caused by increasing the airports' security-check result in unsatisfactory levels of customer service, and those airports whose security screening procedures are particularly time-consuming can expect to lose business to their competitors. Besides, the national economy is seriously affected by the operational efficiency of the airports because they play vital roles in transporting passengers and cargos. Hence, the trade-off between aviation safety and efficiency has been an important research issue in decades [2, 4, 5]. Zhang et al. [6] studied a two-stage security-check system and examined the trade-off between maximizing the security screening level and minimizing the expected customer delay at the U.S.-Canadian border crossings. Wang et al. [4] developed a modelling framework to understand the economic trade-offs embedded in container-inspection decisions, where two important measures at a security-check system were addressed, i.e., the security screening effectiveness and the efficiency.

J. Pan et al. (eds.), *Genetic and Evolutionary Computing*, Advances in Intelligent Systems and Computing 536, DOI 10.1007/978-3-319-48490-7_12

Aviation security operations utilize a prescreening system to help quantify the perceived risk of passengers [7, 8]. This risk information is then used to perform decisions on how each individual passenger and his/her baggage should be screened within the security checkpoints. Due to the limited budget and screening devices resources, only a fraction of the passengers may be screened at the highest security levels [8]. Thus, passengers deemed to pose a higher risk of carrying a threat are screened in security classes containing specialized detection devices and procedures of a more time consuming and operationally expensive nature. With an automated pre-screening system used to determine passengers' perceived risk levels, McLay et al. [9] proposed a methodology to sequentially assign passengers to aviation security resources. Wang [10] studied the trade-off between system security and congestion in a risk-based checkpoint screening system with two kinds of inspection queues, i.e., Selectee Lanes and Non-selectee Lanes. Nie et al. [5] assumed that passengers are classified into several risk classes via some passenger prescreening system, and proposed a simulation-based Selectee Lane queueing design framework to study how to effectively utilize the Selectee Lane resource. McLay et al. [11] studied a sequential stochastic passenger screening problem for aviation security. A fuzzy inference system was developed in [12] to manage the process of passenger security control at an airport. Emerging simulation research is relevant for today's operations managers, and simulation modelling tools can help airport administrations to understand the system behavior and improve the decision process [13].

In this paper, an Arena simulation model will be developed to study the queueing analysis and optimal design of security-check systems for screening aviation passengers in airports. The Arena modeling system from Systems Modeling Corporation is a flexible and powerful tool that allows analysts to create animated simulation models that accurately represent virtually any system [14]. We consider two types of inspection stations in the aviation security-check systems. According to the relevant information about passengers and their risk values, we assign passengers to the appropriate type of security-check stations. That is, when an incoming passenger's risk value is larger than (or equal to) a fixed risk threshold, he/she should go for strict security-check. We take both the waiting time and budget constraint into consideration, and study the optimal design of the aviation security-check systems. Through the analysis of our simulation models, we would like to determine the risk threshold for further inspection and the optimal number of screening devices under the budget constraint.

The paper is organized as follows: In Sect. 2, we introduce the system description and mechanism for the aviation passenger security-check systems. The design of Arena simulation model is proposed in Sect. 3. Section 4 shows the sensitivity analysis of the proposed simulation model on the performance evaluations. Finally, the conclusions are summarized in Sect. 5.

2 System Descriptions

When aviation passengers arrive at the airports, they must pass through the security-check systems for inspection procedures of the Customs. In most airports, there is an automated prescreening system, which can quantify the risk associated with the characteristics of the incoming arrivals. Before crossing the security-check systems, passengers are classified into two risk classes via the automated prescreening system. Each passenger is characterized by an assessed risk value (between zero and one), determined by an automated prescreening system with information and data available from the Customs [10].

There are two kinds of inspection queues in this aviation passenger security-check systems: a Selectee Lane with enhanced scrutiny and a Non-selectee Lanes. Passengers receive one of these two types of security-check procedures based on their attributes (lower risk or higher risk screening). With a risk threshold, passengers are differentiated by their perceived risk levels and are assigned to either the Selectee Lanes or Non-selectee Lanes. For example, when an incoming passenger's risk value is larger than or equal to the risk threshold, he/she should go to Selectee Lane for strict security-check. In contrast, when an incoming passenger's risk value is smaller than a fixed risk threshold, he/she will go to Non-selectee Lanes.

The security-check procedure consists of two distinct operations: inspecting the passenger's carry-on bags and inspecting the passenger himself. There exist two possible errors in the security-check systems, that is, false alarm and false clear. Whatever which error occurs, it will result in a considerable loss of the airport. Given that there is an arrival who is a threat, the total security can be interpreted as the conditional probability that a threat is detected in this security-check system [10]. The total time required to screen a passenger is a function not only of the continuous throughput rate of the electronic equipment (or manual search), but also of the false-alarm rate of the electronic equipment (or manual searchers), and the walking distance required of the passenger through the security checkpoint as well [15].

We assume passengers are screened based on a First-Come-First-Served order. The information about the purchase and overhead costs for the inspection devices would be available from published articles on security checkpoints in the public domain before these screening devices are incorporated into security-check systems.

The model parameters of the security screening operation include:

(1) the passenger screening or service rate (per inspection station),
(2) the passenger arrival rate at the security screening checkpoint, and
(3) the number of screening (service) devices.

The screening rate would depend on whether the inspections are manual/visual or use electronic equipment for automation, the speed of the equipment used, and the operator's skill. The passenger arrival rate would depend on the airlines' departure schedules and the distribution of passenger arrival times in the terminal. The number of security screening devices that the airport administration decides to install is the main controllable variable which can cope with congestion in the security screening operation.

The decision variables include the number of screening devices in Selectee Lanes, the number of screening devices in Non-selectee Lanes, and the risk threshold used to differentiate between selectees and non-selectees. The goal of this work is to analyze the security level and average system congestion of aviation passenger security-check systems in airports. To understand the system behavior of the security-check systems, we develop Arena simulation models to represent components in real-world processes.

3 Design of Arena Simulation Model

In this section, we develop a simulation model with the Arena simulation software to represent components of the aviation passenger security-check system. Our Arena simulation model is a collection of modules, including data modules, logic modules and process modules. Each module contains all of the model parameters, logic, and animation necessary to describe its specific portion of the aviation passenger security-check system. Through supplying values in the module dialogue box, we can build a small model of that component in the real security-check process at an airport.

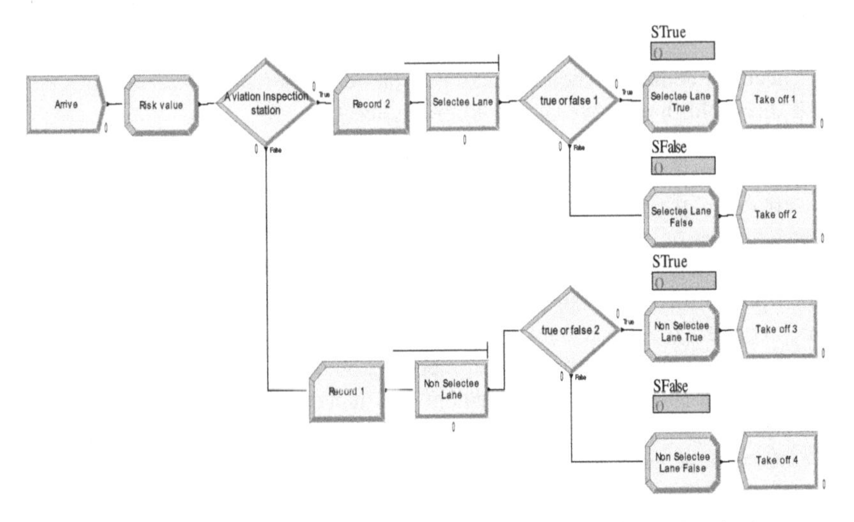

Fig. 1. Flowchart of Arena simulation model for an aviation security-check system.

The flowchart of our Arena simulation model is illustrated in Fig. 1. The aviation passenger security-check system is constructed in the Arena modelling environment by selecting data, logic, and process modules and defining them in the context of the system. The Arrive module generates passengers (randomly) to the security-check system. The Risk Value module assigns a (different) risk value to each incoming passenger. The Aviation Inspection Station module represents the automated prescreening system and differentiates passengers with a risk threshold. The Record 1 module represents the number of passengers in the Non-selectee Lanes, and the Record 2 module represents the number of passengers in the Selectee Lanes. The Selectee Lane module represents the more stringent security-check stations, and the Non-selectee Lane module represents the normal inspection stations. The True or False module distinguishes the true dangerous passengers with threat and the safe passengers. The Selectee Lane True and Non-selectee Lane True modules record the number of dangerous passengers; on the other hand, the Selectee Lane False and Non-selectee Lane False modules record the number of safe passengers. The Take Off modules represent the passengers exist from the security-check system.

When passengers arrive, they must pass through the screening system in aviation security-check system. The customs check the passenger's identity, and assign the risk value to each passenger with an automated prescreening system. Passengers are differentiated into different groups according to a given risk threshold, and then are assigned to appropriate type of security-check stations based on their risk values. There are two type of passenger security-check stations in the system, where the ability of screening is higher in Selectee Lanes and the ability of screening is lower in Non-selectee Lanes. When a certain passenger's risk value is smaller than the given risk threshold, he/she should go to the Non-Selectee Lane for normal inspection; on the other hand, when his/her risk value is larger than the risk threshold, he/she should go to the Selectee Lanes for strict inspection.

4 Numerical Experiments

In this section, we determine the security level and average system congestion in the proposed Arena simulation models. In our experiments, we assume the risk value of passengers follows a Truncated exponential distribution, and its expected value is given as 0.0625. The experimental parameters are selected carefully according to the related literature [9, 11]. Our numerical results are determined through averaging 8 replications of Arena simulation runs. Our numerical experiments are run through Arena simulation software on the PC platform with Intel(R) Core(TM) i7-3770 CPU @ 3.40 GHz and 32 GB RAM.

First, we study the effect of varying the risk threshold on the security level and average waiting time for the security-check systems. In this experiment, the average arrival rate of passengers is given as 1 passenger per unit time. We set the number of screening devices for Selectee Lanes as 2, and the number of screening devices for Non-selectee Lanes is also fixed as 2. The average service rate of Selectee Lanes is given as 1.5 passengers per unit time, and the average service rate of Non-selectee Lanes is 2 passengers per unit time. The numerical results are summarized in Table 1.

Table 1. Numerical results of varying the risk threshold on the security level and average waiting times of security-check systems.

Risk threshold	Proportion of strict security-check	Average waiting time for non-selectee lanes	Average waiting time for selectee lanes	Average waiting time for the system	Security level
0.00	100 %	0.0000	4.7736	4.7736	0.9800
0.01	85.07 %	0.0272	3.6812	3.1357	0.9382
0.02	72.50 %	0.0928	2.6090	1.9170	0.9030
0.03	61.79 %	0.2562	1.9539	1.3051	0.8730
0.04	57.50 %	0.4718	1.3449	0.9738	0.8610
0.05	44.71 %	0.5682	1.1332	0.8208	0.8252
0.10	20.14 %	1.2371	0.1663	1.0214	0.7564

(*continued*)

Table 1. *(continued)*

Risk threshold	Proportion of strict security-check	Average waiting time for non-selectee lanes	Average waiting time for selectee lanes	Average waiting time for the system	Security level
0.15	8.79 %	1.5533	< 0.0001	1.4167	0.7246
0.20	4.00 %	1.8198	< 0.0001	1.7470	0.7112
0.25	1.57 %	1.9172	< 0.0001	1.8871	0.7044
0.30	0.71 %	1.9256	< 0.0001	1.9118	0.7020
0.35	0.36 %	1.9775	< 0.0001	1.9704	0.7010
0.40	0.14 %	1.9849	< 0.0001	1.9821	0.7004
0.50	0.07 %	2.0020	< 0.0001	2.0006	0.7002
0.60	0.00 %	2.0023	0.0000	2.0023	0.7000
0.70	0.00 %	2.0023	0.0000	2.0023	0.7000
0.80	0.00 %	2.0023	0.0000	2.0023	0.7000
0.90	0.00 %	2.0023	0.0000	2.0023	0.7000
1.00	0.00 %	2.0023	0.0000	2.0023	0.7000

In Table 1, it can be found that the passengers' average waiting time for the whole security-check system will decrease to the minimum and then increases when we increase the risk threshold. As the risk threshold is around 0.05, it approaches to the minimum value of the passengers' average waiting time. It also shows that the security level of the security-check system is decreasing when increasing the risk threshold.

Next, we study the effect of varying the number of screening devices on the average waiting time for two types of inspection lanes. The numerical results are summarized in Table 2. Here, we fix the value of risk threshold as 0.05, and the average arrival rate is given as 3 passengers per unit time. Other parameters of our Arena simulation model are unchanged. From the numerical results in Table 2, it shows that the average waiting time can be reduced by increasing the number of screening devices for each type of inspection lanes. However, the reduction is getting smaller due to the fixed average arrival rate.

Table 2. Numerical results of varying the number of inspection devices for two types of inspection lanes on the average waiting times.

Number of screening devices		Proportion of strict security-check	Average waiting time for non-selectee lanes	Average waiting time for selectee lanes	Average waiting time for the system
Non-selectee Lane	Selectee Lane				
1	2	0.4692	320.2200	18.6933	178.7298
1	3	0.4611	269.1600	9.1978	149.3014
1	4	0.4763	265.9600	6.1848	142.2278
1	5	0.4975	260.2300	4.5458	133.0220

(continued)

Table 2. *(continued)*

Number of screening devices		Proportion of strict security-check	Average waiting time for non-selectee lanes	Average waiting time for selectee lanes	Average waiting time for the system
Non-selectee Lane	Selectee Lane				
1	6	0.4692	259.4200	3.6028	139.3788
2	1	0.4630	23.2771	548.5700	266.5117
3	1	0.4683	11.0027	453.9300	218.4052
4	1	0.4630	6.8299	450.9400	212.4731
5	1	0.4630	5.0143	452.5100	212.2252
6	1	0.4630	3.9432	473.8300	221.5222
2	2	0.4688	24.8265	30.1819	27.3368
3	3	0.4705	11.3257	13.9342	12.5530
4	4	0.4653	7.0961	8.3817	7.6943
5	5	0.4658	5.0867	6.0130	5.5181
6	6	0.4683	3.8719	4.6591	4.2405

5 Conclusions

In this paper, we construct a simulation model of the aviation security-check systems for screening passengers in airports. In the sensitivity analysis, we analyze the relation between the model parameters in the proposed model and the observation values of simulation system. We find that both the system congestion and total security are affected by the proportion of passengers arriving to each type of inspection stations. Through studying the characteristics of the proposed simulation model, we determine the optimal design of the aviation security-check systems under the budget constraint. The application of this study would be helpful in designing passenger security-check systems for airports. It is an important research contribution in the academia and managerial practice to deal with the increasing of terrorist attacks these days.

In the future works, more efficient simulation optimization algorithms or heuristics could be designed through understanding the characteristics of the proposed simulation model. Another research direction would focus on testing the other objective functions, such as different inspection costs or security policies.

Acknowledgments. This work was supported by Fujian Provincial Department of Science and Technology, China, under Grant 2016J01330.

References

1. Lee, A.J., Nikolaev, A.G., Jacobson, S.H.: Protecting air transportation: a survey of operations research applications to aviation security. J. Transp. Secur. **1**, 160–184 (2008)
2. Wong, S., Brooks, N.: Evolving risk-based security: a review of current issues and emerging trends impacting security screening in the aviation industry. J. Air Transp. Manage. **48**, 60–64 (2015)

3. Gillen, D., Morrison, W.G.: Aviation security: costing, pricing, finance and performance. J. Air Transp. Manage. **48**, 1–12 (2015)
4. Wang, C.-H., Wu, M.-E., Chen, C.-M.: Inspection risk and delay for screening cargo containers at security checkpoints. In: 11th International Conference on Intelligent Information Hiding and Multimedia Signal Processing (IIH-MSP-2015), pp. 211–214. IEEE Press, New York (2015)
5. Nie, X., Parab, G., Batta, R., Lin, L.: Simulation-based selectee lane queuing design for passenger checkpoint screening. Eur. J. Oper. Res. **219**, 146–155 (2012)
6. Zhang, Z.G., Luh, H.P., Wang, C.-H.: Modeling security-check queues. Manage. Sci. **57**, 1979–1995 (2011)
7. Sewell, E.C., Lee, A.J., Jacobson, S.H.: Optimal allocation of aviation security screening devices. J. Transp. Secur. **6**, 103–116 (2013)
8. Lee, A.J., Jacobson, S.H.: Addressing passenger risk uncertainty for aviation security screening. Transp. Sci. **46**, 189–203 (2012)
9. McLay, L.A., Lee, A.J., Jacobson, S.H.: Risk-based policies for airport security checkpoint screening. Transp. Sci. **44**, 333–349 (2010)
10. Wang, C.-H.: A modelling framework for managing risk-based checkpoint screening systems with two-type inspection queues. In: 3rd International Conference on Robot, Vision and Signal Processing (RVSP 2015), pp. 220–223. IEEE Press, New York (2015)
11. McLay, L.A., Jacobson, S.H., Nikolaev, A.G.: A sequential stochastic passenger screening problem for aviation security. IIE Trans. **41**, 575–591 (2009)
12. Skorupski, J., Uchroński, P.: Managing the process of passenger security control at an airport using the fuzzy inference system. Expert Syst. Appl. **54**, 284–293 (2016)
13. Snowdon, J.L., MacNair, E., Montevecchi, M., Callery, C.A., El-Taji, S., Miller, S.: IBM journey management library: an arena system for airport simulations. J. Oper. Res. Soc. **51**, 449–456 (2000)
14. Kelton, W.D., Sadowski, R.P., Sadowski, D.A.: Simulation with Arena, 2nd edn. McGraw-Hill, New York (2002)
15. Gilliam, R.R.: An application of queueing theory to airport passenger security screening. Interfaces **9**, 117–122 (1979)

Simulation Model and Optimal Design for Call Center Staffing Problems

Chia-Hung Wang[1,2(✉)] and Mao-Hsiung Hung[1]

[1] College of Information Science and Engineering, Fujian University of Technology, Fuzhou 350118, Fujian, China
jhwang728@hotmail.com, mhhung@fjut.edu.cn
[2] Fujian Provincial Key Laboratory of Big Data Mining and Applications, Fuzhou 350118, Fujian, China

Abstract. This paper studies the optimal design for staffing problems in call centers with the use of Arena simulation software. Call center is a labor-intensive business model and spends the major cost in hiring staff, so this research on the staffing optimization can help in reducing the operation cost of call centers. We provide managerial schemes to reduce staffing costs and to improve utilization of personnel at a certain guaranteed service level for the call centers. Through our analysis of the proposed Arena simulation model, several optimal solutions to staffing problems are obtained in our numerical experiments with computer simulation.

Keywords: Simulation · Call center · Queueing system · Arena software · Staffing

1 Introduction

Call centers play an important role in communication with customers, which has become an indispensable part of modern business [1, 2]. According to the survey, the cost of hiring staff in a call center is about 60 % to 70 % of the total operation costs. The saving of few percentage of labor payment often means the reduction of the cost of several million dollars in a large enterprise. Therefore, our research goals are to study how to guarantee the service quality at a certain service level and how to reach the minimal staffing cost in a call center.

In this paper, we use a computer simulation system to model a queuing system of a call center. For a queuing system, traditional mathematics methods are difficult to solve the problem and even cannot obtain a solution [3]. Fortunately, modern computer simulation technologies can be developed to solve the problem effectively, e.g., [4–6]. Combining customer satisfaction and customer service costs, the simulation model is built and executed to obtain some performance parameters in a queuing system, such as customer's waiting time and stay time, service staff's working efficiency and so on. Through the analysis of these parameters, we can achieve more reasonably optimized configuration to meet some requirements of the increments of the system's operation efficiency, the reduction of business cost, and the maintaining of customer satisfaction.

J. Pan et al. (eds.), *Genetic and Evolutionary Computing*, Advances in Intelligent Systems and Computing 536, DOI 10.1007/978-3-319-48490-7_13

For the kind of queuing systems of call centers, service modes and configuration are somewhat different in call centers of different companies [7, 8]. Our proposed simulation model needs few adjustments to adapt different call centers according to the analysis of different system structures and operation rules. As a result, various kinds of optimization solutions can be carry out by the proposed model. The application of this research work could help enterprises largely reduce the customers' waiting time, and then take advantages to enhance their completion. Through the simulation experiments, enterprises can reduce unnecessary expenses and raise the usage of staffing.

2 Design of Arena Simulation Model

In this section, we develop our simulation models with Arena software for studying the queuing management in a call center. The Arena simulation models operate and process the flow optimization based on queuing analysis. The elements of our Arena simulation model are demonstrated as follows.

The Entities are dynamic objects created by simulation system. These dynamic objects can represent objects to be served or to be processed. These objects walk through specific operation and then disappear after they leave the system, such as customers in our study. An attribute is added to an entity to make it characterized. Different entities use different attributes to distinguish each other. In other words, an attributes is a labels to distinguish an entity from other entities. A variable reflects one of system's features which is used to calculate statistics such as average waiting time, maximum waiting time, resource utilization, and so on. Entities usually finish service under a resource. A resource often presents staffs, equipment and spaces. When a resource is seized, it will be released after finishing service. When an entity calls for a resource, if the resource is seized, the entity needs to stay a place (queue) to wait, i.e., a customer stays in a queue to wait for service. An event means a happening of something in the system such as customer's arrival and leaving. A clock means a system's clock in the simulation.

The Create module is a starting point of a model. For a model of call center, the customer arrival is the starting point. We use a Create module to generate entities presenting customers. The Process module is used to store resource, queue and entity. In our model, we set the number of staffs and customer types in Process modules. The Dispose module is used to delete entities of completed service. The Decide module is used to separate entities to different queues depending on their attributes. The expression can be set to operate complex selection in Decide modules. The Assign module is used to assign a value to user defined variables in the system. The Record module is used to record passing entities and attributes. It equips some statistical functions. The Entity module is used to define an entity and modify its attributes. The Resource module is used to set resource in the system, for example the staffs in our model. The Queue module is used to define queuing rules in the system.

3 Arena Modeling of Call Centers

3.1 Modeling of a Call Center with a Single Function

We first conduct a model of a simple call center whose staffs with only one skill. The call center contains a randomly generated arrival flow for calling and a desk of call handling. When a customer's calling arrives at the desk, it will be dealt with by a staff, and then leaves the call center. We build an Arena model of the simple call center, as shown in Fig. 1. Our model contains a Create module to generate arrival customer, a Process module to install a queue and a desk, and a Dispose module to let entities leave the call center after finishing service. This model connects these three modules sequentially.

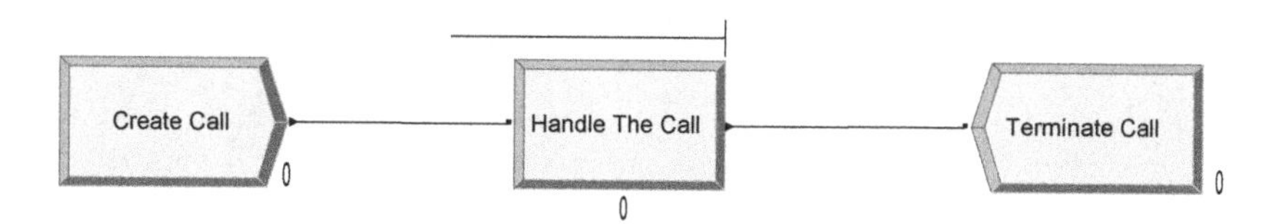

Fig. 1. The Arena simulation model of a simple call center with a single function.

In the setting of model parameters, we assume the inter-arrival time follows an exponential distribution with a mean value of 0.7 min. The call handling time follows a triangular distribution with the shortest/longest processing time of 0.3/1.1 min. Most of the processing time is about 0.75 min.

3.2 Modeling of a Call Center with Double Functions

The double functions mean that two different services are provided in a call center. In a call center of double functions, customers enter calling processes, and then select service types they need. And, then they enter a queue and wait a desk to handle their calling. We set a service circle of 4 h and obtain experimental data to analyze. According to the above flow, we build an Arena model of the call center of double functions, as shown in Fig. 2.

In the proposed model, the Create module generates a customer model. We set the proportion of selection of two functions to be 6:4. The Decide module lets customers choose a service type, and then customers enter a queue to wait to be served. After a service finishes, a customer enter another Decide module. Two Decide modules are used in the last stage. And we add a parameter to a Decide module to judge whether waiting time is greater than 0.5 min or not. We use the judgment of waiting time to obtain two customer counts. The two counts can be used to represent a server level of a call center, namely customer's satisfaction. After services ends, the customer's waiting time during the system, is used to judge whether the customer satisfies. For example, the customer's waiting time is greater than 0.5 min that means it is too long for the customer and he or she is dissatisfied with the service. Based on the judgment, we obtain the two customer counts of satisfaction and dissatisfaction, and then the satisfaction of the system yields.

In our experiments, the customer's arrival time is a random distribution with a mean value of 0.25 min. The call handling time of the two queues of double functions is triangular distribution. The shortest/longest processing time are 0.3/1.9 min for the first queue and 0.3/2.8 min for the second queue. Most of the processing time is about 1.75 min. The two service types are different in two desk of the double functions. Therefore, the two costs of service staffs should be different. The staffing costs of the first and second class are respectively 10 and 11 dollars/person-hour. We set a service circle of 4 h and obtain experimental data to analyze.

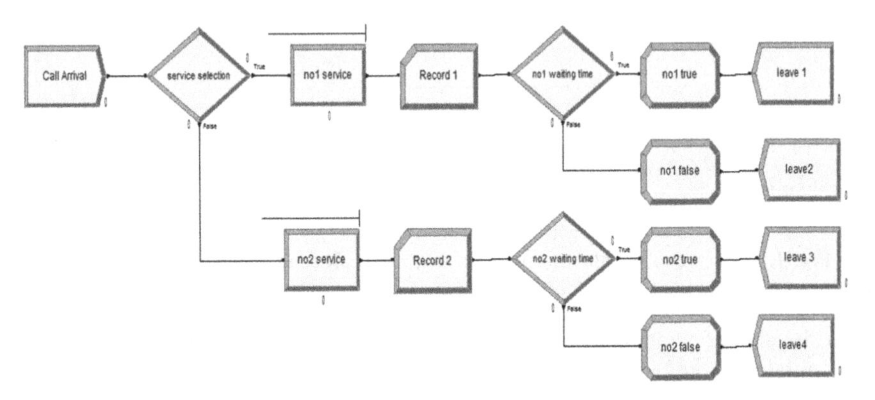

Fig. 2. The Arena simulation model of a call center with double functions.

4 Simulation Results

4.1 Optimization of a Call Center with a Single Function

Arena software simulates the flow of call center and obtain data to optimize. The operation time of simulation is an hour and the number of staffing is three. During an hour of operation, 76 customer's services finished. The longest waiting time in the queue is 28.099 s, and the average time is 1.1336 s. The maximum of customers waiting in the queue was two. The maximum of staffs simultaneously busy was three. The average of busy staff was 0.9451, and the resource utilization was 31.5 %.

For the optimization of our current solution, we chose staffing that means how to hire less staff to reach customer's satisfaction. We create the number of staffs from 1 to 5, and record the average number of busy staffs, resource utilization rate, average waiting time, and maximal waiting time. The numerical results are summarized in Table 1. Based on Table 1, it can be found that the utilization rate and average waiting time decrease as exponential decay when increasing the number of staffs. That means staff's increment generates gradually less effects.

We analyzed the five cases to find the best solution. Only one staff works in the call center and the resource utilization is 96 %. However, the average of waiting time is

Table 1. Simulation results of a call center with a single function.

Number of staffs	Average number of busy staffs	Utilization rate	Average waiting time (sec)	Maximal waiting time (sec)
1	0.9678	96.78 %	131.82	341.8
2	0.8976	44.88 %	8.6476	69.3703
3	0.9451	31.50 %	1.1336	28.099
4	0.9280	23.20 %	0.2043	11.1237
5	0.9280	18.56 %	0.0	0.0

131 s and the maximum waiting time is 341 s. In this case, it is obvious that the solution is infeasible. Not only the staff is always busy, and also many customers always wait and cannot obtain satisfied services. When the staff increases to two persons, the resource utilization is 44 %, the average of waiting time is 8 s. However, the maximum waiting time is more than 1 min and that excess 20 s of our requirement. When three staffs work in the call center, the resource utilization is 33 %, the average of waiting time is about 1 s and the maximum waiting time is 28 s. The few customers need to wait more than 20 s. The customer satisfaction keeps higher level and fair resource utilization yields. When the staff increases to four persons, the resource utilization is merely 23 %, the average of waiting time is less than 1 s, and the maximum waiting time is 11 s. Although the waiting time decreases dramatically, an additional staff makes the cost increased a lot and the resource utilization is quite low. When five staffs work in the call center, all customers do not wait. The maximum waiting time is 0. The resource utilization is only 18 %. Although the customer's satisfaction is exceptionally high, the staffing cost is too much. As a result, this is not a feasible solution. We summarize the above analysis to conclude that three staffs is the best solution having fair resource utilization and low waiting time.

4.2 Optimization of a Call Center with Double Functions

We first defined some symbols which will be used later. The numbers of customers of two service types are S_1 and S_2. Average of busy staff of two queues are F_1 and F_2. The resource utilizations of two queues are P_1 and P_2. We yield two relationships as

$$P_1 = \frac{F_1}{S_1} \text{ and } P_2 = \frac{F_2}{S_2} \tag{1}$$

The overall system's resource utilization is P, which can be calculated through

$$P = P_1 + P_2 \tag{2}$$

The number of customers of completed service is N. The numbers of customers passing two queues are N_1 and N_2. The numbers of satisfied customers of two queues are N_{m1} and N_{m2}. The satisfication of two queues are M_1 and M_2. We derive two relationships as follows:

$$M_1 = \frac{N_{m1}}{N_1} \text{ and } M_2 = \frac{M_{m2}}{N_2}. \quad (3)$$

The total satisfication is calculated with the following formula

$$M = \frac{N_{m1} + N_{m2}}{N_1 + N_2}. \quad (4)$$

Table 2. The hiring cost and waiting times for various combination of staffing numbers.

Number of staffs		Average cost per hour	Average waiting time for no. 1	Average waiting time for no. 2	Maximal waiting time for no. 1	Maximal waiting time for no. 2
S_1	S_2	C	W_1(min)	W_2(min)	W_{max1}(min)	W_{max2}(min)
3	2	52	3.9068	21.3757	14.4696	37.1362
3	3	63	2.6669	1.4077	8.9851	6.1526
3	4	74	3.0169	0.3478	8.5147	4.3168
3	5	85	2.1372	0.0418	5.5188	1.1405
3	6	96	2.3461	0.0130	6.9439	0.8124
4	2	62	0.2818	31.1146	2.5195	64.1823
4	3	73	0.2688	1.5343	2.4185	5.3308
4	4	84	0.3869	0.1879	3.7054	2.1935
4	5	95	0.3712	0.0609	3.7054	1.5080
4	6	106	0.3560	0.0001	3.7054	0.0295
5	2	72	0.0448	25.7889	0.8925	61.6516
5	3	83	0.0777	1.6714	1.2858	5.6565
5	4	94	0.0385	0.1808	0.8548	1.8380
5	5	105	0.0415	0.0723	0.7412	1.5080
5	6	116	0.0366	0.0001	0.6064	0.0001

The averages of waiting time of two queues are W_1 and W_2. The maximun of waiting time of two queues are W_{max1} and W_{max2}. The service cost per hour is C. We build the Arena model for a Call Center with Double Functions, set the combination of staff's numbers and then execute the simulation system. We analyze the obtained simulation data, and compute the hiring cost for every combination of staff‘s numbers, as listed in Table 2. We conducted several experiments to simulate call center's queuing. We yielded large amounts of experimental data and selected data to list in Tables 2 and 3. Customer's satisfaction is calculated based on the waiting time. The waiting time of more than 0.5 min is labelled as dissatisfaction. Therefore, to evaluate service level, we computed customer's satisfaction according to Eqs. (3) and (4). The resource utilization is computed by the average of busy staff during call center's operation, according to Eqs. (1) and (2).

For studying the optimal design, we first consider the customer's satisfaction because customers are always needed to pay most attentions in business management.

Table 3. The satisfaction and utilization rates for various combination of staffing numbers.

Number of staffs		Satisfaction for queue 1	Satisfaction for queue 2	System satisfaction	Utilization rate for queue 1	Utilization rate for queue 2	Total utilization rate
S_1	S_2	M_1	M_2	M	P_1	P_2	P
3	2	21.612 %	2.007 %	14.966 %	96.35 %	98.56 %	97.46 %
3	3	16.696 %	40.053 %	25.945 %	94.19 %	84.08 %	89.13 %
3	4	20.139 %	77.233 %	41.603 %	94.21 %	57.97 %	76.09 %
3	5	16.609 %	96.317 %	46.831 %	94.76 %	46.47 %	70.61 %
3	6	12.095 %	99.215 %	46.440 %	96.55 %	42.19 %	69.37 %
4	2	79.698 %	0.669 %	53.296 %	72.58 %	99.06 %	85.82 %
4	3	79.160 %	26.866 %	58.074 %	74.34 %	87.85 %	81.10 %
4	4	75.541 %	85.224 %	79.286 %	74.57 %	62.26 %	68.42 %
4	5	76.142 %	94.231 %	83.037 %	73.01 %	47.50 %	60.26 %
4	6	77.479 %	100.000 %	85.954 %	73.26 %	39.06 %	56.16 %
5	2	97.789 %	0.673 %	65.198 %	57.77 %	99.06 %	78.41 %
5	3	94.000 %	29.798 %	68.474 %	60.26 %	88.44 %	74.35 %
5	4	99.169 %	85.638 %	93.967 %	59.36 %	61.09 %	60.22 %
5	5	99.154 %	92.837 %	96.751 %	57.55 %	48.11 %	52.83 %
5	6	99.490 %	100.000 %	99.679 %	57.56 %	38.15 %	47.85 %

Therefore, most of companies often choose customer's satisfaction to work for, and manage the solution of service centers based on it. Many researches recommend the solution reaching 80 % of customer's satisfaction, and the resource utilization should be above 50 %. We found out that customer's satisfaction increases with staff number's increment. The customer's satisfaction increases smaller and smaller until it reaches 100 %. There are five solutions of (4, 5), (4, 6), (5, 4), (5, 5) and (5, 6) having above 80 % customer's satisfaction. The solution of (4, 4) has 79.28 % satisfaction closed to the 80 % standard. Therefore, it can be listed in the feasible solutions. Comparing with the above mentioned solutions, customer's satisfaction of other cases are far below our required level of 80 %. The customer's waiting time of these low satisfaction cases are very long, and that is very possible to loss customers. Their average and maximum of waiting time are both very long. Some of them have five more min of waiting time. That indicates that these cases are infeasible solution.

It can be found that the resource utilization (staff's usage) decreases with the staff's increment, and the delta decrement is less and less. The solution of (5, 6) is only one case having 50 % less of resource utilization that causes too much staff's idle time and increases the staffing cost. Other cases of feasible solutions have 50 % more of resource utilization. We reviewed these eliminated solutions and their resource utilizations are quite high. The staffs of these cases are always in busy situation, but the service level cannot serve the customer arrival. The waiting time of customer is very long and low satisfaction results.

After the above mentioned elimination, the five solutions of (4, 4), (4, 5), (4, 6), (5, 4) and (5, 5) remained, and their staffing costs are 84, 95, 106, 94 and 105 dollars/hour, respectively. Therefore, the best solution is (4, 4) in our system. This optimal design reaches the lowest cost and can meet the given service level.

5 Conclusions

In this paper, we present an Arena simulation model for optimal design of staffing problems in call centers. The purpose of the proposed simulation model is to find the feasible solutions with both a guaranteed 80 % service level and minimal staffing costs. We use Arena software to study the simulation model of staffing in call centers under the consideration of customer satisfaction, resource utilization and staffing cost. We analyze the flows of call centers and explore the staffing schemes to obtain optimal solutions in the proposed simulation models with a single function and double functions. This research work could contribute as a good tool for analyzing management decisions in designing and operating call centers. In the future works, we would extend double functions to multiple functions to meet the possible demands of different companies.

Acknowledgments. The authors are thankful for partial support from Fujian Provincial Department of Science and Technology, China, under Grant 2016J01330, and the Education Department of Fujian Province, China, under Grant No. JAT160337.

References

1. Luo, J., Zhang, J.: Staffing and control of instant messaging contact centers. Oper. Res. **61**, 328–343 (2013)
2. Pang, G., Perry, O.: A logarithmic safety staffing rule for contact centers with call blending. Manage. Sci. **61**, 73–91 (2015)
3. Kwan, S.K., Davis, M.M., Greenwood, A.G.: A simulation model for determining variable worker requirements in a service operation with time-dependent customer demand. Queueing Syst. **3**, 265–276 (1988)
4. Altiok, T., Melamed, B.: Simulation Modeling and Analysis with ARENA. Academic Press, Burlington (2010)
5. Atlason, J., Epelman, M.A.: Call center staffing with simulation and cutting plane methods. Ann. Oper. Res. **127**, 333–358 (2004)
6. Bouzada, M.A.C.B.: Dimensioning a call center: simulation or queue theory? J. Oper. Supply Chain Manage. **2**, 34–46 (2009)
7. Koole, G., Mandelbaum, A.: Queueing models of call centers: an introduction. Ann. Oper. Res. **113**, 41–59 (2002)
8. Borst, S., Mandelbaum, A., Reiman, M.I.: Dimensioning large call centers. Oper. Res. **52**, 17–34 (2004)

Knowledge-Based Intelligent Engineering and Its Applications

On the Security of a Certificateless Searchable Public Key Encryption Scheme

Tsu-Yang Wu[1], Fanya Meng[1], Chien-Ming Chen[1(✉)], Shuai Liu[1], and Jeng-Shyang Pan[1,2]

[1] School of Computer Science and Technology, Shenzhen Graduate School of Harbin Institute of Technology, Shenzhen University Town, Xili, Nanshan District, Shenzhen 518055, People's Republic of China
wutsuyang@gmail.com, 237935023@qq.com, chienming.taiwan@gmail.com, liush000@hotmail.com, jengshyangpan@gmail.com

[2] Fujian Provincial Key Lab of Big Data Mining and Apllications, Fujian University of Technology, No. 3, Xueyuan Road, University Town, Fuzhou 350118, People's Republic of China

Abstract. Searchable public key encryption is a cryptographic mechanism which provides an efficient way to search an encrypted keyword. In this paper, we show that Peng et al.'s certificateless searchable public key encryption scheme suffered from a malicious PKG attack and an off-line keyword guessing attack. In the first attack, a malicious PKG can obtain a part of an authorized receiver's private key. In the second attack, the malicious PKG can guess a keyword related to the authorized receiver's trapdoor using the part of receiver's private.

Keywords: Searchable public key encryption · Certificateless · Malicious PKG attack · Off-line keyword guessing attack

1 Introduction

Searchable public key encryption (SPKE for short) (or called public key encryption with keyword search) was proposed by Boneh et al. [5] in 2004. It is a cryptographic mechanism which provides an efficient way to search an encrypted keyword. This method can be applied to an email system which helps user to search some urgent email with specific keywords.

Afterward, virous versions for the SPKE are proposed such as supporting conjunctive keywords [17], supporting multi-user [13], focus on a designated server [4,11,19], proposing enhanced security models [4,19,20], based on the pairing-based cryptosystem [4,11,19,20], based on the ID-based cryptosystem [21], and based on the certificateless cryptosystem [18]. Furthermore, several SPKE schemes were suffered from off-line keyword guessing attacks [4,5,17]. Note that off-line keyword guessing attack means that attacker can select an appropriate keyword and then verify whether the guessed keyword is related to the captured trapdoor.

J. Pan et al. (eds.), *Genetic and Evolutionary Computing*, Advances in Intelligent Systems and Computing 536, DOI 10.1007/978-3-319-48490-7_14

Certificateless cryptosystem was proposed by Al-Riyami and Paterson [1] in 2003. This system can be used to solve the key escrow problem in the ID-based cryptosystem [6,23]. Afterwards, several certificateless-based cryptographic schemes and protocols were proposed such as encryption [2,8–10,15], signature [12,14,24,25], and two party key agreement [16]. Furthermore, Au et al. [3] proposed an attack approach called malicious PKG (private key generator) attack. They pointed out some literatures [1,12,14] suffered from this attack.

In this paper, we demonstrate Peng et al.'s certificateless SPKE (CLSPKE for short) have some security weakness. At first, we adopt a malicious PKG attack approach to show a malicious PKG can obtain a part of an authorized receiver's private key. Then, using the part of receiver's private the malicious PKG can launch an off-line keyword guessing attack successfully.

2 Review of Peng et al.'s Scheme

2.1 Bilinear Pairings

Let G_1 be an additive cyclic group of an elliptic curve $E(F_p)$ over a finite field F_p and G_2 be a multiplicative cyclic group of F_p. The two groups G_1 and G_2 have the same large prime order q.

A bilinear pairing e can be viewed as a map defined by $e : G_1 \times G_1 \rightarrow G_2$. e satisfies the following three properties:

- Bilinear: $\forall$ P, $Q \in G_1$ and a, $b \in \mathbb{Z}_q^*$, $e(aP, bQ) = e(P, Q)^{ab}$.
- Non-degenerate: $\forall$ $P \in G_1$, $\exists$ $Q \in G_1$ such that $e(P, Q) = 1_{G_2}$.
- Computable: $\forall$ P, $Q \in G_1$, there exist efficient algorithms to compute $e(P, Q)$.

For the full descriptions about bilinear pairings, readers can refer to [7,22].

2.2 A Brief Description of Peng et al.'s Scheme

Peng et al.'s scheme [18] consists of following eight phases (algorithms): Setup, Partial private key extract, Set secret value, Set private key, Set public key, CLSPKE, Trapdoor generation, and Test.

Setup Phase. In this phase, PKG generates all needed public parameters and its master/public key pair. PKG runs the Setup algorithm to execute the following steps:

1. Inputting a secure parameter k, to select a bilinear pairing $e : G_1 \times G_1 \rightarrow G_2$, where G_1 and G_2 have the same prime order $q \geq 2^k$.
2. To choose a generator $P \in G_1$ and a master key $s \in \mathbb{Z}_q^*$.
3. To compute the PKG's public key $P_{pub} = s \cdot P$.
4. To select four needed cryptographic hash functions: $H_1 : \{0,1\}^* \rightarrow G_1$, $H_2 : \{0,1\}^* \rightarrow \mathbb{Z}_q^*$, $H_3 : \{0,1\}^* \rightarrow G_1$, and $H_4 : G_2 \rightarrow \{0,1\}^l$.

The public parameters *param* are defined as $param = \{k, e, G_1, G_2, q, P, P_{pub}, H_1, H_2, H_3, H_4\}$.

Partial Private Key Extract Phase. In this phase, PKG generates server S's and receiver R's partial private keys. PKG runs the Partial private key extract algorithm to execute the following steps:

1. Inputting an S's identity $ID_S \in \{0,1\}$, to compute $Q_S = H_1(ID_S)$.
2. The S's partial private key D_S is computed by $s \cdot Q_S$.
3. Inputting an R's identity $ID_R \in \{0,1\}$, to compute $Q_R = H_1(ID_R)$.
4. The R's partial private key D_R is computed by $s \cdot Q_R$.

Set Secret Value Phase. In this phase, server S and receiver R generate its secret values. S runs the Set secret value algorithm to generate its secret value $x \in \mathbb{Z}_q^*$. Similarly, the R's secret value is defined by $y \in \mathbb{Z}_q^*$.

Set Private Key Phase. In this phase, server S and receiver R generate its private keys. Inputting public parameters $param$, S's secret value x, and S's partial private key D_S, S runs the Set private key algorithm to generate its private key $SK_S = (SK_{S_1}, SK_{S_2})$, where $SK_{S_1} = x$ and $SK_{S_2} = x \cdot D_S = x \cdot s \cdot Q_S = x \cdot s \cdot H_1(ID_S)$. Similarly, the R's private key is defined by $SK_R = (SK_{R_1}, SK_{R_2})$, where $SK_{R_1} = y$ and $SK_{R_2} = y \cdot D_R = y \cdot s \cdot Q_R = y \cdot s \cdot H_1(ID_R)$.

Set Public Key Phase. In this phase, server S and receiver R generate its public keys. Inputting public parameters $param$ and S's secret value x, S runs the Set public key algorithm to generate its public key $PK_S = (PK_{S_1}, PK_{S_2})$, where $PK_{S_1} = x \cdot P$ and $PK_{S_2} = x \cdot P_{pub}$. Similarly, the R's public key is defined by $PK_R = (PK_{R_1}, PK_{R_2})$, where $PK_{R_1} = y \cdot P$ and $PK_{R_2} = y \cdot P_{pub}$.

CLSPKE Phase. In this phase, anyone can encrypt a set of keywords $W = \{w_i | i = 1, 2, \ldots, n\}$. Inputting public parameters $param$, R's identity ID_R, R's public key PK_R, S's identity ID_S, and S's public key PK_S, she/he runs the CLPEKS algorithm to execute the following steps:

1. To verify $e(PK_{S_1}, P_{pub}) = e(PK_{S_2}, P)$ and $e(PK_{R_1}, P_{pub}) = e(PK_{R_2}, P)$ hold or not. If not, it aborts and returns false.
2. To compute $Q_R = H_1(ID_R)$ and $Q_S = H_1(ID_S)$.
3. For each keyword w_i,
   - to select a random value $r_i \in \mathbb{Z}_q^*$.
   - to compute $t_i = e(r_i \cdot H_2(w_i) \cdot Q_R, PK_{R_2}) \cdot e(r_i \cdot Q_S, PK_{S_2}) \cdot e(r_i \cdot H_3(w_i), P)$.
   - to computing $U_i = r_i \cdot P$ and $V_i = H_4(t_i)$.

   The resulted ciphertext $C = \{C_1, C_2, \ldots, C_n\}$, where $C_i = (U_i, V_i)$.

Trapdoor Generation Phase. In this phase, receiver R generates a trapdoor of keyword. Inputting public parameters $param$, keyword w_i, R's private key SK_R, and S's public key PK_S, R runs the Trapdoor generation algorithm to execute the following steps:

1. To select a random value $a_i \in \mathbb{Z}_q^*$.
2. To compute
   - $T_1 = a_i \cdot P$.
   - $T_2 = H_2(w_i) \cdot SK_{R_2} \oplus a_i \cdot PK_{S_1}$.
   - $T_3 = H_3(w_i) \oplus a_i \cdot PK_{S_1}$.

   The resulted trapdoor T_w is defined by (T_1, T_2, T_3).

Test Phase. In this phase, server S verifies keyword w_i of trapdoor T_w whether contained in ciphertext C. Inputting public parameters $param$, trapdoor T_w, S's private key, and ciphertext C, S runs the Test algorithm to execute the following steps:

1. To compute $T_2' = T_2 \oplus (SK_{S_1} \cdot T_1)$.
2. To compute $T_3' = T_3 \oplus (SK_{S_1} \cdot T_1)$.
3. To verify $H_4(e(T_2' + SK_{S_2} + T_3', U_i))$ whether equals to V_i.

   If the verification is true, S rcturns accept. Otherwise, S returns reject.

3 Drawbacks of Peng et al.'s Scheme

3.1 Malicious PKG Attack for Receiver

For an authorized receiver R with identity ID_R, a malicious PKG selects $\alpha \in \mathbb{Z}_q^*$ and then returns a generator $P' = \alpha \cdot H_1(ID_R)$ and PKG's public key $P'_{pub} = \alpha \cdot s \cdot H_1(ID_R)$, where s is the master key of PKG.

Then, R runs the Set private key and the Set public key algorithms to generate she/he's private key $SK_R = (SK_{R_1}, SK_{R_2}) = (y, y \cdot s \cdot H_1(ID_R))$ and public key $PK_R = (PK_{R_1}, PK_{R_2}) = (y \cdot P', y \cdot P'_{pub})$, where y is the secret value of R.

Now, the malicious PKG can recover SK_{R_2} using α and R's PK_{R_2}, i.e.

$$\alpha^{-1} \cdot PK_{R_2} = \alpha^{-1} \cdot y \cdot P'_{pub} = \alpha^{-1} \cdot y \cdot \alpha \cdot s \cdot H_1(ID_R) = SK_{R_2}.$$

It will lead an off-line keyword guessing attack on R's trapdoor T_w described below. Note that by the similar method the malicious PKG also can recover server's SK_{S_2}.

3.2 Off-Line Keyword Guessing Attack

The malicious PKG with SK_{R_2} can launch an off-line keyword guessing attack while it captures R's trapdoor $T_w = (T_1, T_2, T_3)$. The malicious PKG executes the following steps:

1. To guess an appropriate keyword w'.
2. To compute $A = H_2(w') \cdot SK_{R_2} \oplus T_2$.
3. To verify $e(A, P'_{pub})$ whether equals to $e(T_1, PK_{s_2})$.

If the verification is true, it means that the guessed keyword w' is related to the trapdoor T_w, the attack success. Otherwise, the malicious PKG goes back to step 1 and continues to execute the steps.

The correctness are described below. Assume that w' is the success guessed keyword. Then, it implies

$$A = H_2(w') \cdot SK_{R_2} \oplus T_2 = a_i \cdot PK_{S_1}.$$

Thus,

$$e(A, P'_{pub}) = e(a_i \cdot x \cdot P', P'_{pub}) = e(a_i \cdot P', x \cdot P'_{pub}) = e(T_1, PK_{s_2}).$$

4 Conclusions

In this paper, we have demonstrated two security weaknesses in Peng et al.'s CLSPKE scheme. In the future, we will propose an improvement based on their scheme with formal security proof.

Acknowledgments. We thank anonymous referees for the comments and suggestions. This work of Tsu-Yang Wu was supported by Natural Scientific Research Innovation Foundation in Harbin Institute of Technology (No. HIT.NSRIF. 2015089) and the work of Chien-Ming Chen was supported in part by the Project NSFC (National Natural Science Foundation of China) under Grant number 61402135 and in part by Shenzhen Technical Project under Grant number JCYJ20150513151706574.

References

1. Al-Riyami, S.S., Paterson, K.G.: Certificateless public key cryptography. In: Laih, C.-S. (ed.) ASIACRYPT 2003. LNCS, vol. 2894, pp. 452–473. Springer, Heidelberg (2003). doi:10.1007/978-3-540-40061-5_29
2. Al-Riyami, S.S., Paterson, K.G.: CBE from CL-PKE: a generic construction and efficient schemes. In: Vaudenay, S. (ed.) PKC 2005. LNCS, vol. 3386, pp. 398–415. Springer, Heidelberg (2005). doi:10.1007/978-3-540-30580-4_27
3. Au, M.H., Mu, Y., Chen, J., Wong, D.S., Liu, J.K., Yang, G.: Malicious KGC attacks in certificateless cryptography. In: Proceedings of the 2nd ACM Symposium on Information, Computer and Communications Security, pp. 302–311. ACM (2007)

4. Baek, J., Safavi-Naini, R., Susilo, W.: Public key encryption with keyword search revisited. In: Gervasi, O., Murgante, B., Laganà, A., Taniar, D., Mun, Y., Gavrilova, M.L. (eds.) ICCSA 2008. LNCS, vol. 5072, pp. 1249–1259. Springer, Heidelberg (2008). doi:10.1007/978-3-540-69839-5_96
5. Boneh, D., Crescenzo, G., Ostrovsky, R., Persiano, G.: Public key encryption with keyword search. In: Cachin, C., Camenisch, J.L. (eds.) EUROCRYPT 2004. LNCS, vol. 3027, pp. 506–522. Springer, Heidelberg (2004). doi:10.1007/978-3-540-24676-3_30
6. Boneh, D., Franklin, M.: Identity-based encryption from the Weil Pairing. In: Kilian, J. (ed.) CRYPTO 2001. LNCS, vol. 2139, pp. 213–229. Springer, Heidelberg (2001). doi:10.1007/3-540-44647-8_13
7. Chen, L., Cheng, Z., Smart, N.P.: Identity-based key agreement protocols from pairings. Int. J. Inf. Secur. **6**(4), 213–241 (2007)
8. Cheng, Z., Chen, L., Ling, L., Comley, R.: General and efficient certificateless public key encryption constructions. In: Takagi, T., Okamoto, T., Okamoto, E., Okamoto, T. (eds.) Pairing 2007. LNCS, vol. 4575, pp. 83–107. Springer, Heidelberg (2007). doi:10.1007/978-3-540-73489-5_6
9. Dent, A.W.: A survey of certificateless encryption schemes and security models. Int. J. Inf. Secur. **7**(5), 349–377 (2008)
10. Dent, A.W., Libert, B., Paterson, K.G.: Certificateless encryption schemes strongly secure in the standard model. In: Cramer, R. (ed.) PKC 2008. LNCS, vol. 4939, pp. 344–359. Springer, Heidelberg (2008). doi:10.1007/978-3-540-78440-1_20
11. Hu, C., Liu, P.: An enhanced searchable public key encryption scheme with a designated tester and its extensions. J. Comput. **7**(3), 716–723 (2012)
12. Huang, X., Susilo, W., Mu, Y., Zhang, F.: On the security of certificateless signature schemes from Asiacrypt 2003. In: Desmedt, Y.G., Wang, H., Mu, Y., Li, Y. (eds.) CANS 2005. LNCS, vol. 3810, pp. 13–25. Springer, Heidelberg (2005). doi:10.1007/11599371_2
13. Hwang, Y.H., Lee, P.J.: Public key encryption with conjunctive keyword search and its extension to a multi-user system. In: Takagi, T., Okamoto, T., Okamoto, E., Okamoto, T. (eds.) Pairing 2007. LNCS, vol. 4575, pp. 2–22. Springer, Heidelberg (2007). doi:10.1007/978-3-540-73489-5_2
14. Li, X.X., Chen, K.F., Sun, L.: Certificateless signature and proxy signature schemes from bilinear pairings. Lith. Math. J. **45**(1), 76–83 (2005)
15. Libert, B., Quisquater, J.-J.: On constructing certificateless cryptosystems from identity based encryption. In: Yung, M., Dodis, Y., Kiayias, A., Malkin, T. (eds.) PKC 2006. LNCS, vol. 3958, pp. 474–490. Springer, Heidelberg (2006). doi:10.1007/11745853_31
16. Lippold, G., Boyd, C., Gonzalez Nieto, J.: Strongly secure certificateless key agreement. In: Shacham, H., Waters, B. (eds.) Pairing 2009. LNCS, vol. 5671, pp. 206–230. Springer, Heidelberg (2009). doi:10.1007/978-3-642-03298-1_14
17. Park, D.J., Kim, K., Lee, P.J.: Public key encryption with conjunctive field keyword search. In: Lim, C.H., Yung, M. (eds.) WISA 2004. LNCS, vol. 3325, pp. 73–86. Springer, Heidelberg (2005). doi:10.1007/978-3-540-31815-6_7
18. Peng, Y., Cui, J., Peng, C., Ying, Z.: Certificateless public key encryption with keyword search. China Commun. **11**(11), 100–113 (2014)
19. Rhee, H.S., Park, J.H., Susilo, W., Lee, D.H.: Improved searchable public key encryption with designated tester. In: Proceedings of the 4th International Symposium on Information, Computer, and Communications Security, pp. 376–379. ACM (2009)

20. Rhee, H.S., Park, J.H., Susilo, W., Lee, D.H.: Trapdoor security in a searchable public-key encryption scheme with a designated tester. J. Syst. Softw. **83**(5), 763–771 (2010)
21. Wu, T.Y., Tsai, T.T., Tseng, Y.M.: Efficient searchable ID-based encryption with a designated server. Ann. Telecommun.-annales des télécommunications **69**(7–8), 391–402 (2014)
22. Wu, T.Y., Tseng, Y.M.: An ID-based mutual authentication and key exchange protocol for low-power mobile devices. Comput. J. **53**(7), 1062–1070 (2010)
23. Wu, T.Y., Tseng, Y.M., Tsai, T.T.: A revocable ID-based authenticated group key exchange protocol with resistant to malicious participants. Comput. Netw. **56**(12), 2994–3006 (2012)
24. Yum, D.H., Lee, P.J.: Generic construction of certificateless signature. In: Wang, H., Pieprzyk, J., Varadharajan, V. (eds.) ACISP 2004. LNCS, vol. 3108, pp. 200–211. Springer, Heidelberg (2004). doi:10.1007/978-3-540-27800-9_18
25. Zhang, Z., Wong, D.S., Xu, J., Feng, D.: Certificateless public-key signature: security model and efficient construction. In: Zhou, J., Yung, M., Bao, F. (eds.) ACNS 2006. LNCS, vol. 3989, pp. 293–308. Springer, Heidelberg (2006). doi:10.1007/11767480_20

On the Study of Trading Strategies Within Limited Arbitrage Based on SVM

Hui-Huang Tsai[1], Mu-En Wu[2(✉)], Wei-Ho Chung[3], and Cheng-Yu Lu[4]

[1] Department of Finance, National United University, Miaoli, Taiwan
hhtsai@nuu.edu.tw
[2] Department of Mathematics, Soochow University, Taipei, Taiwan
mn@scu.edu.tw
[3] Research Center for Information Technology Innovation, Academia Sinica, Taipei, Taiwan
whc@citi.sinica.edu.tw
[4] PIXNET Media, Taipei, Taiwan
rick@pixnet.tw

Abstract. Limited arbitrage will impede the operation of market, then confuses the well-known "Efficient Market Hypothesis" in theory and investment decision of market participants in practice. We develop a contrarian trading strategy, tailored to this kind of situation, with the trading signals derived the technical indicator: BIAS, to indirectly verify the existence of limited arbitrage by testing all listed stocks in Taiwan. Further, we use the well-known machine learning, SVM, to confirm the classification method in this study being free of subjective discretion. Thus we have a robust evidence to support the usefulness of this trading strategy.

Keywords: Limited arbitrage · BIAS · Contrarian · SVM

1 Introduction

LIMITS of ARBITRAGE, e.g. the arbitrageurs are capital or time constrained as described in Shleifer and Vishny (1997), can explain the persistence of mispricing, which could not support the well-known efficient-market hypothesis (EMH) proposed by the Eugene Fama, the 2013 Nobel laureate in Economics, but accords to the stylized facts in financial markets: short-run momentum and long-run reversal. An extensive body of finance literature about the limits of arbitrage concerns the predictability of asset returns. As Gromb and Vayanos (2010) provides a comprehensive survey on this issue, however, the distinction in terms of duration between short-run and long-run remains elusive. This observation motivates us to develop a trading strategy utilizing the mispricing. Once it brings some return, we will have not only the empirical evidence to support that the limits of arbitrage impedes the efficient operation of market but also the suggestion to disadvantaged retail investors.

The core insight in the model of Shleifer and Vishny (1997) is that when the pricing error is not significant or just the beginning, arbitrageurs will not engage in correcting

J. Pan et al. (eds.), *Genetic and Evolutionary Computing*, Advances in Intelligent Systems and Computing 536, DOI 10.1007/978-3-319-48490-7_15

the mispricing eagerly. The deeper is the mispricing, the stronger the motive of arbitrageurs to fix will be. This implies that the optimal trading strategy in the initial stage is the momentum one and that in the later stage is the contrarian one. The solution we think is that using a simple technical trading rule based on a proxy for mispricing, which we choose the technical indicator: BIAS ratio. It derives from the concept of moving averages and measures the divergence of the price from the line of moving average. Moving averages of prices are commonly used in Finance to represent the investment cost during a period. If the price cannot reflect the information completely and timely, moving average it should alleviate this problem.

The technical trading rule in this study is simple: applies the contrarian strategy when the absolute value of the BIAS is big enough. We use a series of market situations, similar to the empirical study of Brock et al. (1992) to test whether this trading rule is suitable to the sample comprised of all listed stocks in Taiwan, then confirm the relation between BIAS and Contrarian strategy with Support Vector Machine (SVM). SVM is a kind of machine learning designed for classification and regression analysis, so we use it to classify the market situation suitable for contrarian strategy or not.

2 Data and the Technical Trading Rule: Contrarian

2.1 Data

The data used in this study is daily prices, including open, high, low and close ones, of 685 listed stocks in Taiwan from the July first in 2009 to the last day in 2015 for avoiding the Financial Tsunami. Every round of trade once initiated, no matter of long or short some stock, will start at the open price in next day, and be closed at the close price at the end of holding period to calculate the rate of return in terms of logarithm.

2.2 The Technical Trading Rule

The formula of *BIAS* ratio is as following:

$$BIAS = \frac{MA(short) - MA(long)}{MA(long)} * 100\ \% \tag{1}$$

Where MAs are the moving averages of typical prices, and short and long mean the numbers of days used to calculate the MAs, where the number of long should be always bigger than that of short. We choose the numbers of short and long among 1, 5, 10, 20, 60 for the sake of market convention.

The trading signal depends on the value of the BIAS. When the BIAS of one stock is big enough, we sell it at the next day with its open price. On the contrary, when the number of one stock is negative and small enough, we will buy it at the next day with its open price. All trade of the stock will be liquidated in some kind of fixed holding days. This kind of trading rule is so-called Contrarian strategy. The condition of 'enough' is described as the following:

1. buy it if the stock's BIAS locates in the first q-Quantiles
2. sell it if the stock's BIAS locates in the last q-Quantiles

We choose the numbers of q among 3, 4, 5, 10 and 20. By the theory of Shleifer and Vishny (1997), we can expect that the bigger the q, the more useful the contrarian strategy would be. With the help of BIAS, we don't need to investigate the fundamental value of the stock and the related mispricing level. To our limited knowledgement, this study is the first one using such a simple technical trading rule to verify the existence of limited arbitrage.

3 The Empirical Results

3.1 The Results from Standard T-Tests

We apply the contrarian strategy to the time-series of each stock to calculate its return individually and pool the returns of all stocks to compute the t-test statistics. If the estimation value of t-test is significantly positive, then we can say the related trade rule is useful and further to reject the EMH, this means the existence of limits of arbitrage. So we propose the null hypothesis like:

$$H_0 : \mu = 0$$

Then the alternative hypothesis is 'two.sided': if the mean of the estimated value is significantly positive, we can say that the contrarian strategy does work; in the opposite, we should apply a different strategy like momentum. Anyway, we expect to see the significantly positive value of estimation term to support the argument of limited arbitrage.

The first empirical result is summarized in Table 1. In that table, the number of days used in the short-term MA is 1. The result of bigger q, e.g. 20 and 10, are located in the left side and that of smaller q, e.g. 5 and 3, in the right side. It is obvious that the majority of the estimated values in the left side are significant positive. This show that the more far the current stock price deviates from its moving average, similar to the more deepening its mispricing is, the more possibly the arbitrageurs actively take actions to correct the stock price. In the contrary, when the mispricing is not obvious, i.e. the BIAS is small, adopting the momentum strategy to follow the forming trend, not contrarian one, is optimal.

Tables 2 and 3 show the results of MA with the days of short-term being 5 and 10 for the stabilizing the performance of the contrarian strategy. The results of both are similar to that of Table 1, and the ratio of significantly positive estimation values in the bigger q is bigger than that in Table 1. This means that the smoothing of the current price will improve the performance of the strategy because that it can alleviate the noise trading.

Table 1. Standard test results for the contrarian strategy with MA (short): one day

Qu	MA.L	Hold	Est	t & p	winRate	Qu	MA.L	Hold	Est	t & p	winRate
20	5	5	**0.088**	(3.949)***	0.609	5	5	5	**0.054**	(1.579)	0.556
20	5	10	−0.026	(−1.063)	0.514	5	5	10	−0.173	(−5.26)***	0.423
20	5	20	−0.103	(−3.506)***	0.501	5	5	20	−0.274	(−7.879)***	0.387
20	5	60	−0.011	(−0.337)	0.514	5	5	60	−0.159	(−4.828)***	0.422
20	10	5	**0.106**	(5.67)***	0.62	5	10	5	−0.049	(−1.513)	0.502
20	10	10	−0.027	(−1.252)	0.493	5	10	10	−0.259	(−7.824)***	0.397
20	10	20	−0.08	(−3.065)***	0.505	5	10	20	−0.346	(−10.316)***	0.326
20	10	60	**0.007**	(0.215)	0.536	5	10	60	−0.06	(−1.728)*	0.498
20	20	5	**0.133**	(7.837)***	0.651	5	20	5	−0.225	(−7.029)***	0.368
20	20	10	**0.038**	(2.053)**	0.539	5	20	10	−0.354	(−11.22)***	0.327
20	20	20	−0.002	(−0.101)	0.523	5	20	20	−0.343	(−10.428)***	0.374
20	20	60	**0.114**	(4.175)***	0.584	5	20	60	−0.079	(−2.338)**	0.458
20	60	5	**0.217**	(15.792)***	0.766	5	60	5	**0.021**	(0.838)	0.542
20	60	10	**0.192**	(12.972)***	0.712	5	60	10	−0.021	(−0.816)	0.499
20	60	20	**0.18**	(10.879)***	0.693	5	60	20	−0.008	(−0.299)	0.514
20	60	60	**0.231**	(10.936)***	0.654	5	60	60	−0.025	(−0.84)	0.504
10	5	5	**0.089**	(3.125)***	0.594	3	5	5	**0.09**	(2.337)**	0.543
10	5	10	−0.081	(−2.805)***	0.482	3	5	10	−0.18	(−4.9)***	0.434
10	5	20	−0.199	(−6.118)***	0.42	3	5	20	−0.202	(−5.507)***	0.428
10	5	60	−0.074	(−2.104)**	0.488	3	5	60	−0.259	(−7.509)***	0.371
10	10	5	**0.032**	(1.298)	0.559	3	10	5	−0.12	(−3.314)***	0.455
10	10	10	−0.116	(−4.215)***	0.436	3	10	10	−0.278	(−7.917)***	0.399
10	10	20	−0.197	(−6.585)***	0.419	3	10	20	−0.373	(−10.301)***	0.358
10	10	60	−0.068	(−1.985)**	0.493	3	10	60	−0.142	(−4.121)***	0.445
10	20	5	**0.015**	(0.659)	0.545	3	20	5	−0.396	(−10.557)***	0.334
10	20	10	−0.089	(−3.6)***	0.455	3	20	10	−0.487	(−13.968)***	0.293
10	20	20	-0.157	(−5.278)***	0.445	3	20	20	−0.386	(−11.029)***	0.334
10	20	60	**0.002**	(0.072)	0.515	3	20	60	−0.294	(−8.721)***	0.391
10	60	5	**0.161**	(8.561)***	0.677	3	60	5	−0.087	(−2.832)***	0.479
10	60	10	**0.114**	(5.896)***	0.629	3	60	10	−0.134	(−4.423)***	0.434
10	60	20	**0.121**	(5.669)***	0.619	3	60	20	−0.016	(−0.539)	0.499
10	60	60	**0.12**	(4.746)***	0.599	3	60	60	−0.016	(−0.515)	0.504

Results for daily data from 2009/07/01 to 2015/12/31. Logarithmatic cumulative return are calculated after signals for fixed periods (in the column named as Hold) of following days: 5, 10, 20, 60. The days of MA (short) in BIAS formula is 1 in this table and that of the MA (long) is as following: 5,10, 20 and 60, in the MA.L column. The Bias ratios are furtherly divided by the time series of the stock individually into 4 kind of quantiles (Qu for short): 20, 10, 5 and 3. The trading signals will be initiated when its BIAS touchs the first quantile or the last quantile. The Est column reports the estimated values of t-test and the t & p column reports the related t-statistics and p-values. The number of asterisk represents the p-vlaue is smaller than 0.1, 0.05 and 0.01. Finally, the column of winRate is the winning rate.

Table 2. Standard test results for the contrarian strategy with MA (short): five days

Qu	MA. L	Hold	Est	t & p	winRate	Qu	MA. L	Hold	Est	t & p	winRate
20	10	5	**0.167**	(10.483)***	0.69	5	10	5	**0.032**	(1.191)	0.525
20	10	10	**0.028**	(1.483)	0.557	5	10	10	−0.266	(−8.827)***	0.376
20	10	20	**0.01**	(0.397)	0.558	5	10	20	−0.265	(−8.335)***	0.38
20	10	60	**0.11**	(3.817)***	0.585	5	10	60	−0.092	(−2.761)***	0.471
20	20	5	**0.128**	(9.339)***	0.682	5	20	5	−0.132	(−4.921)***	0.436
20	20	10	**0.056**	(3.488)***	0.563	5	20	10	−0.257	(−8.816)***	0.347

(continued)

Table 2. (*continued*)

Qu	MA. L	Hold	Est	t & p	winRate	Qu	MA. L	Hold	Est	t & p	winRate
20	20	20	**0.083**	(4.272)***	0.589	5	20	20	−0.201	(−6.313)***	0.426
20	20	60	**0.167**	(6.743)***	0.614	5	20	60	−0.04	(−1.214)	0.487
20	60	5	**0.225**	(18.583)***	0.768	5	60	5	**0.09**	(3.925)***	0.592
20	60	10	**0.201**	(15.683)***	0.742	5	60	10	**0.064**	(2.718)***	0.56
20	60	20	**0.217**	(14.602)***	0.723	5	60	20	**0.106**	(4.191)***	0.601
20	60	60	**0.246**	(12.2)***	0.681	5	60	60	**0.033**	(1.14)	0.538
10	10	5	**0.148**	(7.129)***	0.622	3	10	5	−0.12	(−3.766)***	0.449
10	10	10	−0.067	(-2.935)***	0.465	3	10	10	−0.396	(−11.526)***	0.308
10	10	20	−0.109	(-3.72)***	0.456	3	10	20	−0.348	(−10.45)***	0.354
10	10	60	−0.03	(-0.93)	0.499	3	10	60	−0.239	(−7.327)***	0.401
10	20	5	**0.094**	(4.715)***	0.605	3	20	5	−0.378	(−11.348)***	0.3
10	20	10	−0.001	(-0.055)	0.506	3	20	10	−0.455	(−13.98)***	0.283
10	20	20	−0.021	(-0.811)	0.517	3	20	20	−0.343	(−10.085)***	0.345
10	20	60	**0.089**	(2.909)***	0.564	3	20	60	−0.236	(−6.881)***	0.413
10	60	5	**0.186**	(11.099)***	0.69	3	60	5	−0.066	(−2.356)**	0.494
10	60	10	**0.168**	(9.398)***	0.668	3	60	10	−0.076	(−2.675)***	0.49
10	60	20	**0.205**	(10.489)***	0.671	3	60	20	**0.01**	(0.346)	0.526
10	60	60	**0.179**	(7.257)***	0.624	3	60	60	−0.056	(−1.803)*	0.475

Results for daily data from 2009/07/01 to 2015/12/31. Logarithmatic cumulative return are calculated after signals for fixed periods (in the column named as Hold) of following days: 5, 10, 20, 60. The days of MA (short) in BIAS formula is 5 in this table and that of the MA (long) is as following: 10, 20 and 60, in the MA.L column. The Bias ratios are furtherly divided by the time series of the stock individually into 4 kind of quantiles (Qu for short): 20, 10, 5 and 3. The trading signals will be initiated when its BIAS touchs the first quantile or the last quantile. The Est column reports the estimated values of t-test and the t & p column reports the related t-statistics and p-values. The number of asterisk represents the p-vlaue is smaller than 0.1, 0.05 and 0.01. Finally, the column of winRate is the winning rate.

3.2 The Robustness of Empirical Results

To verify the combination of the days of the long-term MA and of the holding period feasible, we use the SVM to classify the data by the C-classification and the radial kernel. Although the parameters of the days are numeric, they should be treated in this study as categorical attributes. So we adopt the recommendation of Chang and Lin (2011) to use the same numbers to represent a category attribute for the stabilization of SVM. The predict results extending the abovementioned tables are in Table 4 and we can see easily that the error of classification is very rare, thus it supports the robustness of the above results.

Table 3. Standard test results for the contrarian strategy with MA (short): ten days

Qu	MA	Hold	Est	t & p	winRate	Qu	MA	Hold	Est	t & p	winRate
20	20	5	**0.036**	(2.862)***	0.537	5	20	5	−0.199	(−7.558)***	0.406
20	20	10	**0.025**	(1.638)	0.526	5	20	10	−0.225	(−8.073)***	0.369
20	20	20	**0.064**	(3.555)***	0.566	5	20	20	−0.136	(−4.371)***	0.453
20	20	60	**0.145**	(6.201)***	0.577	5	20	60	−0.044	(−1.349)	0.474
20	60	5	**0.177**	(14.817)***	0.731	5	60	5	**0.077**	(3.479)***	0.591
20	60	10	**0.172**	(14.286)***	0.721	5	60	10	**0.091**	(3.964)***	0.587

(*continued*)

Table 3. *(continued)*

Qu	MA	Hold	Est	t & p	winRate	Qu	MA	Hold	Est	t & p	winRate
20	60	20	**0.209**	(15.509)***	0.724	5	60	20	**0.168**	(6.937)***	0.625
20	60	60	**0.213**	(10.884)***	0.654	5	60	60	**0.036**	(1.32)	0.533
10	20	5	−0.03	(−1.572)	0.483	3	20	5	−0.437	(−13.491)***	0.286
10	20	10	−0.053	(−2.45)**	0.467	3	20	10	−0.456	(−13.526)***	0.292
10	20	20	−0.016	(−0.651)	0.517	3	20	20	−0.261	(−7.787)***	0.391
10	20	60	**0.093**	(3.241)***	0.578	3	20	60	−0.209	(−6.103)***	0.409
10	60	5	**0.164**	(10.081)***	0.658	3	60	5	−0.032	(−1.118)	0.509
10	60	10	**0.167**	(9.795)***	0.658	3	60	10	−0.004	(−0.151)	0.542
10	60	20	**0.217**	(11.251)***	0.673	3	60	20	**0.074**	(2.479)**	0.569
10	60	60	**0.164**	(7.046)***	0.612	3	60	60	−0.035	(−1.102)	0.493

Results for daily data from 2009/07/01 to 2015/12/31. Logarithmatic cumulative return are calculated after signals for fixed periods (in the column named as Hold) of following days: 5, 10, 20, 60. The days of MA (short) in BIAS formula is 10 in this table and that of the MA (long) is as following: 20 and 60, in the MA.L column. The Bias ratios are furtherly divided by the time series of the stock individually into 4 kind of quantiles (Qu for short): 20, 10, 5 and 3. The trading signals will be initiated when its BIAS touchs the first quantile or the last quantile. The Est column reports the estimated values of t-test and the t & p column reports the related t-statistics and p-values. The number of asterisk represents the p-vlaue is smaller than 0.1, 0.05 and 0.01. Finally, the column of winRate is the winning rate.

Table 4. The predictive results on the above three tables by SVM

Panel A **MA1**

pred	y: Contrarian	Momentum
Contrarian	20	0
Momentum	2	58

Panel B **MA5**

pred	y: Contrarian	Momentum
Contrarian	27	0
Momentum	1	32

Panel C **MA10**

pred	y: Contrarian	Momentum
Contrarian	21	0
Momentum	1	18

Results of prediction by training the above 3 tables by SVM with C-classification and the radial kernel. The error sample will be located out of the diagonal line.

4 Conclusion

Mispricing deepening, implying the existence of limited arbitrage, violates the well-known efficient market hypothesis, and most of the related researches focus on the momentum strategy to show the existence of profit opportunity in short term. This coincides the old saying in Wall Street that “trend is your friend;” However, the duration of a trend is elusive. In this study, we propose a new contrarian strategy, which depends on the signals derived from the BIAS ratio, to alleviate this problem. Every sample for standard t-test is calculated from the historical prices of a stock listed in Taiwan market, and the empirical results support the limited arbitrage theory of

Shleifer and Vishny (1997). Further, we use the well-known machine learning, support vector machine (SVM), to show the key of a successful classification is on the parameters of BIAS ratio. The result shows that our trading rule is free of subjective discretion.

References

Chang, C.-C., Lin, C.-J.: LIBSVM: a library for support vector machines. ACM Trans. Intell. Syst. Technol. **2**, 27:1–27:27 (2011)

Brock, W., Lakonishok, J., LeBaron, B.: Simple technical trading rules and the stochastic properties of stock returns. J. Finance **47**, 1731–1764 (1992)

Gromb, D., Vayanos, D.: Limits of arbitrage. Ann. Rev. Finance Econ. **2**(1), 251–275 (2010)

Shleifer, A., Vishny, R.W.: The limits of arbitrage. J. Finance **52**, 35–55 (1997)

A More Efficient Algorithm to Mine Skyline Frequent-Utility Patterns

Jerry Chun-Wei Lin[1(✉)], Lu Yang[1], Philippe Fournier-Viger[2], Siddharth Dawar[3], Vikram Goyal[3], Ashish Sureka[4], and Bay Vo[5]

[1] School of Computer Science and Technology, Harbin Institute of Technology Shenzhen Graduate School, Shenzhen, China
jerrylin@ieee.org, luyang@ikelab.net
[2] School of Natural Sciences and Humanities, Harbin Institute of Technology Shenzhen Graduate School, Shenzhen, China
philfv@hitsz.edu.cn
[3] Indraprastha Institute of Information Technology, Delhi, India
{siddharthd,vikram}@iiitd.ac.in
[4] ABB Corporate Research, Bangalore, India
ashish.sureka@in.abb.com
[5] Faculty of Information Technology, Ho Chi Minh City University of Technology, Ho Chi Minh City, Vietnam
bayvodinh@gmail.com

Abstract. In the past, a SKYMINE approach was proposed to both consider the aspects of utility and frequency of the itemsets to mine the skyline frequency-utility skyline patterns (SFUPs). The SKYMINE algorithm requires, however, the amounts of computation to mine the SFUPs based on the utility-pattern (UP)-tree structure performing in a level-wise manner. In this paper, we propose more effective algorithms to mine the SFUPs based on the utility-list structure. Substantial experiments are carried to show that the proposed algorithms outperform the state-of-the-art SKYMINE to mine the SFUPs in terms of runtime and memory usage.

Keywords: Skyline · Utility · Frequent · Umax · Utility-list

1 Introduction

The frequent itemset mining (FIM) is the fundamental task of Knowledge discovery in database, which is used to identify the set of frequent itemsets (FIs) [3,11,12,14,20,22]. In real-life situations, only frequency of the itemsets reveals the insufficient information. To solve the limitation of FIM, high-utility itemset mining (HUIM) [5,7,15,16,25] was proposed to discover the "useful" and "profitable" itemsets from the quantitative database. Lin et al. then presented the high-utility pattern (HUP)-tree algorithm [15] to mine the HUIs. The UP-Growth+ algorithm [24] was further designed to adopt several pruning strategies to speed up mining process based on the developed utility-pattern (UP)-tree.

J. Pan et al. (eds.), *Genetic and Evolutionary Computing*, Advances in Intelligent Systems and Computing 536, DOI 10.1007/978-3-319-48490-7_16

Yeh et al. [26] first proposed the two-phase algorithms for mining high utility and frequency itemsets. Podpecan et al. [21] then proposed a fast algorithm to mine the utility-frequent itemsets. However, it is difficult to define appropriate utility threshold and minimum support threshold for retrieving the required information.

Skyline contains the dominance relationship between tuples based on multi-dimensions. Borzsonyi et al. [6] addressed the first work of skylines in the context of databases and developed several algorithms based on block nested loops, divide-and-conquer, and index scanning mechanism. Chomicki et al. [9] employs a certain ordering of tuples in the window to increase performance of [6]. Tan et al. [23] proposed progressive (or on-line) algorithms that can progressively output skyline points without scanning the entire dataset. Other related works of skyline are still developed in progress [2,8,13,19].

Although FIM and HUIM have widely range in real-world applications, both of them can only focus on one aspect by respectively considering the occurrence frequency or the utility of the itemsets. Goyal et al. [10] first defined the skyline frequent-utility pattern (SFUP) and designed a SKYMINE algorithm to mine the itemsets with high occurrence frequency and high utility. However, the numerous candidates are required to be generated of the SKYMINE, which is a time-consuming task. The problem of memory leakage also happens in the SKYMINE since it is necessary to generate the numerous candidates for mining SFUPs. To speed up the mining process and reduce the memory usage, we design new algorithms to efficient mine the SFUPs. An efficient utility-list structure is adopted in this paper to efficiently mine the SFUPs without candidate generation. Besides, an *umax* array is further developed to keep the maximal utility under the occurrence frequency. A pruning strategy is also developed to reduce the search space for mining SFUPs. Extensive experiments on various databases were conducted and the results showed that the proposed algorithm has better performance than that of the SKYMINE for mining SFUPs.

2 Preliminaries and Problem Statement

Let $I = \{i_1, i_2, \ldots, i_m\}$ be a finite set of m distinct items. A quantitative database is a set of transactions $D = \{T_1, T_2, \ldots, T_n\}$, where each transaction $T_q \in D$ $(1 \leq q \leq n)$ is a subset of I and has a unique identifier q, called its TID. Besides, each item i_j in a transaction T_q has its purchase quantity (internal utility) and denoted as $q(i_j, T_q)$. A profit table ptable $= \{pr(i_1), pr(i_2), \ldots, pr(i_m)\}$ indicates the profit value of each item i_j. A set of k distinct items $X = \{i_1, i_2, \ldots, i_k\}$ such that $X \subseteq I$ is said to be a k-itemset, where k is the length of the itemset. An itemset X is said to be contained in a transaction T_q if $X \subseteq T_q$.

Definition 1. The occurrence frequency of an itemset X in D is denoted as $f(X)$, where X is a set of items and $f(X)$ is defined as the number of transactions T_q in D containing X as:

$$f(X) = |\{X \subseteq T_q \wedge T_q \in D\}|. \quad (1)$$

Definition 2. The utility of an item i_j in a transaction T_q is denoted as $u(i_j, T_q)$ and defined as:

$$u(i_j, T_q) = q(i_j, T_q) \times pr(i_j). \tag{2}$$

Definition 3. The utility of an itemset X in a transaction T_q is denoted as $u(X, T_q)$ and defined as:

$$u(X, T_q) = \sum_{i_j \subseteq X \wedge X \subseteq T_q} u(i_j, T_q). \tag{3}$$

Definition 4. The utility of an itemset X in a database D is denoted as $u(X)$, and defined as:

$$u(X) = \sum_{X \subseteq T_q \wedge T_q \in D} u(X, T_q). \tag{4}$$

Definition 5. The transaction utility of a transaction T_q is denoted as $tu(T_q)$ and defined as:

$$tu(T_q) = \sum_{X \subseteq T_q} u(X, T_q). \tag{5}$$

Definition 6. The transaction-weighted utility of an itemset X in D is denoted as $twu(X)$ and defined as:

$$twu(X) = \sum_{X \subseteq T_q \wedge T_q \in D} tu(T_q). \tag{6}$$

The above definitions are used to find whether the FIs or HUIs. The state-of-the-art algorithms whether in FIM or HUIM cannot consider both frequency and utility together. To obtain the skyline frequent-utility patterns (SFUPs), the definitions are given below.

Definition 7. An itemset X dominates another itemset Y in D, denoted as $X \succ Y$ iff $f(X) \geq f(Y)$ and $u(X) \geq u(Y)$.

Definition 8. An itemset X in a database D is a skyline frequent-utility pattern (SFUP) iff it is not dominated by any other itemset in the database by considering both the frequency and utility factors.

Problem Statement: Based on the above definitions, we define the problem of skyline frequent-utility pattern mining (SFUPM) as discovering the set of non-dominated itemsets in the database by considering both the frequency and utility factors.

3 Proposed Algorithms for Mining SFUPs

In this section, two efficient algorithms are designed for mining the set of skyline frequent-utility patterns (SFUPs). It is based on the well-known utility-list structure to fast combine the itemsets by simple join operation. Details are given as follows.

3.1 Utility-List Structure

Let $\triangleright$ be an ascending order on the items in the databases. The utility-list [17] of an itemset X in a database D is a set of tuples, in which each tuple consists of three fields as $(tid, iutil, rutil)$. The tid is the transaction ID containing the itemset X. The $iutil$ and $rutil$ elements of a tuple respectively are the utility of X in tid; i.e., $u(X, T_q)$ and the resting utility of the items except the itemset X in tid, which can be defined as: $\sum_{i_j \in T_q \wedge i_j \notin X} u(i_j, T_q)$. Here is an example to illustrate the construed utility-list shown in Fig. 1.

E		
tid	iutil	rutil
2	4	14
4	4	2
5	8	14

C		
tid	iutil	rutil
1	2	7
2	3	11
3	2	9
4	2	0
6	1	11

B		
tid	iutil	rutil
1	2	5
2	2	9
5	4	10
6	8	3

A		
tid	iutil	rutil
2	4	5
3	4	5
5	5	5
6	3	0

D		
Tid	Iutil	Rutil
1	5	0
2	5	0
3	5	0
5	5	0
7	5	0

Fig. 1. The constructed utility-lists.

In addition to keep the maximal utility of the frequency value, an utility-max ($umax$) array is then set in the beginning of the developed algorithm, which is used to keep the maximal utility of the frequency value.

Definition 9. An $umax$ array keeps the maximal utility of the frequency value r, which is defined as $umax(r)$.

Definition 10. An itemset X is considered as a potential SFUP (PSFUP) if its frequency is equal to r and non-itemset having higher utility than $u(X)$.

3.2 Pruning Strategy

To mine the SFUPs, the SKYMINE algorithm generates numerous candidates from the UP-tree structure, and the upper bound of the itemsets is overestimated based on the two-phase model. To solve above problems, we define a strategy to speed up mining process of the SFUPs.

Pruning Strategy: Let X be an itemset, and let the extensions of X by appending an item Y to X as $(X \cup Y)$ such that $X \triangleright Y$. If the sum of $iutil$ and $rutil$ values in the utility-list of X is less than $umax(r)$, $r = f(X)$, then all the extensions of X are not SFUPs.

For each itemset X, it can be known that the frequency of its extensions is higher than or equals to itself. Based on above pruning strategy, it can be found that there must be an itemset Y dominates X' having the same frequency with X and same utility with $umax(r)$, $r = f(X)$. It indicates that X' is not a SFUP.

3.3 Proposed Algorithms

In the designed algorithms, the utility-list structure is first constructed. The transaction-weighted utility (twu) of all 1-itemsets is discovered and the 1-itemsets in the database are also sorted in twu-ascending order. The sorted database is then used to construct the initial utility-list of 1-itemsets. After that, the P-Miner algorithm is used to find the potential SFUPs. The pseudo-code of the **P-Miner** algorithm is described in Algorithm 1, which is used to find the potential SFUPs.

Algorithm 1. P-Miner

```
Input: P.UL, the utility-list of the itemset P; P'ULs, the set of utility-lists of
       P's extensions; umax, an array to keep the maximum utility of the
       varied frequencies.
Output: PSFUIs; the set of P's potential skyline frequent-utility itemsets.
1  for each X in P'ULs do
2      if sum(X.iutil) ≥ umax(f(X)) then
3          umax(f(X)) ← sum(X.iutil);
4          PSFUIs ← X;
5          remove Y from PSFUIs if (f(Y) == f(X));
6      if sum(X.iutil) + sum(X.rutil) ≥ umax(f(X)) then
7          exULs := null;
8          for each utilit-list Y after X in P'ULs do
9              exULs := exULs + construct(P.UL.X, Y);
10         P-Miner(X, exULs, umax, PSFUPs);
```

After all PSFUPs are discovered, the proposed mining algorithm is then executed to find the actual SFUPs from PSFUPs. The proposed mining algorithm is shown in Algorithm 2.

Algorithm 2. Proposed mining algorithm

```
Input: PSFUPs, the set of potential SFUPs.
Output: SFUPs, the set of skyline frequent-utility itemsets.
1  for each X ∈ PSFUPs do
2      for each Y ∈ PSFUPs do
3          if u(X) ≥ u(Y) ∧ f(X) > f(Y)||u(X) > u(Y) ∧ f(X) ≥ f(Y) then
4              SFUPs ← X ∪ SFUPs;
5              remove Y from PSFUPs;
6  return SFUPs;
```

4 Experimental Results

In this section, substantial experiments were conducted to evaluate the proposed algorithm for mining SFUPs on several datasets. Note that only one existing algorithm called SKYMINE [10] was proposed to mine the SFUPs by considering both the frequency and utility of the itemsets. Four real-world datasets called chess [1], mushroom [1], foodmart [18] and retail [1] and one synthetic T10I4N4KDXK dataset [4] were used in the experiments to evaluate the performance of the proposed algorithm. In the experiments, the program is terminated if the runtime exceeds 2×10^2 s or the memory leakage occurred.

4.1 Runtime

The proposed algorithm was compared with the state-of-the-art SKYMINE algorithm [10] on five datasets and the results are shown in Table 1.

Table 1. Runtime of the compared algorithms.

	Proposed mining algorithm	SKYMINE
Chess	202.69 s	-
Mushroom	10.57 s	202.82 s
Foodmart	2.64 s	98.59 s
Retail	117.57 s	-
T10I4N4KD100K	57.32 s	346.45 s

From Table 1, it can be observed that there are no results of the SKYMINE algorithm on chess and retail datasets. The reason is that the memory leakage occurred for those two datasets and the algorithm is terminated. We also can observe that the proposed algorithm outperforms the SKYMINE algorithm and generally up to almost one or two orders of magnitude faster than the SKYMINE algorithm. The proposed algorithm always has better results than that of SKYMINE algorithm since the proposed algorithm can directly exact the actual utility of the itemsets with only two scans of dataset. The SKYMINE algorithm generates, however, many redundant candidates with the overestimated value of the itemsets. Thus, the SKYMINE algorithm requires more time for generating the candidates and determining the actual SFUPs.

4.2 Memory Usage

In this section, the memory usage of the proposed algorithm and the SKYMINE algorithm were also compared. The results on all datasets are shown in Table 2.

From Table 2, there are no results for the memory usage of the SKYMINE algorithm on chess and retail datasets since the memory leakage occurred.

Table 2. Memory usage of the compared algorithms.

	Proposed mining algorithm	SKYMINE
Chess	137.00 M	-
Mushroom	188.87 M	423.80 M
Foodmart	32.13 M	667.46 M
Retail	143.68 M	-
T10I4N4KD100K	234.30 M	446.69 M

It can be clearly seen that the proposed algorithm requires less memory compared to the SKYMINE algorithm on all datasets. The reason is that the SKYMINE algorithm mines, however, the SFUPs based on UP-Growth algorithm, which requires to generate the numerous candidates and it is not an efficient way to mine the SFPUs.

5 Conclusion

In this paper, more efficient algorithms are proposed to mine a set of skyline frequent-utility itemsets without candidate generation by considering the frequency and utility factors. The designed algorithms rely on the utility-list structure and the *umax* array for mining SFUPs. A pruning strategy are used to early prune the unpromising candidates for deriving the SFUPs. Based on the designed algorithms, it is unnecessary to pre-defined the minimum support or utility thresholds but the set of useful and meaning information can be returned and discovered. Substantial experiments were conducted on both real-life and synthetic datasets to asses the performance of the proposed algorithm in terms of runtime and memory usages.

Acknowledgment. This research was partially supported by the National Natural Science Foundation of China (NSFC) under grant No. 6150309.

References

1. Frequent itemset mining dataset repository (2012). http://fimi.ua.ac.be/data/
2. Afrati, F.N., Koutris, P., Suciu, D., Ullman, J.D.: Parallel skyline queries. Theory Comput. Syst. **57**(4), 1008–1037 (2015)
3. Agrawal, R., Srikant, R.: Fast algorithm for mining association rules. In: International Conference on Very Large Data Bases, pp. 487–499 (1994)
4. Agrawal, R., Srikant, R.: Quest synthetic data generator (1994). http://www.Almaden.ibm.com/cs/quest/syndata.html
5. Ahmed, C.F., Tanbeer, S.K., Jeong, B.S., Le, Y.K.: Efficient tree structures for high utility pattern mining in incremental databases. IEEE Trans. Knowl. Data Eng. **21**(12), 1708–1721 (2009)

6. Borzsonyi, S., Kossmann, D., Stocker, K.: The skyline operator. In: International Conference on Data Engineering, pp. 421–430 (2001)
7. Chan, R., Yang, Q., Shen, Y.D.: Mining high utility itemsets. In: IEEE International Conference on Data Mining, pp. 19–26 (2003)
8. Chan, C.Y., Jagadish, H.V., Tan, K.L., Tung, A.K.H., Zhang, Z.: Finding k-dominant skylines in high dimensional space. In: ACM SIGMOD International Conference on Management of Data, pp. 503–514 (2006)
9. Chomicki, J., Godfrey, P., Gryz, J., Liang, D.: Skyline with presorting. In: International Conference on Data Engineering, pp. 717–720 (2003)
10. Goyal, V., Sureka, A., Patel, D.: Efficient skyline itemsets mining. In: The International C* Conference on Computer Science & Software Engineering, pp. 119–124 (2015)
11. Grahne, G., Zhu, J.: Efficiently using prefix-trees in mining frequent itemsets. In: IEEE ICDM Workshop on Frequent Itemset Mining Implementations (2003)
12. Han, J., Pei, J., Yin, Y.: Mining frequent patterns without candidate generation. In: ACM SIGKDD International Conference on Management of Data, pp. 1–12 (2000)
13. Kossmann, D., Ramsak, F., Rost, S.: Shooting stars in the sky: an online algorithm for skyline queries. In: International Conference on Very Large Data Bases, pp. 275–286 (2002)
14. Lin, C.W., Hong, T.P., Lu, W.H.: The pre-FUFP algorithm for incremental mining. Expert Syst. Appl. **36**(5), 9498–9505 (2009)
15. Lin, C.W., Hong, T.P., Lu, W.H.: An effective tree structure for mining high utility itemsets. Expert Syst. Appl. **38**(6), 7419–7424 (2011)
16. Liu, Y., Liao, W., Choudhary, A.: A two-phase algorithm for fast discovery of high utility itemsets. In: Ho, T.B., Cheung, D., Liu, H. (eds.) PAKDD 2005. LNCS (LNAI), vol. 3518, pp. 689–695. Springer, Heidelberg (2005). doi:10.1007/11430919_79
17. Liu, M., Qu, J.: Mining high utility itemsets without candidate generation. In: ACM International Conference on Information and Knowledge Management, pp. 55–64 (2012)
18. Microsoft, Example database foodmart of Microsoft analysis services. http://msdn.microsoft.com/en-us/library/aa217032(SQL.80).aspx
19. Papadias, D., Tao, Y., Seeger, B.: Progressive skyline computation in database systems. ACM Trans. Database Syst. **30**(1), 41–82 (2005)
20. Park, J.S., Chen, M.S., Yu, P.S.: An effective hash based algorithm for mining association rules. In: ACM SIGMOD International Conference on Management of Data, pp. 175–186 (1995)
21. Podpecan, V., Lavrac, N., Kononenko, I.: A fast algorithm for mining utility-frequent itemsets. In: International workshop on Constraint-based Mining and Learning, pp. 9–20 (2007)
22. Savasere, A., Omiecinski, E., Navathe, S.: An efficient algorithm for mining association rules in large databases. In: International Conference on Very Large Databases, pp. 432–444 (1995)
23. Tan, K.L., Eng, P.K., Ooi, B.C.: Efficient progressive skyline computation. In: International Conference on Very Large Data Bases, pp. 301–310 (2001)
24. Tseng, V.S., Shie, B.E., Wu, C.W., Yu, P.S.: Efficient algorithms for mining high utility itemsets from transactional databases. IEEE Trans. Knowl. Data Eng. **25**(8), 1772–1786 (2012)

25. Yao, H., Hamilton, H.J., Geng, L.: A unified framework for utility-based measures for mining itemsets. In: ACM SIGKDD International Conference on Utility-Based Data Mining, pp. 28–37 (2006)
26. Yeh, J.-S., Li, Y.-C., Chang, C.-C.: Two-phase algorithms for a novel utility-frequent mining model. In: Washio, T., et al. (eds.) PAKDD 2007. LNCS (LNAI), vol. 4819, pp. 433–444. Springer, Heidelberg (2007). doi:10.1007/978-3-540-77018-3_43

Recent Advances on Evolutionary Optimization Technologies

Research on Expressway Emergency Vehicle Allocation Based on Improved Particle Swarm Optimization

Lieyang Wu(✉)

Jiangxi Expressway Networking Management Center, Nanchang 330036, China
lieyang_wu@163.com

Abstract. Firstly, an improved particle swarm optimization algorithm is proposed to solve the allocation of expressway emergency vehicles. Compared with the standard PSO algorithm, the particle population number of the improved PSO algorithm is increased, due to particle flight behavior of different populations is different and particle information between different populations is exchanged, so the swarm population diversity of the improved PSO algorithm is increased, and its ability to jump out of local optimum is improved. Moreover, the improved algorithm is applied to the allocation of emergency vehicles, that is, the mathematical model is established to solve the shortest travel distance of the emergency vehicle, and the mathematical model is optimized by the proposed algorithm to obtain the optimal solution. The experimental results show that the improved algorithm proposed in this paper is feasible and effective to solve the expressway emergency vehicle allocation problem.

Keywords: Particle swarm optimization · Population diversity · Expressway · The allocation of emergency vehicles

1 Introduction

With the continuous develop of the transportation industry, there are more and more expressway emergency events, and allocation problem of emergency vehicle is becoming more and more complex. On the other hand, a reasonable allocation scheme of emergency vehicle can reduce traveling distance of the emergency vehicle and shorten the time of emergency rescue; in addition, it also can improve the utilization rate of emergency vehicles, lower freight costs, and enhance transport efficiency.

With the continuous construction of expressway network and the rapid develop of transportation, expressway network structure is more and more large and complicated. The allocation problem complexity of emergency vehicle mainly depends on the expressway network structure complexity and the dynamic changes of whole expressway network emergency, so solving allocation problem of emergency vehicle with traditional mathematical methods becomes more and more difficult.

Through simulating the process of birds finding food, a new optimization algorithm is obtained which is named particle swarm optimization algorithm (PSO) [1, 2]. The PSO algorithm is simple to design, easy to realize and has high calculation speed

J. Pan et al. (eds.), *Genetic and Evolutionary Computing*, Advances in Intelligent Systems and Computing 536, DOI 10.1007/978-3-319-48490-7_17

and other advantages, so that it is a good method to solve the optimization problem. However, the PSO algorithm is prone to premature convergence when solving complex problems, so the optimal problem usually can't be effectively solved [3].

Against the premature convergence of particle swarm algorithm, an improved particle swarm optimization algorithm is proposed in this paper. By increasing populations with different flight behavior and exchanging particle information between different populations, population diversity is increased, which can make the particles jump out of local optimum, thereby global searching ability of the algorithm is greatly improved. The improved algorithm proposed in this paper is applied to the allocation problem of expressway emergency vehicle, and the optimal solution of the problem can be obtained.

2 Mathematical Model of the Allocation Problem of Emergency Vehicle

Problem description: Supposed that there are n sending stations of emergency vehicle (source station) S_i, $i \in [1, n]$, S_i station has a_i emergency vehicles; There are m arrival stations of emergency vehicle (destination) $D_j, j \in [1, m]$, D_j station demands b_j vehicle emergency vehicles; The distance from the transmission station S_i to destination station D_j (transport costs or time) is l_{ij}. An allocation set of emergency vehicle flow x_{ij} from the sending station to the destination station will be determined, which makes the total traveling distances (transportation cost or time) of all emergency vehicles being minimal under satisfying certain constraints [4, 5].

A mathematical model to meet the above requirements is set up:

$$\min F = \sum_{i=1}^{n} \sum_{j=1}^{m} x_{ij} l_{ij} \tag{1}$$

Constraint conditions are as follows:

$$\sum_{i=1}^{m} x_{ij} \leq a_i \tag{2}$$

$$\sum_{j=1}^{n} x_{ij} \leq b_i \tag{3}$$

$$x_{ij} \in \{0, 1, 2, \cdots\} \quad i = 1, 2, \cdots, n; \; j = 1, 2, \cdots, m \tag{4}$$

Formula (1) is the objective function, which describes the optimal solutions of the problem, namely, the minimum distance; Formula (2) indicates that the number of emergency vehicles from the sending stations to the destination stations must not be greater than total number of emergency vehicles of sending station; Formula (3) indicates that the number of emergency vehicles arriving destination station must be not greater than the demand number; Formula (4) shows the constraint that the number of the emergency vehicle x_{ij} must being integer.

If sending emergency vehicles and receiving emergency vehicles is balance, namely, the total number of emergency vehicles of sending stations is equal to the total number of emergency vehicles of destination stations; Formula (2) is equal to formula (3). On the contrary, if sending emergency vehicle and receiving emergency vehicle isn't balance, virtual sending stations or virtual destination stations will be set up, which can make the unbalance problem become the balance problem, that is, supposed that the number of emergency vehicles of virtual stations being $\sum_{j=1}^{m} a_i - \sum_{i=1}^{n} b_j$, and the traveling distance of the emergency vehicle of virtual stations being 0, so the original problem is transformed into balance problem of sending and receiving.

3 Particle Swarm Optimization Algorithm and Its Improving Algorithm

3.1 The Principles of Particle Swarm Optimization Algorithm

The basic idea of the particle swarm optimization (PSO) algorithm is derived to simulate birds finding food behavior. The birds are abstracted as the particles with no quality and no volume in solution space, which fly closer to the optimal solution step by step with the help of their own historical optimum and the population optimum.

Firstly, the particle swarm is initialized, that is, the initial position and initial velocity of the particle are determined. Particle population size is N, and the position and the velocity of the i th particle in the D dimensional space can be respectively represented as $X_i = (x_{i1}, x_{i2}, \ldots\ldots, x_{iD})$ and $V_i = (v_{i1}, v_{i2}, \ldots\ldots, v_{D1})$.

The optimal position of each particle marked as *pbest* $P_i = (p_{i1}, p_{i2}, \ldots\ldots, p_{iD})$ and the optimal position of group optimal location can be (*gbest*) $P_g = (p_{g1}, p_{g2}, \ldots\ldots\ldots, p_{gD})$, determined by comparing the fitness value of each particle when they are flying.

$$v_{id}^{t+1} = \omega v_{id}^{t} + c_1 r_1 (p_{id}^{t} - x_{id}^{t}) + c_2 r_2 (p_{gd}^{t} - x_{id}^{t}) \tag{5}$$

$$x_{id}^{t+1} = x_{id}^{t} + v_{id}^{t+1} \tag{6}$$

The inertia weight (ω) controls the influence of a particle's velocity, resulting in a memory effect. The use of the inertia weight can improve performance in a number of applications. Originally, it was linearly decreased during a run, providing a balance between exploration and exploitation.

Accelerating constants c_1, c_2 play a role in adjusting a particle's social swarm experience and its own experience. r_1, r_2 represents a vector of random variables following the uniform distribution between 0 and 1. According to specific problem, generally the maximum iteration number or minimum error is selected as the termination condition.

3.2 The Improved Particle Swarm Optimization Algorithm

Unlike standard PSO algorithm which use only one swarm to search the optimal solution in the search space, in multi-swarm co-evolutionary algorithm, the swarm is divided into several sub-swarm, each sub-swarm evolved independently while sharing information through migration to each other.

The proposed PSO algorithm in this paper divides the particle swarm into two groups with the same size. One adopts the standard PSO model, the other corresponding to the reverse flight model, and its iteration equations are as follows:

$$v_{id}^{t+1} = \omega v_{id}^{t} + c_1 r_1 (p_{id}^{t} - x_{id}^{t}) + c_2 r_2 (p_{gd}^{t} - x_{id}^{t}) \tag{7}$$

$$x_{id}^{t+1} = x_{id}^{t} - v_{id}^{t+1} \tag{8}$$

In the search process, the worst adaptation of particles in the first swarm (s0) will be exchanged with the optimal adaptation of the particles in the second swarm (s1). In the entire search process, the inferior particles in the first swarm will be continuously exchanged with the superior particles in the second swarm in this way. During the evolution process, with the different models the particles falling to local extreme value can get new vitality. In a word, the advanced algorithm improves the population diversity, promote the information communication between particles and can effectively avoid local convergence.

Thc flowchart of the improved PSO algorithm is shown in Fig. 1.

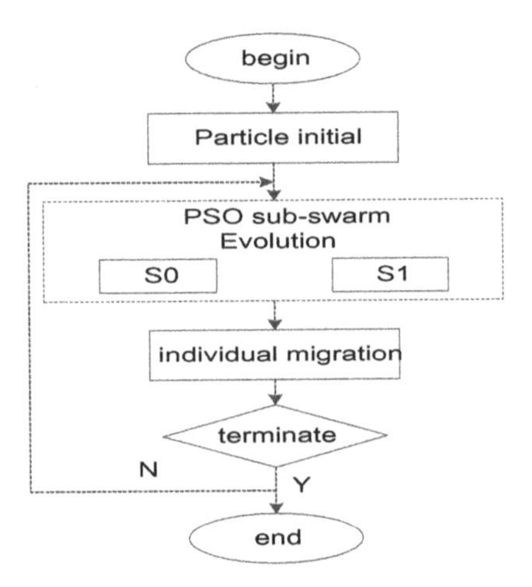

Fig. 1. Flow chart of the algorithm

4 Application of Improved Algorithm in the Allocation of Emergency Vehicles

4.1 Problem Description

Supposed that a expressway network has 7 emergency stations, at some time point, there are 3 sending stations of emergency vehicles (sending station) and 4 demand stations (destination stations), moreover, the number of sending emergency vehicles and the number of receiving emergency vehicles is equal, as well as there are 240 emergency vehicles. The allocation of emergency vehicles is shown in Table 1, and transport distance as shown in Table 2.

Table 1. Emergency vehicle allocation balance sheet

Station	B_1	B_2	B_3	B_4	Send
A_1	x_{11}	x_{12}	x_{13}	x_{14}	80
A_2	x_{21}	x_{22}	x_{23}	x_{24}	50
A_3	x_{31}	x_{32}	x_{33}	x_{34}	110
Receive	40	70	60	70	240

Table 2. Transport distance (km)

Station	B_1	B_2	B_3	B_4
A_1	$l_{11} = 200$	$l_{12} = 600$	$l_{13} = 200$	$l_{14} = 550$
A_2	$l_{21} = 100$	$l_{22} = 500$	$l_{23} = 150$	$l_{24} = 450$
A_3	$l_{31} = 400$	$l_{32} = 250$	$l_{33} = 550$	$l_{34} = 300$

4.2 Parameter Setting

The maximum iteration number of the algorithm is 200 times. The total number of particles is 60. The each particles number of the two populations is 30 in this paper. For the standard PSO, the inertia weight ω is linearly reduced from 0.95 to 0.4. However, in this paper, the inertia weight of the first group is linearly reduced from 0.95 to 0.4, and the inertia weight of the second group is linearly increased from 0.4 to 0.95. The learning factors c_1 and c_2 is 2, that is $c_1 = c_2 = 2$.

4.3 Result Analysis

When using the standard PSO and the improved PSO respectively to solve the allocation problem of emergency vehicles, the corresponding allocation scheme and the optimization results are shown in Tables 3 and 4.

From the operation results, it can be seen that the number of times of finding the global optimal solution using the algorithm in this paper are more than standard PSO, as well as the average value is better, the solution of the allocation problem being more reasonable, in addition, it can save more time and cost. In this paper, the improved

Table 3. Optimization results (optimum solution: 61000)

Station	B_1	B_2	B_3	B_4	Send
A_1			60	20	80
A_2	40			10	50
A_3		70		40	110
Receive	40	70	60	70	240

Table 4. Comparison of optimization performances between two optimization algorithms (Independently run 30 times)

Algorithm	Mean of fitness	Maximum value	Minimum value	Times of finding optimal solution
Standard PSO	67869	93000	61000	8
Improved PSO	61255	61500	61000	17

algorithm is not easy to fall into the local optimum, which increases the population diversity and makes the particle jump out of the local extreme value, so as to improve the global optimization performance of the algorithm.

5 Conclusions

With the continuous construction of expressway network and the rapid develop of transportation, expressway network structure is more and more large and complicated, how to allocate the emergency vehicle problem to be solved urgently. According to the standard PSO algorithm, it is easy to premature convergence. In this paper, by increasing the number of particle population and improving the diversity of the algorithm, the improved PSO algorithm has a good global optimization ability. The improved algorithm is applied to solve the problem of emergency vehicle allocations, the experimental results show that the improved algorithm can find the optimal allocation scheme, which can solve the high speed and high precision.

References

1. Kennedy, J., Eberhart, R.: Particle swarm optimization. In: IEEE International Conference on Neural Networks, Perth, Australia, IEEE Service Center Piscataway NJ, pp. 1942–1948 (1995)
2. Eberhart, R., Kennedy, J.: A new optimizer using particle swarm theory. In: Proceedings of the 6th International Symposium on Micro Machine and Human Science, Nagoya, Japan, IEEE Service Center Piscataway NJ, pp. 39–43 (1995)

3. Shi, Y., Eberhart, R.: A modified particle swarm optimizer. In: IEEE World Congress on Computational Intelligence, pp. 69–73 (1998)
4. Jing, A.L., Stephen, C.H.L.: Allocation of empty containers between multi-ports. Eur. J. Oper. Res. **182**, 400–412 (2007)
5. Dong, L., Boliang, L.: Research on the multi-stage optimization model of empty railcar distribution. J. China Railway Soc. **29**, 1–6 (2007)

Local-Learning and Reverse-Learning Firefly Algorithm

Zhifeng Xie[1], Jia Zhao[1,2(✉)], Hui Sun[1,2], Hui Wang[1,2], and Kun Wang[1]

[1] School of Information Engineering, Nanchang Institute of Technology, Nanchang 330099, China
13803506291@163.com, zhaojia925@163.com, wk880819@163.com, sunhui@nit.edu.cn, huiwang@whu.edu.cn

[2] Provincial Key Laboratory for Water Information Cooperative Sensing and Intelligent Processing, Nanchang 330099, China

Abstract. Firefly algorithm is a nature-inspired, which has shown an effective performance on many optimization problems. However, it may suffer from premature convergence and trap in local optimum easily on many optimization problems. Therefore, we propose a new FA variant, called LRRSFA, which aims to solve the problem of premature convergence and local optimum. LRFA mainly has three changed points. First, the fixed original attractiveness is replaced by random variable attractiveness. Second, it is neighborhood search of global optimal particle. Third, some particle move towards given particles which are chosen while initialing population in specific condition. Results tested on eleven standard benchmark function are better than standard firefly algorithm.

Keywords: Firefly algorithm · Premature convergence · Local optimum

1 Introduction

Firefly Algorithm (FA) [1] is one of the swarm intelligence [2–5] for hard optimization problem. It has been developed by mating and social behavior of fireflies. The algorithm is a research hotspot of scholars due to its simple concept, easy implementation, and effectiveness. With in-depth research, it obtains good effect in the field of each application, such as complex networks developed by Amiri et al. [6] in 2013, energy conservation proposed by Coello and Mariani [7] in 2013, image compression inspired by Horng [8] in 2012, and so on.

Although the algorithm is widely used in various fields, the algorithm itself also has many defects. It has low convergence rate during the search process, traps into local optimum easily and gets the poor accuracy of value, and so on. A few researches have improved it. Lukasik et al. [9] in 2009 researched the parameters of the FA and improved it. It greatly improved the accuracy of the solution but it slowed down the convergence speed. The FA successfully applied to solve multi-objective optimization

J. Pan et al. (eds.), *Genetic and Evolutionary Computing*, Advances in Intelligent Systems and Computing 536, DOI 10.1007/978-3-319-48490-7_18

problems in the economic emission load dispatch problem while expanding the field of application of FA by Apostolopoulos et al. [10] in 2001. ChaiEad et al. [11] used FA to solve the nonlinear function optimization problems with noise, and simplified the setting of the algorithm parameters.

In FA, the intensity of light of firefly is associated to the fitness of objective function. The brighter firefly, the better fitness. With a fully attraction model, a particle in the swarm only moves toward other brighter firefly. In other words, a firefly will move many times in each generation. If swarm size is set to N, the FA process executes $N * (N-1)/2$ operations while most of bio-inspired algorithm conducts N. Therefore, the FA has high computational time complexity. A particle only move toward brighter fireflies, this way of the search results in lack of population diversity. Results show that the algorithm is easy to trap into local optimum. Aiming at the problems above, in this paper, we propose a new FA called local-learning and reverse-learning FA (LRFA). The LRFA mainly changes the search process of particle and makes particle mobile with more diversity. Tests are conducted on eleven benchmark functions and results are promising compared to standard FA. Hence, the LRFA outperforms the FA.

2 Firefly Algorithm

2.1 The Introduction of Firefly

Firefly is a small beetle, because its tail can emit fluorescence. There are 2000 species fireflies in the world. Fireflies light emitting device has a light emitting cells that contain a phosphorus chemicals called fluorescein. Light occurs when the fluorescent reacts with oxygen under the catalysis of luciferase, with consuming ATP. There are a variety of biological significance of firefly's flash. Most of the fireflies attract the opposite sex bugs in order to mate and breed; some prey on other insects by flashing; other emit warning light for protecting oneself. The purpose of firefly luminescence is worthy of further exploration and research.

By observing the flying process of firefly, the phenomenon is similar to particle random search in swarm intelligence optimization algorithm heavily. Firefly algorithm is inspired by the flying behavior of firefly.

2.2 Firefly Algorithm

The standard firefly algorithm inspired by the behavior of fireflies was developed by Yang. Its implementation must meet three rules:

(1) All fireflies are unisex. Therefore, one firefly will move toward to other fireflies regardless of their sex.
(2) The brightness decides to the direction of the firefly mobile. One firefly will be attracted to the brighter one. The brightness is affected or determined by values of objective function. For a minimization problem, the smaller the value of function, the better the firefly.

(3) Attractiveness decides to the degree of distance of the firefly mobile. The attractiveness and brightness decrease as the distance of any two fireflies increase.

In FA, the relative intensity of the light of firefly can be calculated as follows:

$$I = I_0 e^{-\gamma r_{ij}} \tag{1}$$

where I_0 is the intensity of the light at, $r_{ij} = 0$, r_{ij} is the distance between two fireflies. The parameter γ is the light absorption coefficient, which is usually set to 1.

The attractiveness equation is proposed as follows:

$$\beta = \beta_0 e^{-\gamma r_{ij}^2} \tag{2}$$

where β_0 is the attractiveness at $r_{ij} = 0$, and r_{ij} is the distance between two fireflies. The parameter γ is the light absorption coefficient, which is usually set to 1.

From two fireflies x_i and x_j, their distance r_{ij} can be defined by:

$$r_{ij} = \left\| x_i - x_j \right\| = \sqrt{\sum_{d=1}^{D} \left(x_{id} - x_{jd} \right)^2} \tag{3}$$

where D is the problem dimension.

Location update formula of firefly, which is attracted to another brighter firefly, is determined by:

$$x_{id}(t+1) = x_{id}(t) + \beta_0 e^{-\gamma r_{ij}^2} \left(x_{jd}(t) - x_{id}(t) \right) + \alpha \varepsilon \tag{4}$$

where x_{id} and x_{jd} are the d th dimension value of firefly x_i and x_j, respectively. The parameter t represents the number of each iteration. The parameter α is the step factor and it is distributed in the range [0, 1]. The parameter ε is a random number and it is uniformly distributed in the range [0,1].

The pseudo-code of the standard FA is described in Algorithm 1, where N is the population scale, MAX_GEN is the maximum number of iteration and it is the condition of the decision to terminate the program. The paper describes minimization problems. Hence, $f(x_j) < f(x_i)$ indicates firefly should move towards to firefly.

Algorithm 1: The Standard Firefly Algorithm

1	Initialize a population of fireflies and a set of parameter.
2	Calculate the fitness values of each firefly.
3	**while** (*t*<*MAX_GEN*) **do**
4	**for** $i = 1$ **to** N **do**
5	**for** $j = 1$ **to** N **do**
6	**if** $f(x_j) < f(x_i)$ **do**
7	Move firefly x_i towards x_j according to Eq.(4)
8	Calculate the fitness values of the new solution
9	**end if**
10	**end for**
11	**end for**
12	Gen++
13	**end while**

3 Proposed Approach

In this section, we present a new FA variant, called local-learning and reverse-learning firefly algorithm (LSFA). The LSFA is improved in three space. First, original attractiveness is no longer a fixed value. Second, the best particle conducts local search. Third, some particle executes reverse-learning.

3.1 Original Attractiveness Randomization

In standard FA, the original attractiveness is a constant. We know that the attractiveness is proportional to the brightness of firefly. In other words, the brighter firefly should have higher the attractiveness. Therefore, the original attractiveness should be different for any firefly in all population. The attractiveness is associated with the brightness and brightness is determined by value of objective function. So, the objective function value of every firefly should be a measure standard of the original attractive. The equation of the original attractive is proposed as follows:

$$\beta_0 = I_j / \left(I_i + I_j\right) \tag{5}$$

where I is the intensity of the brightness of firefly. I_i and I_j are the intensity of the brightness of firefly x_i and x_j, respectively.

3.2 The Local Search of the Optimal Particle

In order to avoid premature convergence, the best particle in every iteration is conducted by local search. The degree of local search of the optimal particle is controlled by two parameters of the value of the optimal and suboptimal particle. The updated equation of the best particle is proposed as follows:

$$P'_{g1} = r_1 P_{g1} + r_2 (P_{g1} - P_{g2}) \tag{6}$$

where the parameter r_1 and r_2 are uniformly distributed in the range [−1,1], the range be able to control the direction of the local search of the optimal firefly particle efficiently. The parameter P_{g1} and P_{g2} are the optimal and suboptimal particle in last generation, respectively. The parameter P'_{g1} is updated value in current generation.

3.3 The Reverse-Learning of Particle

The reverse-learning enhances population diversity and makes particle escape from local optimum. Reverse-learning is conditional constraint. When the optimal particle is not update in certain iterations, the program conducts reverse-learning. Therefore, the population is divided into two parts. The reverse-learning is conducted on some particle. The rest keep the original way of updating.

At present reverse-learning particles trap into local optimum. Hence, the some particle should not update according to Eq. (4). Otherwise, they still keep in local optimum position. Reverse-learning particle must get bigger step factor for escaping local optimum point. All particle in population are close each other while global particle is not still updated. So, the reverse-learning particle should move towards to the particle which is far from them. The particle that we initialize the population randomly for the first time should be object of reverse-learning particle mobile. The updated equation of the reverse-learning particle is proposed as follows:

$$P'_{Ri} = r_3 P_{Ri} + r_4 (P_{Ri} - P_{Ij}) \tag{7}$$

where the parameter r_3 and r_4 are uniformly distributed in the range [−1,1], the range is able to control the direction of the local search of the reverse-learning particle and make them escape the local position efficiently. The parameter P_{Ri} is the ith reverse-learning particle. The parameter P_{Ij} is the jth random initialized particle. The parameter is updated value of the ith reverse-learning particle.

Algorithm 2. The LRFA

```
1   Initialize a population of fireflies and a set of parameter.
     The current initialized particle are set to object of reverse-learning.
2   Calculate the fitness values of each firefly.
3   while (t < MAX_GEN) do
4       for i = 1 to N do
5           for j= 1 to N do
6               if f(x_j) < f(x_i) do
7                   Move firefly x_i towards x_j according to Eq.(4)
8                   Calculate the fitness values of the new solution
9               end if
10          end for
11      end for
12   update the best particle Eq.(6)
13   if (reverse-learning condition)
14           some particles update according to Eq.(7)
15           the rest of particles update according to Eq.(4)
16   end if
17   Gen++
18   end while
```

4 Experiment

4.1 Benchmark Functions

Tests are conducted on set of eleven well-known benchmark functions. The benchmark functions are listed in Table 1.

4.2 Test and Results

The comparison is conducted between the standard FA and the LRFA. The experiment uses the control variable method. Such is fixed population and variable dimension. The population size and the maximum iterator are set to 30 and 1000, respectively. The parameter β_0 is equal to 1.0. The dimension for every candidate solution are set to 5,25,50 and 100 respectively. Each of the experiments is executed 30 times using different random seeds. Results of the experiments are listed in the following Table 2.

From the data in the Table 2, experimental results of the LSFA are better than them of the FA on the whole. When the dimension is set to 5 and 25, the results of are close to the optimal values function. However, it is set to 50 and 100, the results show that

Table 1. Eleven benchmark function

Name	Function	Range	Optimum
Sphere	$f_1(x) = \sum_{i=1}^{D} x_i^2$	[−100.100]	0
Schwefel 2.22	$f_2(x) = \sum_{i=1}^{D} \lvert x_i \rvert + \prod_{i=1}^{D} x_i$	[−10, 10]	0
Schwefel 1.2	$f_3(x) = \sum_{i=1}^{D} \left(\sum_{j=1}^{i} x_j \right)^2$	[−100, 100]	0
Schwefel 2.21	$f_4(x) = max\{\lvert x_i \rvert, 1 \le i \le D\}$	[−100, 100]	0
Rosebrock	$f_5(x) = \sum_{i=1}^{D} \left[100(x_{i+1} - x_i^2)^2 + (1 - x_i^2)^2 \right]$	[−30, 30]	0
Step	$f_6(x) = \sum_{i=1}^{D} \lfloor x_i + 0.5 \rfloor$	[−100, 100]	0
Quartic	$f_7(x) = \sum_{i=1}^{D} ix_i^4 + random[0, 1)$	[−1.28, 1.28]	0
Schwefel 2.26	$f_8(x) = \sum_{i=1}^{D} -x_i \sin \sqrt{\lvert x_i \rvert}$	[−500, 500]	−12569.5
Rastrigin	$f_9(x) = \sum_{i=1}^{D} [x_i^2 - 10 \cos 2\pi x_i + 10]$	[−5.12, 5.12]	0
Ackley	$f_{10}(x) = -20exp\left(-0.2\sqrt{\frac{1}{D}\sum_{i=1}^{D} x_i^2} \right) - exp\left(\frac{1}{D}\sum_{i=1}^{D} \cos 2\pi x \right) + 20 + e$	[−32, 32]	0
Griewank	$f_{11}(x) = \frac{1}{4000}\sum_{i=1}^{D} (x_i)^2 - \prod_{i=1}^{D} \cos\left(\frac{x_i}{\sqrt{i}}\right) + 1$	[−600, 600]	0

there are big gap from the optimal fitness of function. Hence, the performance of the LSFA is better when the optimization problem is a low dimension.

The Table 3 is Friedman detection result. From the data in the tables, all the mean rank of the LRFA in four forms are smaller than that of the FA. Therefore, it is the conclusion that the performance of the LRFA is better than that of the FA.

Table 2. Results of test of the standard FA and the LRFA

Functions		D = 5		D = 25		D = 50		D = 100	
		FA	LRFA	FA	LRFA	FA	LRFA	FA	LRFA
F1	Mean	1.81e-04	**4.20e-05**	3.29e-02	**1.21e-02**	1.23e-01	**4.94e-02**	4.17e-01	**1.69e-01**
	Std	1.87e-04	**9.60e-05**	1.50e–02	**8.74e-03**	5.26e-02	**3.48e-02**	1.75e-01	**8.11e-02**
F2	Mean	1.97e-02	**8.83e-03**	7.63e-01	**4.74e-01**	4.89e+08	**1.13e+02**	1.17e+09	**5.44e+06**
	Std	3.09e-02	**1.11e-02**	3.07e-01	**2.17e-01**	3.01e+08	**7.02e+01**	1.30e+09	**1.32e+08**
F3	Mean	1.98e-02	**8.84e-03**	8.19e-01	**5.29e-01**	4.89e+08	**1.13e+02**	1.17e+09	**5.44e+06**
	Std	1.12e-01	**4.96e-02**	4.19e+00	**2.61e+00**	2.70e+09	**6.23e+02**	6.52e+09	**1.36e+08**
F4	Mean	2.79e-02	**1.29e-02**	8.90e-01	**5.78e-01**	4.89e+08	**1.13e+02**	1.17e+09	**5.44e+06**
	Std	1.56e-01	**6.93e-02**	6.14e+00	**3.90e+00**	3.80e+08	**8.79e+02**	9.13e+09	**1.39e+08**
F5	Mean	8.13e-02	**1.10e+00**	3.31e+01	**2.54e+01**	9.31e+08	**1.67e+02**	2.34e+09	**6.37e+06**
	Std	2.76e-01	**1.56e+00**	1.07e+02	**5.18e+00**	4.70e+09	**1.08e+03**	1.11e+10	**1.45e+08**

(continued)

Table 2. (*continued*)

Functions		D = 5		D = 25		D = 50		D = 100	
		FA	LRFA	FA	LRFA	FA	LRFA	FA	LRFA
F6	Mean	5.58e+01	**1.10e+00**	4.84e+02	**2.54e+01**	9.31e+08	**1.67e+02**	2.34e+09	**6.37e+06**
	Std	2.32e+02	**6.25e+00**	9.01e+02	**1.39e+02**	6.93e+09	**1.42e+03**	1.70e+10	**1.49e+08**
F7	Mean	5.58e+01	**1.10e+00**	4.84e+02	**2.54e+01**	9.31e+08	**1.98e+02**	2.34e+09	**6.37e+06**
	Std	3.83e+02	**8.70e+00**	2.80e+03	**1.96e+02**	8.61e+09	**1.62e+03**	2.13e+10	**1.53e+08**
F8	Mean	2.08e+03	**1.86e+03**	1.07e+04	**1.03e+04**	9.31e+08	**2.08e+04**	2.34e+09	**6.42e+06**
	Std	5.02e+02	**5.12e+02**	3.87e+03	**4.75e+02**	1.00e+10	**2.02e+03**	2.48e+10	**1.57e+08**
F9	Mean	2.09e+03	**1.86e+03**	1.08e+04	**1.03e+04**	9.31e+08	**2.14e+04**	2.34e+09	**6.42e+06**
	Std	1.62e+04	**1.02e+04**	5.87e+04	**5.63e+04**	1.12e+10	**1.14e+05**	2.79e+10	**1.61e+08**
F10	Mean	2.10e+03	**1.86e+03**	**1.08e+04**	**1.30e+04**	9.31e+08	**2.14e+04**	2.34e+09	**6.42e+06**
	Std	1.62e+04	**1.44e+04**	8.32e+04	**7.95e+04**	1.23e+10	**1.64e+05**	3.07e+10	**1.64e+08**
F11	Mean	2.10e+03	**1.86e+03**	1.08e+04	**1.03e+04**	9.31e+08	**2.14e+04**	2.34e+09	**6.42e+06**
	Std	1.98e+04	**1.76e+04**	1.02e+-5	**9.74e+04**	1.33e+10	**2.01e+05**	3.33e+10	**1.68e+08**
w/e/l		11/0/0		11/0/0		11/0/0		11/0/0	

Table 3. Mean ranks achieved by Friedman for the standard FA and the LRFA

D = 5 Friedman Test

Algorithm	Mean rank
LRFA	1.09
FA	1.91

D = 25 Friedman Test

Algorithm	Mean rank
LRFA	1.09
FA	1.91

D = 50 Friedman Test

Algorithm	Mean rank
LRFA	1.00
FA	2.00

D = 100 Friedman Test

Algorithm	Mean rank
LRFA	1.00
FA	2.00

5 Conclusion

The paper presents an improved FA variant called the local-leaning and reverse-learning of firefly algorithm (LRFA). The LRFA is improved from three aspects. First, original attractiveness is no longer a fixed value. Second, the best particle conducts local search. Third, some particle executes reverse-learning. Results tested on eleven standard benchmark functions are better than standard firefly algorithm. The firefly algorithm is novel optimization and there are improved points. The parameter of step factor α is the improved key point in next research for firefly algorithm.

Acknowledgment. This work is supported by the National Natural Science Foundation of China under Grant (Nos. 51669014, 61663029), Science Foundation of Jiangxi Province under Grant (No. 20161BAB212037), Jiangxi Province Department of Education Science and Technology Project under Grant (No. GJJ151133).

References

1. Yang, X.S.: Nature-Inspired Metaheuristic Algorithms. Luniver Press, UK (2010)
2. Al-Rifaie, M.M., Bishop, J.M.: Swarmic sketches and attention mechanism. In: Machado, P., McDermott, J., Carballal, A. (eds.) EvoMUSART 2013. LNCS, vol. 7834, pp. 85–96. Springer, Heidelberg (2013)
3. Zhao, J., Lv, L.: Two-phases learning shuffled frog leaping algorithm. Int. J. Hybrid Inf. Technol. **8**, 195–206 (2015)
4. Hui, S., Xiaolu, S., Jia, Z., Hui, W.: Hybrid algorithm of particle swarm optimization and artificial bee colony with its application in wireless sensor networks. Sens. Lett. **12**, 392–397 (2014)
5. Yuan, Z., Montes de Oca, M.A., Birattari, M., Stützle, T.: Continuous optimization algorithms for tuning real and integer parameters of swarm intelligence algorithms. Swarm Intell. **6**, 49–75 (2012)
6. Amiri, B., Hossain, L., Crawford, J.W., Wigand, R.T.: Community detection in complex networks: multi-objective enhanced firefly algorithm. Knowl. Based Syst. **46**, 1–11 (2013)
7. Dos Santos Coelho, L., Mariani, V.C.: Improved firefly algorithm approach applied to chiller loading for energy conservation. Energy Build. **59**, 273–278 (2013)
8. Horng, M.H.: Vector quantization using the firefly algorithm for image compression. Expert Syst. Appl. Int. J. **39**, 1078–1091 (2012)
9. Łukasik, S., Żak, S.: Firefly algorithm for continuous constrained optimization tasks. In: Nguyen, N.T., Kowalczyk, R., Chen, S.-M. (eds.) ICCCI 2009. LNCS, vol. 5796, pp. 97–106. Springer, Heidelberg (2009)
10. Apostolopoulos, T., Vlachos, A.: Application of the firefly algorithm for solving the economic emissions load dispatch problem. Int. J. Comb. **23**, 1687–9163 (2011)
11. Chaiead, N., Aungkulanon, P., Luangpaiboon, P.: Bees and firefly algorithms for noisy non-linear optimisation problems. In: International Multi conference of Engineers and Computer Scientists (2011)

Swarm Intelligence and Its Applications

Using Parallel Compact Evolutionary Algorithm for Optimizing Ontology Alignment

Xingsi Xue[1,2(✉)], Pei-Wei Tsai[1,2], and Li-Li Zhang[3]

[1] College of Information Science and Engineering, Fujian University of Technology, Fuzhou 350118, Fujian, China
jack8375@gmail.com
[2] Fujian Provincial Key Laboratory of Big Data Mining and Applications, Fujian University of Technology, Fuzhou 350118, Fujian, China
[3] Harbin University of Commerce, Harbin 150028, Heilongjiang, China

Abstract. On the basis of our former work based on Compact Evolutionary Algorithm (CEA), in this paper, we introduce parallel technology into Compact Evolutionary Algorithm (CEA), and design an Parallel Compact Evolutionary Algorithm (PCEA) based ontology matching technology to further improve the efficiency of solving the ontology meta-matching problem. Comparing with CEA based approach, our approach is able to further reduce the time and memory consumption while at the same time ensures the correctness and completeness of the alignments. The Experiment is carried out on the OAEI 2015 benchmark, and the results show that our approach is able to reduce the executing time and main memory consumption of the tuning process while at the same time ensures the quality of the alignment.

Keywords: Ontology meta-matching problem · Parallel Compact Evolutionary Algorithm · OAEI 2015

1 Introduction

With the development of Semantic Web, more and more ontologies with different terms or different taxonomies are being developed and many of them describe similar domain. For this reason, a key challenge is enabling the interoperability among different ontologies. In fact, ontologies can interoperate only if correspondences between their elements have been identified and established. As such, ontology engineers face the problem of how to map various different ontologies to enable a common understanding in order to support communication among existing and new domains. This process is commonly known as ontology alignment which can be described as follows: given two ontologies, each describing a set of discrete entities (which can be classes, properties, predicates, etc.), find the relationships (e.g., equivalence or subsumption) that hold between these entities [1].

It is highly impractical to align the ontologies manually when the size of ontologies is considerable large. Thus, numerous alignment systems have arisen

J. Pan et al. (eds.), *Genetic and Evolutionary Computing*, Advances in Intelligent Systems and Computing 536, DOI 10.1007/978-3-319-48490-7_19

over the years. Each of them could provide, in a fully automatic or semi-automatic way, a numerical value of similarity between elements from separate ontologies that can be used to decide whether those elements are semantically similar or not. Since none of the similar measures could provide the satisfactory result independently, most ontology alignment systems combine a set of different similar measures together by aggregating their aligning results. How to select the appropriate similar measures, weights and thresholds in ontology aligning process in order to obtain a satisfactory alignment is called meta-matching which can be viewed as an optimization problem and be addressed by techniques like Evolutionary Algorithm (EA). However, for dynamic applications, it is necessary to perform the similarity measures combination and system self-tuning at run time, and thus, beside quality (correctness and completeness) of the aligning results, the efficiency (execution time and main memory) of the aligning process is of prime importance especially when a user cannot wait too long for the system to respond or when memory is limited. Therefore, state-of-the-art ontology meta-matching systems tend to adopt different strategies within the same infrastructure to improve the efficiency of aligning process, and even though, the intelligent aggregation of multiple aligning results is still an open problem. Compact Evolutionary Algorithm (CEA) [2] belong to the class of Estimation of Distribution Algorithm (EDA) as the explicit representation of the population is replaced with a probability distribution [3]. On the basis of our former work based on CEA [10], in this paper, we introduce parallel technology into CEA, and design an Parallel Compact Evolutionary Algorithm (PCEA) to further improve the efficiency of solving the ontology meta-matching problem. Comparing with the approach based on CEA, our proposal is able to further reduce the time and memory consumption while at the same time ensures the correctness and completeness of the alignments.

2 PCEA for Optimizing Ontology Alignment

2.1 The Optimal Model for Ontology Alignment Problem

$$\begin{cases} max \quad f(X) = f - measure(X) \\ s.t. \quad X = (x_1, x_2, ..., x_n)^T \\ \qquad\quad x_i \in [0,1], i = 1...n \end{cases} \tag{1}$$

In our work, we take maximizing values of f-measure [8] as the goal we expect to achieve, and X represents the parameter set, i.e. the weights for aggregating various alignments and a threshold for filtering the aggregated alignment, used to obtain the final alignment.

2.2 The Detailed Procedure of PCEA for Optimizing Ontology Alignment

The CEA consists of the following: a binary vector of length n is randomly generated by assigning a 0.5 probability to each gene to take eighter the value 0 or the value 1. This description of the probabilities, initialized with n values all equal to 0.5, is named as Probability Vector (PV). By means of the PV two individuals are sampled and their fitness values are calculated. The winner solution, i.e. the solution characterized by a higher performance, biases the PV on the basis of a parameter N_p called virtual population. More specifically, if the winner solution in correspondence to it ith gene display a 1 while the loser solution displays a 0 the probability value in position ith of the PV is augmented by a quantity $1/N_p$. On the contrary, if the winner solution in correspondence to its ith gene displays a 0 while the loser solution displays a 1 the probability value in position ith of the PV is reduced by a quantity $1/N_p$. If the genes in position ith display the same value for both the winner and loser solutions, the ith probability of PV is not modified. This scheme is equivalent to (steady-state) pair-wise tournament selection. With the function *compete*() we simply mean the fitness-based comparison, i.e. we regard the individual with higher f-measure as the one with better quality. For the sake of clarity, the pseudo-code of CEA is displayed in Table 1:

Table 1. Pseudo code of Compact Evolutionary Algorithm

```
t = 0;
// PV initialization
for i = 1 : n do
   initialize PV[i] = 0.5;
end for
generate elite by means of PV;
while termination condition is not met do
   generate an individuals a by means of PV;
   [winner, loser] = compete(a, elite);
   // PV Update
   for i = 1 : n do
      if winner[i]! = loser[i] then
         if winner[i] == 1 then
            compute[i]=PV[i] + 1/N_p;
         end if
         else
            compute PV[i] = PV[i] − 1/N_p;
         end if
      end if
   end for
   if winner == a then
      elite = a;
   end if
   t = t + 1;
end while
```

Table 2. Pseudo code of Parallel Compact Genetic Algorithm

```
int tid = CurrentThread.getID();
// PV initialization
PV[tid] = generateInitPV();
elite[tid] = generateAnIndividual(tid, PV[tid]);
a[tid] = generateAnIndividual(tid, PV[tid]);
syncThreads();
// Fitness evaluation
fitness_elite = getFitness(elite);
fitness_a = getFitness(elite);
syncThreads();
// Competition
[winner, loser] = compete(a, elite);
// PV Update
PV[tid] = updatePV(winner, tid, PV[tid], elite[tid], a[tid]);
if winner == a then
    elite[tid] = a[tid];
end if
syncThreads();
```

We assign the number of threads in PCEA to be equal to the number of variables in PV vector which is determined by the number of binary bits and dimension of the optimizing function. When the PCEA is invoked, the threads start to evaluate the PV values as shown in the Table 1, and each of the binary bit in PV is manipulated by one thread.

3 Experimental Results and Analysis

In the experiments, we use the well-known benchmark provided by the Ontology Alignment Evaluation Initiative (OAEI) 2015 [7]. Table 3 shows a brief description of each test of the benchmark.

3.1 Experiments Configuration

In our work, the following similarity measures are used: (1) Levenstein distance [4] (Syntactic Measure), (2) Jaro distance [9] (Syntactic Measure), (3) Linguistic distance [6] (Linguistic Measure), (4) Taxonomy distance [5] (Taxonomy-based Measure). The aggregation strategy is weighted average approach which can be defined as follows:

$$Sim_{aggregate}(e_1, e_2) = \sum_{i=1}^{n} w_i \cdot Sim_i(e_1, e_2) \quad (2)$$

where: $\sum_{i=1}^{n} w_i = 1, w_i \in [0, 1]$, n is the number of considered similarity measures; w_i is a weight; $Sim_i(e_1, e_2)$ is an instance of the similarity function represented in this section. The hardware configurations used to run the algorithms are provided below: (1) Processor: Intel Core (TM) i7; (2) CPU speed: 2.93 GHz; (3) RAM capacity: 4 GB.

Table 3. Brief description of benchmarks

ID	Brief description
101	Strictly identical ontologies
103	A regular ontology and other with a language generalization
104	A regular ontology and other with a language restriction
201	Ontologies without entity names
203	Ontologies without entity names and comments
204	Ontologies with different naming conventions
205	Ontologies whose labels are synonymous
206	Ontologies whose labels are in different languages
221	A regular ontology and other with no specialization
222	A regular ontology and other with a flattened hierarchy
223	A regular ontology and other with a expanded hierarchy
224	Identical ontologies without instances
225	Identical ontologies without restrictions
228	Identical ontologies without properties
230	Identical ontologies with flattening entities
231	Identical ontologies with multiplying entities
301	A real ontology about bibliography made by MIT
302	A real ontology with different extensions and naming conventions
304	A regular ontology and a real ontology which is close to it

3.2 Results and Analysis

All the values shown in Tables 4, 5 and 6 are the average figures in ten independent runs. Specifically, Table 4, where symbol R and P refer to recall and precision [8], respectively, shows the comparison of the qualities of the alignment obtained by the approach based on CEA with our approach. While Tables 5 and 6 present the comparison of the average executing time and main memory consumption per generation by the approach based on CEA with our way, respectively.

As it can be seen from the second column and third column in Table 4 that, except benchmark 302, all the other benchmarks' alignment quality obtained by two approaches are identical to each other. With respect to benchmark 302, although the recall and precision of two alignments are slightly different, the f-measure obtained by two approaches is the same. Therefore, we may draw the conclusion that, from the aspect of the quality of alignment, our proposal is effective.

Table 4. Comparison of the qualities of the alignments obtained by CEA based approach with our approach

ID	$f-measure\ (R,\ P)$ (CEA)	$f-measure\ (R,\ P)$ (our approach)
101	1.00 (1.00, 1.00)	1.00 (1.00, 1.00)
103	1.00 (1.00, 1.00)	1.00 (1.00, 1.00)
104	1.00 (1.00, 1.00)	1.00 (1.00, 1.00)
201	0.94 (0.91, 0.97)	0.94 (0.91, 0.97)
203	0.99 (0.98, 1.00)	0.99 (0.98, 1.00)
204	0.98 (0.98, 0.99)	0.98 (0.98, 0.99)
205	0.93 (0.91, 0.97)	0.93 (0.91, 0.97)
206	0.70 (0.65, 0.75)	0.70 (0.65, 0.75)
221	1.00 (1.00, 1.00)	1.00 (1.00, 1.00)
222	1.00 (1.00, 1.00)	1.00 (1.00, 1.00)
223	0.99 (0.99, 1.00)	0.99 (0.99, 1.00)
224	1.00 (1.00, 1.00)	1.00 (1.00, 1.00)
225	1.00 (1.00, 1.00)	1.00 (1.00, 1.00)
228	1.00 (1.00, 1.00)	1.00 (1.00, 1.00)
230	1.00 (1.00, 1.00)	1.00 (1.00, 1.00)
231	1.00 (1.00, 1.00)	1.00 (1.00, 1.00)
301	0.75 (0.70, 0.80)	0.75 (0.70, 0.80)
302	0.74 (0.62, 0.90)	0.74 (0.63, 0.89)
304	0.93 (0.92, 0.95)	0.93 (0.92, 0.95)

Table 5. Comparison of the executing time taken per generation by the approach based on CEA with our approach

ID	Time (ns) (CEA)	Time (ns) (our approach)	Improvement ratio (%)
101	513,052,637	213,014,237	58.48
103	491,086,716	223,671,309	54.45
104	1,124,264,677	750,359,110	33.25
201	1,501,238,286	945,347,900	37.02
203	5,736,132,665	4,536,778,237	20.90
204	5,757,619,474	3,256,387,212	43.44
205	5,636,454,913	4,560,212,332	19.09
206	5,676,888,543	4,457,654,229	21.47
221	5,909,427,595	4,332,347,833	26.68
222	6,527,802,835	5,677,343,232	13.02
223	7,419,041,247	5,909,443,342	20.34
224	531,648,017	231,856,238	56.38
225	4,284,469,953	2,334,679,346	45.50
228	2,900,399,632	1,334,562,300	53.98
230	4,921,782,238	2,611,236,398	46.94
231	5,801,469,182	2,342,637,239	59.61
301	1,486,276,329	934,239,202	37.14
302	5,760,455,309	2,348,623,330	59.22
304	4,852,889,962	3,450,436,238	28.89
Average	4,043,810,537	2,655,306,803	34.34

From the Tables 5 and 6, we can see that, in all benchmarks, our approach improves the executing time and the main memory consumption per generation. Specifically, the improvement degree is on average by 34.34 % and 11.53 % respectively.

According to the experiment results showing above, comparing with the approach by using CEA, the utilization of PCEA is able to highly reduce the executing time and main memory consumption of the tuning process while at the same time ensures the correctness and completeness of the alignments.

Table 6. Comparison of the main memory consumed per generation by the approach based on CEA with our approach

ID	Memory (byte) (CEA)	Memory (byte) (our approach)	Improvement ratio (%)
101	12,135,256	10,233,452	15.67
103	28,130,568	26,425,990	6.05
104	33,429,416	29,342,220	12.22
201	71,847,624	69,326,359	3.50
203	74,940,416	68,334,209	8.81
204	52,807,184	48,406,230	8.33
205	32,311,840	28,228,435	12.63
206	71,956,096	64,981,237	9.69
221	66,327,008	62,980,290	5.04
222	63,017,368	57,439,223	8.85
223	81,903,184	78,320,298	4.37
224	75,580,408	73,231,298	3.10
225	31,199,368	28,341,609	9.15
228	54,376,960	26,521,982	51.22
230	54,754,840	29,342,232	46.41
231	66,127,752	59,348,509	10.25
301	18,888,808	17,368,298	8.04
302	82,554,504	76,546,230	7.27
304	71,045,104	68,340,225	3.80
Average	54,912,300	48,582,017	11.53

4 Conclusion

Ontology matching is an important step in ontology engineering. Although lots of work have been done to tackle this problem, there are still various challenges left for the researchers to deal with. One of these challenges is the selection of matchers and self-configuration of them. For dynamic applications it is necessary to perform matcher combination and self-tuning at run time, and thus, efficiency of the configuration search strategies becomes critical. To this end, in this paper, we propose to use PCEA to tune the parameters of ontology matching system in order to improve the efficiency. From the aspect of the quality of the alignment, the executing time and main memory consumption, the experiment results show our approach's efficiency by comparing the approach by using CEA with our approach. It turns out that our approach is able to reduce the executing time and main memory consumption of the tuning process while at the same time ensures the quality of the alignment.

Acknowledgment. This work is supported by the National Natural Science Foundation of China (No. 61503082) and Natural Science Foundation of Fujian Province (No. 2016J05145 and 2016J05146).

References

1. Euzenat, J., Valtchev, P.: Similarity-based ontology alignment in owl-lite. In: Proceedings of the 16th European Conference on Artificial Intelligence, Valencia, Spain, pp. 333–337, August 2004
2. Harik, G.R., Lobo, F.G., Goldberg, D.E.: The compact genetic algorithm. IEEE Trans. Evol. Comput. **3**(4), 287–297 (1999)
3. Larranaga, P., Lozano, J.A.: Estimation of Distribution Algorithms: A New Tool for Evolutionary Computation, vol. 2. Springer, New York (2002). IEEE Transactions on Evolutionary Computation
4. Maedche, A., Staab, S.: Measuring similarity between Ontologies. In: Gómez-Pérez, A., Benjamins, V.R. (eds.) EKAW 2002. LNCS (LNAI), vol. 2473, pp. 251–263. Springer, Heidelberg (2002). doi:10.1007/3-540-45810-7_24
5. Melnik, S., Garcia-Molina, H., Rahm, E.: Similarity flooding: a versatile graph matching algorithm and its application to schema matching. In: 18th International Conference on Data Engineering, Bali, Indonesia, pp. 117–128 (2002)
6. Miller, G.A.: Wordnet: a lexical database for English. Commun. ACM **38**, 39–41 (1995)
7. OAEI: Ontology alignment evaluation initiative (oaei) (2016). http://oaei.ontologymatching.org/2015/
8. Rijsbergen, C.J.V.: Foundation of evaluation. J. Documentation **34**, 365–373 (1974)
9. Winkler, W.: The state record linkage and current research problems. Technical report, Statistics of Income Division, Internal Revenue Service Publication (1999)
10. Xue, X., Liu, J., Tsai, P., Zhan, X., Ren, A.: Optimizing ontology alignment by using compact genetic algorithm. In: 2015 11th International Conference on Computational Intelligence and Security, Guangzhou, China, pp. 231–234 (2016)

Robot Path Planning Optimization Based on Multiobjective Grey Wolf Optimizer

Pei-Wei Tsai[1,2], Trong-The Nguyen[3(✉)], and Thi-Kien Dao[3]

[1] College of Information Science and Engineering, Fujian University of Technology, Fuzhou, China
[2] Fujian Provincial Key Laboratory of Big Data Mining and Applications, Fujian University of Technology, Fuzhou, China
[3] Department of Information Technology, Hai-Phong Private University, Haiphong, Vietnam
vnthe@hpu.edu.vn

Abstract. For the environment of robot motion, workspace consisted of the positions and shapes of obstacles, optimization for robot operations requires not only one criteria but also several criteria. In this paper, a novel multi-objective method for optimal robot path planning is proposed based on Grey wolf optimizer (GWO). Two criteria of distance and smooth path of the robot path planning issue are transformed into a minimization one for fitness function. The position of the globally best agent in each iterative can be reached by the robot in sequence permutation. Series simulations are implemented in different static environments for the optimal path when the robot reaches its target. The results show that the proposed method provides the robot reaches its target with colliding free obstacles and the alternative method of optimization for robot planning.

Keywords: Grey wolf optimizer · Motion path planning

1 Introduction

In hazardous environments, it is really hard to reach directly the operations for man, but a robot can be an effective tool to do those operations at there. The robots have been paid attention to applying in many fields including industry, agriculture, architecture and military because of their particular abilities that can replace workers [1]. One of the important issues in moving robot navigation is optimizing path planning efficiency according to some parameters such as cost, distance, energy, and time. The most common task for a motion robot is the ability to find an optimal start-to-target path amid obstacles with collisional free [2]. An optimal path can be obtained successfully from one location to another and reaches its goal with avoiding all obstacles. It means that the planning must satisfy some criteria such as length of the planning path is the shortest, and energy consumption of robot is the lowest [3]. Many traditional methods and metaheuristic algorithms could have solved to this problem, but a scale of the problem is large and high degrees of freedom, the traditional methods have to pay expensively for high computational costs and complexities [4]. Metaheuristic methods

J. Pan et al. (eds.), *Genetic and Evolutionary Computing*, Advances in Intelligent Systems and Computing 536, DOI 10.1007/978-3-319-48490-7_20

have been proposed to deal with the above drawbacks of classical approaches because of the NP-completeness nature of path planning problems. Inspired by the natural phenomena or the evolution genetic, or swarm intelligent, metaheuristic methods have been applied successfully solve many problems in many areas included the robotic [5]. Grey wolf optimizer (GWO) is a new swarm intelligence algorithm that developed based on the hunting behavior of grey wolves [6].

As a result of NP- completeness of the robot path planning problem, metaheuristic methods have been developed increasingly to cope with high computational costs and complexities of classic methods, especially for high degrees of freedom. In many real-life situations, the robot needs to keep a certain safe interspace from obstacles to avoid collisions. The most of these optimization methods have solved to the robot planning problem with respect to a single objective function, mainly the path length. However, the obtained paths from these methods are generally non-smooth and their practicality is questionable in most of the cases since mobile robots lose considerable energy and time when turning their course of moving abruptly. Many robot operations required a path plan that is efficient over several parameters.

Therefore, in this paper, we have worked out a planning algorithm with two objectives simultaneously, which are the distance and smooth criteria. This is done through an aggregative weighting multi-objective approach incorporated in a GWO context.

2 Path Planning Problem

The motion path planning is one of the vital issues of robot navigation because it affects the efficiency and accuracy of the robot working performance. The path planning problem is typically stated as: Given a robot and a two-dimensional workplace including obstacles and danger, sources need to find out an optimal collision-free path from start state to target state according to some performance merits, e.g. the length, time, smoothness, and energy. In this paper, the criteria of the shortness length and the smooth path are considered to optimize robot working performance. The path planning problem should be modeled with the workplace of robots and the related criteria as follows.

Robot Workplace Modeling. For simplicity and small sensitivity to obstacles' shapes, the robot workplace can be described as a path in the global coordinates O–XY, the start position and the target position of the robot, the polygon entities and the dashed circles represent obstacles respectively. In order to decrease the dimension of decision variable, a coordinate transformation is first used to locate the new X'-axis to coincide with the line Start – Target when Start – Target intersects the X-axis. The corresponding transformation formula is as follows:

$$\begin{bmatrix} x' \\ y' \end{bmatrix} = \begin{bmatrix} cos\varphi & -\sin\,\varphi \\ sin\varphi & cos\varphi \end{bmatrix} \times \begin{bmatrix} x \\ y \end{bmatrix} + \begin{bmatrix} x_{Start} \\ y_{Targ} \end{bmatrix} \tag{1}$$

where φ is angle of anti-clockwise rotation from the X-axis to the line Start–Target, (x_{Start}, y_{Targ}) is the point Start in the coordinates $O_{x,y}$, and (x', y') is the point (x, y) in the new coordinates $Start_{x_{0,}y_{0,}}$. First, the line Start–Target is divided into $n+1$ equal segments by n points. After drawing n vertical lines through these points in turn, a set of parallel lines denoted $\{l_1, l_2, ..l_n\}$. A complete path $\{p_1, p_2, ..p_n\}$ can be constructed by sampling at random on vertical lines of $l_1, l_2, ..l_n$. Hence, the robot path planning problem is transformed into optimizing the following set of points $\{Start, p_1, p_2, ..p_n, \text{and} Target\}$. This path is collision-free constraint. It means each point in this path is not covered by obstacles, and each line among the set does not intersect with obstacles. The search agents are generated in the beginning with respect to the robot's initial position and regarding its sensing range.

Robot Path Planning Objectives. For length of path, supposing that the start state and the target state are p_0 0 and p_{n+1}, the length of a path can be approximated by:

$$L(p) = \sum\nolimits_{i=0}^{n} d(p_i, p_{i+1}) \tag{2}$$

where $L(p)$ is the length of path and $d(p_i, p_{i+1})$ represents the distance between p_i and p_{i+1}. In the coordinates $Start_{x_{0,}y_{0,}}$, since the line Start –Target is divided into $n+1$ equal segments, the value of $d(p_i, p_{i+1})$ can be calculated as following:

$$d(p_i, p_{i+1}) = \sqrt{\left(\frac{d(p_0, p_{n+1})}{n+1}\right)^2 + \left(y'_{p_i} + y'_{p_{i+1}}\right)^2} \tag{3}$$

Where $\left(x'_{p_i} - x'_{p_{i+1}}\right)^2$ is set to $\left(\frac{d(p_0, p_{n+1})}{n+1}\right)^2$. The initial population is generated such that along each sensing direction, an agent is created at a certain distance from the robot, determined by the range of the used sensor. The first objective function is shortness path that is can be defined as the Euclidean distance between the agent and the goal point in each iteration:

$$F_1(p) = \sum\nolimits_{i=0}^{n-1} d_i \tag{4}$$

If any obstacle point is within the sensing range in that direction, a point near the obstacle's border is selected as the search agent at that direction.

Second objective function is smoothness path that is mathematically expressed as the angle between the two hypothetical lines connecting the goal point to the robot's two successive positions in each iteration, i.e. p_i^* and p_{i+1}^*, in which i is the iteration number.

$$F_2(p) = \sum\nolimits_{i=0}^{n-1} \varphi_i + \delta \times L \tag{5}$$

where $F_1(p)$ and $F_2(p)$ are the objective functions of shortness criterion and smoothness criterion respectively; for φ_i is the angle between the two line segments $(0 \le \varphi \le \pi)$, connecting the point p_i; δ is a positive constant; L is the number of line segments in the path. The total cost of fitness or objective function of feasible path P with n points is obtained by optimization process in later section.

3 Multiobjective GWO for Path Planning

The basic version of the grey wolf optimizer (GWO) [6] is only for single objective optimization. In order to solve multiobjective functions of the motion robot planning, GWO is extended to multiobjective grey wolf optimizer (MOGWO). The basic version of GWO and Pareto optimal front will be first briefly reviewed, and then the motion robot planning will be dealt with it based on MOGWO.

3.1 Meta-Heuristic Grey Wolf Optimizer

GWO was imitated the behavior of searching and attacking prey of Grey wolf [6]. In the leadership hierarchy of grey wolves included four guided types alpha (α), beta (β), delta (δ), and omega (ω). Where α is considered the fittest solution, and then β, and δ are considered the second and the third best solutions respectively. Omega (ω) could be assumed the rest of the candidate solutions. GWO algorithm consists of the constructed mathematical models as follows:

$$\vec{X}(t+1) = \overrightarrow{X_p}(t) - \vec{A} \cdot \left| \vec{C} \cdot \overrightarrow{X_p}(t) - \vec{X}(t) \right| \quad (6)$$

where $\overrightarrow{X_p}(t)$ is the position vector of the prey, and $\vec{X}(t)$ indicates the position vector of a grey wolf. The $\vec{A}$ and $\vec{C}$ are vectors parameters that calculated as $\vec{A} = 2\vec{a} \cdot \vec{r_1} - \vec{a}$ and $\vec{C} = 2 \cdot \vec{r_2}$ respectively. Where components $\vec{a}$ are linearly decreased from 2 to 0 over the course of iterations and r_1, r_2 are random vectors in [0, 1]. A grey wolf in the position of (X, Y) can update its position according to the position of the prey (X*, Y*). The first three best solutions are obtained so far and oblige the other search agents (including the omegas) to update their positions according to the best search agents.

$$\vec{X}(t+1) = \frac{\overrightarrow{X_1} + \overrightarrow{X_2} + \overrightarrow{X_3}}{3} \quad (7)$$

where X_1 is alpha candidate, and calculated as $\overrightarrow{X_1} = \overrightarrow{X_\alpha} - \overrightarrow{A_1} \cdot \left| \overrightarrow{C_1} \cdot \overrightarrow{X_\alpha} - \vec{X} \right|$; X_2 is alpha candidate, and calculated as $\overrightarrow{X_2} = \overrightarrow{X_\beta} - \overrightarrow{A_2} \cdot \left| \overrightarrow{C_2} \cdot \overrightarrow{X_\beta} - \vec{X} \right|$, and X_3 is alpha candidate, and calculated as $\overrightarrow{X_3} = \overrightarrow{X_\delta} - \overrightarrow{A_3} \cdot \left| \overrightarrow{C_3} \cdot \overrightarrow{X_\delta} - \vec{X} \right|$. The position of the prey is estimated by alpha, beta, and delta and other wolves update their positions randomly around the prey during the hunt. The grey wolves finish the hunt by attacking the prey when it stops moving.

3.2 Pareto Optimal Front

The domination of a solution vector $x = (x_1, x_2, .., x_n)^T$ on a vector $y = (y_1, y_2, .., xy_n)^T$ for a minimization problem if and only if $x_i \leq y_i$ for $\forall_i \in \{1, \ldots, n\}$ and

$\exists_i \in \{1,\ldots,n\} : x_i < y_i$. It means that is no component of x is larger than the corresponding component of y, and at least one component is smaller. Similarly, the dominance relationship could be defined by

$$x \preccurlyeq y \Leftrightarrow x \prec y \vee x = y. \tag{8}$$

For maximization problems, the dominance can be defined by replacing symbol of $\prec$ with the symbol of $\succ$. Therefore, a point x_* is called a non-dominated solution if no solution can be found that dominates on it. The Pareto front PF of a multiobjective can be defined as the set of non-dominated solutions as following.

$$PF = \{s \in S | \nexists\, s' \in S : s' \prec s\} \tag{9}$$

where S is the solution set. A good approximation could be obtained from the Pareto front if a diverse range of solutions should be generated using efficient techniques [7].

3.3 Optimal Robot Path Planning Based on MOGWO

The optimal solution of multiobjective optimization can be obtained from the Pareto optimal solution. Multiobjective optimization issue for a minimization problem with d-dimensional decision vectors and h objectives is given by

$$\begin{aligned} &\mathit{Minimize}\ F(\boldsymbol{x}) = (f_1(\boldsymbol{x}), f_2(\boldsymbol{x}), ..\mathrm{f_h}(\boldsymbol{x})) \\ &\text{Subject to} \qquad \boldsymbol{x} \in [\boldsymbol{x}_L, \boldsymbol{x}_U] \end{aligned} \tag{10}$$

where x is a decision vector as a set of $(x_1, x_2, .., x_u) \in X \in R^d$ and $F(x)$ is the objective function with the objective vector as a set of $(f_1, f_2, .., f_u) \in Y \in R^h$. The decision vector x is belonging to the d-dimensional decision space X, which is corresponding to the space d dimensional of search agents in GWO. The objective function $F(x)$ belongs to the $h-$ dimensional objective space Y, in which it is mapping functions from the decision space to the objective space. x_L, andx_U are lower and upper bound constraints of the agent range, respectively. The set of all the search agents meeting the constraints forms the decision space feasible set $\Omega = \{\boldsymbol{x} \in R^d | \boldsymbol{x} \in [\boldsymbol{x}_L, \boldsymbol{x}_U]\}$. The purpose of optimization is to find the Pareto-optimal solution. The decision space includes the dimension d and the objective space h. We begin with a generated population of N_p search agents randomly so that these search agents should distribute among the search space as uniformly as possible. This can be achieved by using sampling techniques via uniform distributions. The model of the path planning problem with the two objective functions are defined by Eqs. (4) and (5) consist of the objective function $F(p)$. Therefore, from Eqs. (4), (5) and (10) can be formulated in the optimum mathematical form in MOGWO as.

$$\begin{aligned} &\text{Minimize}\ \boldsymbol{F}(\boldsymbol{p}) = (f_1(\hat{x}_i, \hat{y}_i), f_2(\hat{x}_i, \hat{y}_i)) \\ &\mathit{Subject\ to}\ (\hat{x}_i, \hat{y}_i) \in (\hat{x}_L, \hat{y}_L), (\hat{x}_U, \hat{y}_U) \\ &i = m + 1, ..n, \end{aligned} \tag{11}$$

where decision vectors $p = (\hat{x}_i, \hat{y}_i)$ are the estimated coordinates corresponding to solutions in GWO. $(\hat{x}_L, \hat{y}_L), (\hat{x}_U, \hat{y}_U)$ are the lower and upper bound constraint values, f_1 is the objective function of the length path constraint, and f_2 is the objective function of the smoothness path constraint. Obtaining the multiobjective Pareto optimal solution is the ultimate goal of building a multiobjective optimal model for robot planning issues, which meets both the shortest path constraint and the smoothest path constraint. Therefore the main essence of MOGWO can be described as determining the dominant relationship according to the decision space feasible set Ω and the Pareto front $\boldsymbol{F}(\boldsymbol{p}^*)$ saving Pareto optimal solution set S in an archive by Eq. (10) and updating the best solution of multiobjective.

The basic steps of the optimization are described as follows:

Step 1: Modeling robot workspace included obstacles' positions and shapes, and the robot's start and target positions

Step 2: Sparse solution is to map the search agents to a model of path planning during optimization

Step 3: Implementing the proposed MOGWO to find optimal paths of the above model

Step 3.1: WHILE the maximum number of iterations has not been reached, i.e., $t <= T_{max}$, DO
If (A < 1) Update the global best position Eq. (7);
Else Update the positions of agent Eq. (6);
Take the best position of agent;
Calculate the objective values and the constraint- violated degree of each agent by Eq. (11)

Step 3.2: Store all non-dominated feasible particles into the feasible archive, and non-dominated infeasible agents into the infeasible archive; Update the feasible archive and the infeasible archive; Increase the loop counter, $t = t + 1$;

Step 3.3: Output optimal results

Step 4: Guide the robot to the target position by the optimal path selected

4 Simulation Results

This section investigates the proposed multiobjective Grey wolf optimizer (MOGWO) for optimal robot path planning. The simulations have been done applying MOGWO with objective functions f_1 and f_2, and Pareto archived evolution strategies. The traditional path planning methods can only generate a feasible path when running once. The proposed method of MOGWO for solving robot path planning problem in this paper can generate multiple feasible paths for robots to choose when running once. A criterion for performance evaluation of multi-objective optimization algorithm is the error rate (ER). The error rate measures the probability whether the obtained non-domination solution is the actual Pareto frontier or not. The calculation method is given in Eq. (12)

$$ER = \frac{\sum_{i=1}^{n'} x_i}{n'} \tag{12}$$

where n' is number of the obtained optimal points in Pareto frontier. If the obtained solution is an actual Pareto frontier elements, then x_i is set to 0, otherwise, x_i is set to 1. A robot path planning in the environment with one concave obstacle is carried out in order to verify the proposed method. Initialization parameters of the algorithm are as follows. The population size is 100. The maximum number of iterations is 500. The environment map of the simulation is set to 300*400 pixels. The number of obstacles, the coordinate of the starting point and the coordinate of the target point of the robot can be set or reset as GUI scheme shown in Fig. 1. The shape of the obstacle can be square, rectangle, or circle, and the position of the obstacle are generated randomly in the environment of the robot working space.

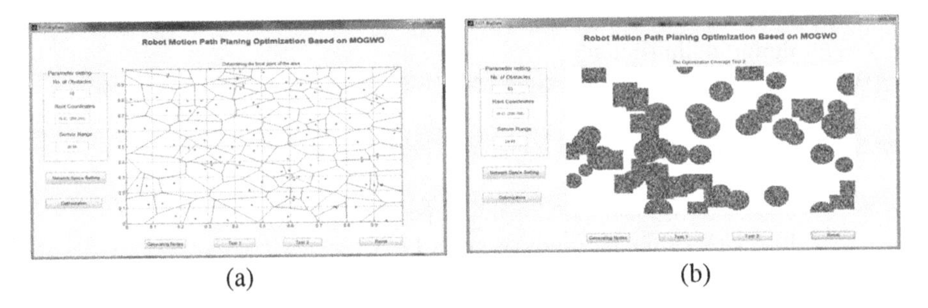

(a) (b)

Fig. 1. Setting environments of robot workspace model: (a) Obtaining center id of sensing areas of obstacles, (b) Generating obstacles randomly

Experiments have been conducted some tests with different obstacles density based on two objective functions which are the path length, and smoothness. The shape of the obstacle is to set concave in order to verify the effectiveness of the paths in complex environments. The path planning results are shown in Fig. 2. As shown in the figures, the robot can well avoid the concave obstacle and find the shortest path from the starting point to the target point.

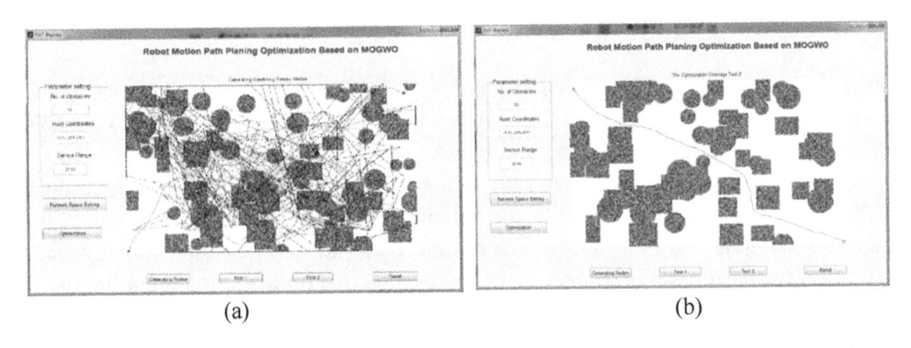

(a) (b)

Fig. 2. Simulation of optimization of robot path planning: (a) Lines setting are not shown, (b) Configuration of parse solutions of robot workspace.

The experiment is to verify the effectiveness of the paths generated by one single robot based on the proposed method and the performance is analyzed by comparison with another method as multi-object genetic algorithm MOGA [8]. The probability of the obtained non-domination solution Pareto frontier is calculated according to Eq. (12), ER = 0.1 for MOGWO, however, this figure for MOGA for robot planning problem ER = 0.2.

5 Conclusion

This paper proposed a novel multi-objective grey wolf optimizer (MOGWO) for optimization the robot path planning problem. The environment of robot workspace consists of the positions and shapes of obstacles, and the robot's start and target positions were modeled and search agents mapped to a sparing solution in each iteration of robot planning during optimization. In the proposed method, MOGWO handles two objectives simultaneously: the shortness path and the smoothness path. The position of the globally best agent is selected in each iterative, and reached by the robot in sequence. In addition, the robot processor updates its information during the moving, and the environment is partially unknown for the robot due to the limit detection range of its sensors. Simulations results show that the proposed method provides effectively complete the robot path planning task with a convincing performance. The robot reaches its target with colliding free obstacles. The simulation results were compared with the obtained of MOGA method, the quality of the proposed method MOGWO is slightly increased and the error rate of the proposed method is less than MOGA method.

References

1. Kramer, J., Scheutz, M.: Development environments for autonomous mobile robots: a survey. Auton. Robot. **22**(2), 101–132 (2007)
2. Choset, H., Lynch, K.M., Hutchinson, S., Kantor, G., Burgard, W., Kavraki, L., Thrun, S.: Principles of robot motion: theory, algorithms, and implementations. IEEE Robot. Autom. Mag. **12**(3), 110 (2005)
3. Russell, R.A., Bab-Hadiashar, A., Shepherd, R.L., Wallace, G.G.: A comparison of reactive robot chemotaxis algorithms. Robot. Auton. Syst. **45**(2), 83–97 (2003)
4. Hussein, A., Mostafa, H., Badrel-din, M., Sultan, O., Khamis, A.: Metaheuristic optimization approach to mobile robot path planning. In: 2012 International Conference on Engineering and Technology (ICET), pp. 1–6 (2012)
5. Goerzen, C., Kong, Z., Mettler, B.: A survey of motion planning algorithms from the perspective of autonomous UAV guidance. J. Intell. Robot. Syst. Theory Appl. **57**(1–4), 65–100 (2010)
6. Mirjalili, S., Mirjalili, S.M., Lewis, A.: Grey wolf optimizer. Adv. Eng. Softw. **69**, 46–61 (2014)
7. Zavala, G.R., Nebro, A.J., Luna, F., Coello, C.A.C.: A survey of multi-objective metaheuristics applied to structural optimization. Struct. Multi. Optim. **49**(4), 537–558 (2014)
8. Castillo, O., Trujillo, L., Melin, P.: Multiple objective genetic algorithms for path-planning optimization in autonomous mobile robots. Sof Comput. **11**, 269–279 (2007)

Enhanced Diversity Herds Grey Wolf Optimizer for Optimal Area Coverage in Wireless Sensor Networks

Chin-Shiuh Shieh[1(✉)], Trong-The Nguyen[2], Hung-Yu Wang[1], and Thi-Kien Dao[1]

[1] Department of Electronics Engineering, National Kaohsiung University of Applied Sciences, Kaohsiung, Taiwan
csshieh@kuas.edu.tw

[2] Department of Information Technology, Hai-Phong Private University, Haiphong, Vietnam

Abstract. Wireless Sensor Networks (WSNs) have been envisioned as the emerging technology and applied widely, but they also faced many practical challenges. One of such challenges is the coverage issue because of a high coverage rate ensures a high quality of service of the WSN. This paper proposes a novel method to optimize sensor coverage based on the enhanced diversity herds grey wolf optimizer (EGWO). In the proposed method, coverage overlaps and holes of deploying WSN are considered to a mathematical model for the objective function of the optimization problem. Quality performance of the proposed method is evaluated through simulation in several scenarios of WSN. The simulation results compared with other methods such as the grey wolf optimizer and the genetic algorithm shows that the proposed algorithm achieves a good coverage and a competitor.

Keywords: Enhanced diversity herds grey wolf optimizer · Area converge · Wireless sensor networks

1 Introduction

The advancement in electronics, communications and information technologies and the Internet have led to the rapid proliferation of WSN. WSNs are envisioned as the future technology and are emerging as an interesting research area among the research organizations, academia, and industries [1]. Sensor-enabled products and their networks are becoming a commonplace, and central to the everyday life, e.g., healthcare, smart homes, object tracking and monitoring, and so on. WSNs consist of a large number of inexpensive sensors that have quite limited resources. Sensors are small in size and are integrated with a sensing unit and wireless communication capabilities. These nodes are being deployed in a wide terrain to perform their intended tasks efficiently. Typically, the heterogeneous sensor networks that are more practical, having better network performance (i.e., multi-hop communication, delay tolerant, etc.) and lifetime, scalability, efficient load-balancing, and are cost-efficient [2].

J. Pan et al. (eds.), *Genetic and Evolutionary Computing*, Advances in Intelligent Systems and Computing 536, DOI 10.1007/978-3-319-48490-7_21

These wireless sensors, however, have several constraints such as restricted sensing and communication range as well as limited battery capacity [5]. The limitations raise several issues including coverage, connectivity, network lifetime, scheduling and data aggregation. In order to prolong the WSN lifetime, energy conservation measures such as scheduling and data aggregation must be taken. Scheduling conserves energy by turning off the sensors whenever possible while data aggregation tries to conserve the energy by reducing the energy used in transmitting the data. Connectivity and coverage problems stem from limited communication and sensing range of the involved sensors. To ensure connectivity, the sensors need to be placed close enough to each other so that they are within the communication range. In the other hand, the coverage problem concerns on how to guarantee that each of the points in the region of interest (RoI) is covered by the sensors. In order to maximize the coverage, the sensors need to be placed not too close to each other so that the sensing capability of the network is fully utilized and at the same time not too far from each other to prevent forming coverage holes (area outside sensing a range of sensors). A sensor's prime function is to sense the environment for any occurrence of the event of interest. Therefore, the coverage is one of the major concerns in WSN, and it is used as a key for quality of service (QoS) evaluation in WSN [3].

Moreover, metaheuristic algorithms have been successfully applied to solve optimization problems in various areas (e.g., engineering, financial, and management fields [4]). In view of the practical utility of optimization problems, there is a need for efficient and robust computational algorithms which can solve optimization problems arising in requirement constraints. A novel metaheuristic algorithm of Enhanced diversity herds grey wolf optimizer (EGWO) [5] was developed based on the frame of grey wolf optimizer.

In this paper, the coverage problem is formulated based on detecting the coverage holes, and the EGWO is used to find an optimum or near optimum solution to it. Optimal placement of the sensors according to a fitness function is deployed by minimizing the coverage holes based on Voronoi diagram. The advantage of using Voronoi diagram over other geometrical structures, for instance, the grid, is that its computational complexity is controlled only by one parameter which is the number of sensors in the network.

2 The Coverage Problem in WSN

Given network with a set of N number of sensors, $S = \{s_1, s_2, \ldots, s_N\}$ and a the region of interest (RoI) coverage problem is how to place the sensors for maximizing coverage percentage in the RoI and minimizing coverage holes. The coverage in WSN falls in three categories including area coverage, point coverage, and barrier coverage [6]. The main concern of the coverage problem is whether to cover an area, boundary surveillance or monitoring a set of points of interest. The area coverage is on how to cover an area with the sensors, while the point coverage deals with the coverage of a set of points of interest, and the barrier coverage is the probability of undetected penetration. This work deals with area coverage, where the objective is to maximize the coverage percentage. For minimization problem can be viewed as where the objective is how to make sure the total area of the coverage holes in the network is as small as possible.

Several factors related to the coverage problem in WSN include as insufficient sensors to cover the whole RoI, limited sensing range, and random deployment. Since the sensors operate under a limited power supply, some of them might die out, resulting in an inadequate number of sensors to fully cover the whole RoI thus causing holes to exist. Another reason is the sensor's restricted sensing range, of course, this issue can be solved by using sensors with larger sensing range, but this type of sensors are more expensive [7]. In several scenarios included in a hostile and unreachable environment i.e., battlefield and a steep terrain, sensors can be dropped from the air. In these cases, WSNs have to be the ability to be randomly deployed without the need for manual interference [7, 8]. However, random deployment could cause some of the sensors to fall too close to each other while others are too far apart. In both situations coverage problem arises; in the first case, the sensing capabilities of the sensors are wasted and the coverage is not maximized, while in the latter case, blind spots are formed.

As stated above, the coverage can be enhanced by using sensors with larger sensing range but this is costly. Thus among the commonly used solutions is to address the problem during the deployment phase. Rather than random deployment, the deployment of WSN can be done using a predetermined plan [9]. In predetermined deployment, the WSN coverage is improved by carefully planning the positions of the sensors in the RoI prior to their deployment. Then, the sensors are placed according to the plan either manually or with the help of a mobile robot. However, this method is costly and suitable only for small WSN. As for random deployment, the initial coverage can be enhanced by manipulating the locomotion capability of the sensors or by using incremental deployment after the initial one. In the mobility approach, the mobile sensors are self-repositioned after early deployment, to achieve a better arrangement and the coverage is maximized. On the other hand, the incremental deployment method involves analyzing the initial coverage and adding sensors at locations with coverage hole. Overall, for both deployment methods, the aim is to solve the coverage problem using sensors' placement. Maximizing area coverage in the given n sensor nodes WSN with 2D domain A can be formulated as follows.

$$F = Maximize\left(area\left(\bigcup_{i=1}^{k} \bigcup_{j=1}^{n_i} c_{r_i}\left(x_{ij}, y_{ij}\right) \cap A\right)\right), \tag{1}$$

where $area(X)$ is the area of the domain X of the deployed network; $c_{r_i}\left(x_{ij}, y_{ij}\right)$ is the circle centered at (x_{ij}, y_{ij}) and the radius is r_i; n is the number of sensors; n_i is the number of sensors for type $i(i = 1, 2, \ldots, k)$, k is the number of sensor types. The area coverage of the solution X is calculated as given:

$$Area(X) = \int_A I_X(x, y)dxdy = \lim_{L \to \infty} \frac{area(A)}{L} \sum_{l=1}^{L} I_X(\widetilde{x_l}, \widetilde{y_l}), \tag{2}$$

where $area(A)$ is the area of domain A with $I_X(x, y)$ is set to 1 if $(x, y) \in X$, otherwise it is set to 0.

3 Enhanced Grey Wolf Optimizer for Area Coverage

The base version of the grey wolf optimizer (GWO) [10] is not much diversity. Enhanced grey wolf optimizer is designed based on original GWO to enhance diversity for avoiding local optimum rather obtained global optimum.

3.1 Meta-Heuristic Grey Wolf Optimizer

GWO was developed based on imitating the behavior of searching and attacking prey of Grey wolf [10]. In the leadership hierarchy of grey wolves included four guided types alpha (α), beta (β), delta (δ), and omega (ω). Where α is considered the fittest solution, and then β, and δ are considered the second and the third best solutions respectively. Omega (ω) could be assumed the rest of the candidate solutions. GWO algorithm consists of the constructed mathematical models as follows:

$$\vec{X}(t+1) = \overrightarrow{X_p}(t) - \vec{A} \cdot \left| \vec{C} \cdot \overrightarrow{X_p}(t) - \vec{X}(t) \right| \tag{3}$$

where $\overrightarrow{X_p}(t)$ is the position vector of the prey, and $\vec{X}(t)$ indicates the position vector of a grey wolf. The $\vec{A}$ and $\vec{C}$ are vectors parameters that calculated as $\vec{A} = 2\vec{a} \cdot \vec{r_1} - \vec{a}$ and $\vec{C} = 2 \cdot \vec{r_2}$ respectively. Where components $\vec{a}$ are linearly decreased from 2 to 0 over the course of iterations and r_1, r_2 are random vectors in [0, 1]. A grey wolf in the position of (X, Y) can update its position according to the position of the prey (X^*, Y^*). The first three best solutions are obtained so far and oblige the other search agents (including the omegas) to update their positions according to the best search agents.

$$\vec{X}(t+1) = \frac{\overrightarrow{X_1} + \overrightarrow{X_2} + \overrightarrow{X_3}}{3} \tag{4}$$

where X_1 is alpha candidate, and calculated as $\overrightarrow{X_1} = \overrightarrow{X_\alpha} - \overrightarrow{A_1} \cdot \left| \overrightarrow{C_1} \cdot \overrightarrow{X_\alpha} - \vec{X} \right|$; X_2 is alpha candidate, and calculated as $\overrightarrow{X_2} = \overrightarrow{X_\beta} - \overrightarrow{A_2} \cdot \left| \overrightarrow{C_2} \cdot \overrightarrow{X_\beta} - \vec{X} \right|$, and X_3 is alpha candidate, and calculated as $\overrightarrow{X_3} = \overrightarrow{X_\delta} - \overrightarrow{A_3} \cdot \left| \overrightarrow{C_3} \cdot \overrightarrow{X_\delta} - \vec{X} \right|$. The position of the prey is estimated by alpha, beta, and delta and other wolves update their positions randomly around the prey during the hunt. The grey wolves finish the hunt by attacking the prey when it stops moving.

3.2 Enhanced Diversity Herds GWO

There are two considered characters in neighborhood structure, small size and communicating. Not as other evolutionary algorithms that prefer larger population, GWO needs a comparatively smaller population size. Especially for simple problems, a population with three to five wolves can achieve satisfactory results. GWO with small neighborhoods performs better on complex problems. In order to increase diversity, the small sized herds are employed by dividing the wolves in GWO into the groups. Each herd uses its own members to search for a better area in the search space. Since the small sized herds are searching using their own best historical information, they are able to converge to a local optimum. So, a randomized regrouping schedule should be set for optimizing by the probability weight setting, and a new configuration of small herds has started the searching the best global target. The exchangeable information is activated between herds whenever the communication strategy is triggered. The benefit of cooperation and exploitation is achieved through the communicating information. The fitness sharing available resource is one common used in the Niching techniques. The herd GWO has its own wolves as known search agent and the finest agents are evaluated according to the fitness function. These best agents among all the wolves in one group will be assigned to the poorer agents based on the fitness evaluation in the other groups, replace them and update agents for each herd after running the exchanging period.

Let Gj be the group, where j is the index of the group, n is a number of groups, $j = 0, 1, 2, \ldots, n-1$; and m be number wolves of a group, called population size of the group. While $t \cap R \neq \theta$, k search agents (where the top k fitness in the group G_j) will be copied to $G_{(j+1)}$ to replace the same number of search agents with the worst fitness. Every R generation, the population is regrouped randomly and starts searching using a new configuration of small herds. In this way, the good information obtained by each herd is exchanged among the herds. Simultaneously the diversity of the population is increased. It is not surprising that it performs better on complex multimodal problems. The steps can be described as follows:

1. Initialization: Initialize a, A, C, generate $m \times n$ search agents and divide population into n groups randomly, with m individuals in each group G. Assign R the exchanging period for executing $Xijt$ solutions, where $i = 0, 1, \ldots, m-1$; $j = 0, 1, \ldots, n-1$; t is the current iteration and set to 1.
2. Evaluation: Evaluate the value of $f(X_{ijt})$ for search agents in j-th group G_j.
3. Update: Update the position of the current search agent by Eqs. (3) and (4), and a, A, and C.
4. Communication Strategy: Migrate k best agents among G_{tj} to the (j + 1)-th group G_{tj+1}, mutate G_{tj+1} by replacing k poorer agents in that group, If $mod(i, R) = 0$, regroup the herds randomly, and update all of the group in each R iterations.
5. Termination: Repeat Step 2 to Step 5 until the predefined value of the function is achieved or the maximum number of iterations has been reached. Record the best value of the function $f(X_{\mathrm{ijt}})$ and the best agent solution among all the agent positions X_{ijt}.

3.3 Optimal Area Coverage in WSN

This subsection presents the proposed method of optimal deployment of the WSN by maximizing area coverage of RoI. The WSN is assumed to be deployed in an area of a two-dimensional square with sensing RoI sensors, knowing their positions, and possessing a capability of locomotor. It means that position of simulated sensor nodes can be able to immigrate and change their locations during the optimization process. Scheme optimizer is to be executed at a base station after an initial random placement. The sensors final optimal positions will be transmitted by the base station to the sensors, based on this information the sensors will move to their optimal positions.

The fitness function evaluates the solution encoded in a search agent of grey wolves. Here coverage problem is considered as a minimization problem where the objective is to minimize the total area of overlaps and coverage holes.

$$F_{fitness} = Minimize \sum_{i=1}^{N} \left(\sum_{j=i+1}^{N} overlap(s_i, s_j) + \sum_{j=i+1}^{N} hole(s_i, s_j) \right) \quad (5)$$

Subject to:

$$overlap(s_i, s_j) = \begin{cases} 0 \quad if\, d(s_i, s_j) \geq r_{s_i} + r_{s_j},\ hole \\ \omega \cdot \Delta d\, if\, |r_{s_i} - r_{s_j}| \leq d(s_i, s_j) \leq r_{s_i} + r_{s_j} \\ \beta \cdot \min(r_{s_i}, r_{s_j})\, if\, d(s_i, s_j) < |r_{s_i} - r_{s_j}| \end{cases} \quad (6)$$

$$hole(s_i, s_j) = \sqrt{d(s_i, s_j) - r_{s_i} + r_{s_j}} + \frac{1}{4}\sqrt{d(s_i, s_j) - r_{s_i} + r_{s_j}} \quad (7)$$

where Δd is set to $r_{s_i} + r_{s_j} - d(s_i, s_j)$, and $d(s_i, s_j)$ is the distance between s_i and s_j. To measure the coverage holes, a set of points, called interest points, are required to be selected. The interest points set consists of the vertices of the Voronoi polygons which obtained from the computed Voronoi diagram [11], and a number of points distributed evenly on the boundary of the polygons. If the interest point is a Voronoi vertex, the hole area is approximated as the circular area around the vertex not covered by the nearest sensor, while if it is a corner point on the boundary, then it is a quarter of the circle.

The basic steps of the optimization are described as follows:

Step 1: Modeling network space including a solution as S is set to $(s_1, s_2, \ldots, s_N)$, where N is number of sensor nodes. The coordinates of s_i positions is set to (x_i, y_i), where $i = 1, 2, ..N$. Calculating the overlap and hole of the network coverage.

Step 2: Parsing solution is as mapping search agents to a model of the area coverage optimization.

Step 3: Implementing the proposed EGWO to find optimal deployment WSN of the above model.

4 Simulation Results

In this section, several simulations are performed to evaluate the correctness and the efficiency of the proposed method for coverage in WSN optimization.

The simulation results are also compared those obtained from similar other methods include WOA and GA [12]. Fitness functions for the experiments are averaged over different random seeds with 25 runs. The goal of the optimization is to minimize the outcome for objective function Eq. (5), and constraints in Eqs. (6) and (7) are handled out in 3.3 step 2 of the parsing solution. The detail of parameter settings of EGWO can be found in [5]. A network is set up with 200 sensors over a region of interest of 200 m × 200 m. The sensing range of nodes is 25 m. Figure 1 shows the GUI result of the proposed method of optimal coverage in WSNs after the phase of generating sensor nodes.

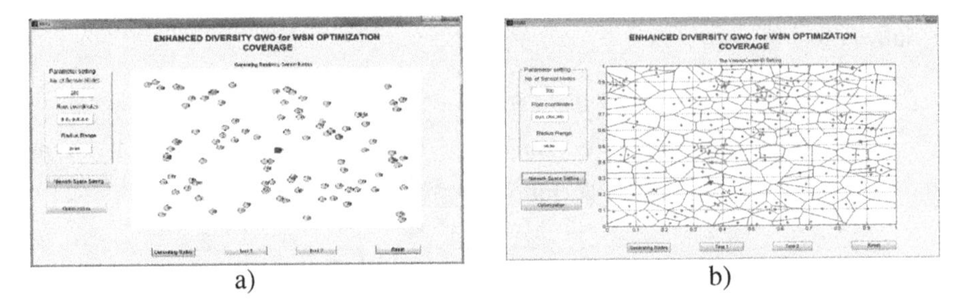

Fig. 1. The GUI of the optimal coverage in WSNs: (a) a scheme for the phase of generating sensor nodes, (b) a scheme for Voronoi diagram phase

Table 1. The comparison the proposed method of EGWO, with the GWO, and the GA methods in terms of quality performance evaluation and speed

Method	Avg. of obtained	Time consumption (s)
GA [12]	1.7019E + 01	13.22
GWO	1.5334E + 01	9.123
EGWO	1.4681E + 01	8.133

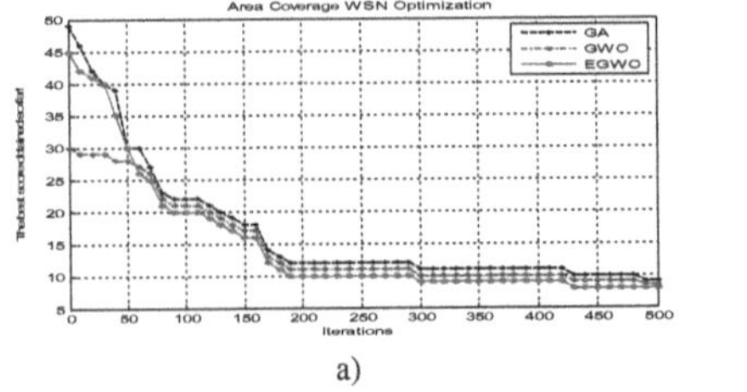

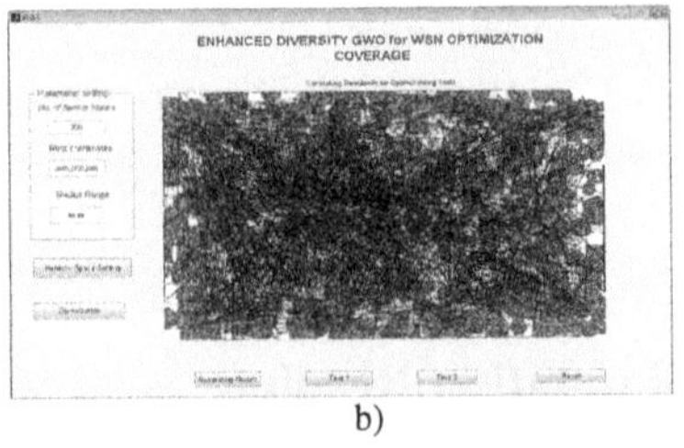

Fig. 2. (a) The comparisons of the proposed method EGWO with GWO and GA approaches in terms of the average of the obtained minimum value for the objective function in 25 times, (b) The GUI result of the optimization of in the proposed method for coverage in WSNs

Table 1 and Fig. 2 (a) compare the quality of performance and time running for deployment optimization of the proposed method of EWOA with the GWO and GA method for area coverage optimization. Clearly, the average the obtained values of the objective functions in EWGO outperforms the other methods of GWO and GA. Figure 2 (b) illustrates the optimization results of the proposed method for coverage in WSNs.

5 Conclusions

In this paper, a novel optimization method was proposed for maximizing the area coverage problem in WSN. The proposed method considers the overlaps of coverage area and coverage holes to formulate the objective function of the coverage problem. Voronoi diagram is used for estimating coverage holes in optimizing the coverage problem, and the overlapping coverage in WSN is modeled based on the Monte Carlo method. In this way, the controllable trajectories of a reduced number of mobile nodes can be exploited in order to improve the coverage rate. Additionally, the enhanced diversity herbs in the grey wolf optimizer can provide the optimization algorithm to avoid dropping optimum local in solving the complex problems like the coverage problem. Several deterministic countermeasures are presented in order to verify the behavior of the proposed method. Simulation results compared with other methods in the library such as WOA and GA show that the proposed method provides good coverage within a reasonable computational time, and be alternatives competitor. As this is an ongoing research project, in the future we will focus more on doing more tests for different conditions and optimizing other problems in WSN.

Acknowledgments. We are grateful to the support from Ministry of Science and Technology, Taiwan, R.O.C. to this study, under the grant numbers MOST 104-2221-E-151-068- and MOST 105-2221-E-151-007-.

References

1. Yick, J., Biswanath Mukherjee, D.G.: Wireless sensor network survey. Elseiver Comput. Netw. **52**, 2292–2330 (2008)
2. Othman, M.F., Shazali, K.: Wireless sensor network applications: a study in environment monitoring system. In: Procedia Engineering, pp. 1204–1210 (2012)
3. Nguyen, T.-T., Shieh, C.-S., Horng, M.-F., Ngo, T.-G., Dao, T.-K.: Unequal clustering formation based on bat algorithm for wireless sensor networks. The Sixth KSE, pp. 667–678 (2015)
4. Boussaïd, I., Lepagnot, J., Siarry, P.: A survey on optimization metaheuristics. Inf. Sci. **237**, 82–117 (2013)
5. Shieh, C.-H., Nguyen, T.-T., Horng, M.-F., Dao, T.-K.: A communication strategy enhanced diversity herds grey wolf optimizer for multimodal optimization. ICIC Express Lett. **7**, 1511–1517 (2016). Part B

6. Ammari, H.M., Mulligan, R., Ammari, H.M., Mulligan, R.: Coverage in wireless sensor networks: a survey. Netw. Protoc. Algorithms **2**, 93–131 (2010)
7. García-hernández, C.F., Ibargüengoytia-gonzález, P.H., García-hernández, J., Pérez-díaz, J.A.: Wireless sensor networks and applications: a survey. J. Comput. Sci. **7**, 264–273 (2007)
8. Nguyen, T.-T., Shieh, C.-S., Dao, T.-K., Wu, J.-S., Hu, W.-C.: Prolonging of the network lifetime of WSN using fuzzy clustering topology. In: Proceedings - 2013 2nd International Conference on Robot, Vision and Signal Processing, RVSP 2013 (2013)
9. Fan, G., Jin, S., Processing, D.: Coverage problem in wireless sensor network : a survey. J. Netw. **5**, 1033–1040 (2010)
10. Mirjalili, S., Mirjalili, S.M., Lewis, A.: Grey wolf optimizer. Adv. Eng. Softw. **69**, 46–61 (2014)
11. Jooyandeh, M., Mohades, A., Mirzakhah, M.: Uncertain voronoi diagram. Inf. Process. Lett. **109**, 709–712 (2009)
12. Gao, W., Chen, Q., Jiang, M., Li, Y., Wang, S.: The optimization of genetic algorithm in wireless sensor network coverage. Int. J. Signal Process. Image Process. Pattern Recognit. **8**, 255–264 (2015)

Utilizing IABC and Time Series Model in Investigating the Influence of Adding Monitoring Indicator for Foreign Exchange Rate Forecasting

Pei-Wei Tsai[1,2], Wen-Ling Wang[3], Jui-Fang Chang[3(✉)], Zhi-Sheng Chen[3], and Yong-Hui Zhang[1,2]

[1] College of Information Science and Engineering, Fujian University of Technology, Fuzhou, China
peri.tsai@gmail.com, zyh@fjut.edu.cn

[2] Fujian Provincial Key Laboratory of Big Data Mining and Applications, Fujian University of Technology, Fuzhou, China

[3] Department of International Business, National Kaohsiung University of Applied Science, Kaohsiung, Taiwan
{1103346120,1104346114}@gm.kuas.edu.tw, rose@kuas.edu.tw

Abstract. This work focuses on the NTD/USD exchange rate and the Monitoring Indicator in years of 2006 to 2010 to forecast the foreign exchange rate via Time-series models including GARCH (1,1) and EGARCH (1,1), and a computational intelligence model called IABC. In order to compare the rate forecasting ability of these models, the MAPE is consecutively applied as the evaluating criterion after the forecasting process. The experimental results indicate that it is effective to enhance the ability of foreign exchange rate forecasting by adding the Monitoring Indicator as a new reference variable in the IABC model. Based on the experimental results, we find that IABC is the most effective one to forecast the foreign exchange rate. Nevertheless, when IABC is suffered from the local optimum in the solution space, the forecasting ability would present a significant drop.

Keywords: GARCH · EGARCH · IABC · Rate forecasting · Monitoring indicator

1 Introduction

As technology develops, international trades among nations become more frequent. Economic globalization makes international finance a more important issue to concern with. Exchange rate forecasting thus gradually become one of the most concerned issues, as it not only influences trades but nations' economies and political stabilities.

Some scholars forecast variables by conducting different models, and Time Series Analysis is one of the most common model used. Engle (1982) [1] developed ARCH that indicates conditional heteroscedastic would be influenced by the square of earlier

J. Pan et al. (eds.), *Genetic and Evolutionary Computing*, Advances in Intelligent Systems and Computing 536, DOI 10.1007/978-3-319-48490-7_22

stage error term, and that meanwhile the distribution of conditional heteroscedastic error term is normal distribution. Bollerslev (1986) [2] took a step further in the expansion ARCH model and introduced the GARCH model. He suggested that conditional heteroscedastic error term would not only be influenced by the square of the error term in the earlier stage, and that it would also be influenced by the conditional heteroscedastic in the earlier stage. GARCH(1,1) is usually employed when adopting the GARCH model. Nelson (1991) [3] developed Exponential GARCH model (EGARCH). This model is used when market is influenced by different issues, it may react to them and make difference reactions. Swarm Intelligence is an algorithm that is based on creatures' collective behavior.

Tsai et al. (2008) [4] improved Artificial Bee Colony by applying Universal Gravitation to Interactive Artificial Bee Colony. Comparing to ABC, IABC expands searching range to solve optimization problems. That makes it able to break the limitation of previous space.

There are several reasons that influence exchange rate. Thus, scholar developed many economic models to investigate the relations among exchange rate and different variables. Tsai et al. (2015) [5] forecasted exchange rate through nine different macro-economic variables and Consumer confidence, and the results were good. However, there are less researches concern economic boom conditions may influence exchange rate or not. Thus, this research adding Monitoring Indicator as a new variable to examine whether it can enhance the forecasting accuracy or not.

2 Experiment Design

Tsai et al. (2015) applied nine macro-economic variables and Consumer Confidence Index to forecast the foreign exchange rates. In our work, we include the Monitoring Indicator as the newly involved variable to make it eleven variables in total in the foreign exchange rate forecasting. The variables involved are listed in Table 1.

Table 1. Variables related to foreign exchange rate forecasting.

Variables	Variables
Rate (NTD/USD)	Stock Return
Consumer Price Index	M1
Commercial Paper Rate	M1B
Federal Fund Rate	Consumer Confidence Index
Balance of Trade	Monitoring Indicator
Foreign Investment	

This research applied eleven variables to forecast foreign exchange rates; all data we use is collected from the source in TEF; the focused period is bounded in January 1st in 2006 to December 31st in 2010. Thus, there are 1246 records in total.

There are four test processes setup in this work including the Pearson's correlation coefficient test for testing the normality; the autocorrelation examination and the unit root test; the final step is feeding the data into the Time-series model or the IABC model. After obtaining the forecasting results, the MAPE is used to evaluate and to trace the foreign exchange rate forecasting error. The experiment design is decomposed as follows:

Step 1. Conducting Pearson's correlation coefficient test by Eq. (1) to examine whether the new variable, Monitoring Indicator has any relevance with exchange rates or not. Two variables, X and Y, are in linear correlation, and the amount are between minus 1 to 1. The result shows that the new variable is relevant to the exchange rate.

$$r_{XY} = \frac{COV_{XY}}{S_X S_Y} \tag{1}$$

Step 2. In Time Series, data is determined whether it is in correspondence with normal distribution by doing JB test. The formula shows as follows:

$$JB = \frac{T-n}{6}\left[S^2 + \frac{1}{4}(K-3)^2\right] \tag{2}$$

where S stands for skewness and K denotes the kurtosis.

Step 3. For correlation problems, this research conducted Lijun-Box Q tests to examine whether variables are self-correlated. The formula shows as follows:

$$Q(p) = n(n+2)\sum_{k=1}^{p}\frac{1}{n-k\rho_k^2} \sim \chi^2(p) \tag{3}$$

where n represents the samples and k stands for lag order.

The statistics are in chi-square distribution, which means the degree of freedom is zero.

Step 4. In Time Series, data would be stationary or non-stationary. If data is non-stationary, the regression analysis will be spurious regression, which means it is less reliable. Thus, unit root test must be done to make sure data are stationary. The unit root tests applied in this research are ADF test and PP test.

Step 5.1. The GARCH model equations displaying this phenomenon are as follows:

$$y_t|\Omega_t \sim N(x_t a, \sigma^2) \tag{4}$$

$$\varepsilon_t = y_t - x_t a \tag{5}$$

$$\sigma_t^2 = \alpha_0 + \sum_{i=1}^{q} a_i \varepsilon_{t-i}^2 + \sum_{i=1}^{p} \beta_i \sigma_{t-i}^2, \alpha_i \geq 0, \beta_i \geq 0 \tag{6}$$

where σ_t^2 stands for function of squared residues in the past q period and the conditional variance in the past p period. Both p and q are GARCH levels.

Step 5.2. The EGARCH model equations displaying this phenomenon are as follows:

$$y_t = x_t b + \varepsilon_t \tag{7}$$

$$\varepsilon_t \middle| \Omega_{t-1} \sim N(0, \sigma^2) \tag{8}$$

$$\ln \sigma_t^2 = \alpha_0 + \sum_{i=1}^{q} \left[\alpha_i \left(\left| \frac{\varepsilon_{t-i}}{\sigma_{t-i}} \right| - E \left| \frac{\varepsilon_{t-i}}{\sigma_{t-i}} \right| + \gamma \frac{\varepsilon_{t-i}}{\sigma_{t-i}} \right) \right] + \sum_{j=1}^{p} \beta_j \ln \sigma_{t-j}^2 \tag{9}$$

where β_j is the function of β_{j-1}, α_i means the parameter that impacted by α_{i-1}, γ represents the parameter that asymmetry deviation for last period to current period, and $\frac{\varepsilon_{t-i}}{\sigma_{t-i}}$ stands for the normalize residual.

Step 6. MAPE is used as the evaluating criterion of the model's forecasting ability. The closer to zero the result is, the better the forecasting ability will be. The formula displays as follows:

$$MAPE = \frac{1}{n} \sum_{t=1}^{n} \frac{\left| \hat{S}_t - S_t \right|}{S_t} \times 100\ \% \tag{10}$$

where $\hat{s}_t$ stands for the forecasted exchange rate in period t, S_t stands for the actual exchange rate in period t, and n stands for the amount of data (Table 2).

Table 2. Typical MAPE values for the model evaluation

MAPE (%)	Forecasting ability
MAPE < 10	Best
10 < MAPE ≤ 20	Better
20 < MAPE ≤ 50	Reasonable
50 < MAPE	Worse

There are six steps in conducting IABC model for testing the normality, the autocorrelation examination, the unit root test, and the fitness allocation:

Step 1. Conducting descriptive statistics and JB test on every variable to make sure data is in correspondence with normal distribution.

Step 2. Conducting Ljung-Box Q test to determine the residues of data should conform to white noise.

Step 3. Conducting ADF test and PP test to make sure data are stationary.

Step 4. Utilizing a set of forecasted values of every variable, based on the weight ratio to forecast errors and determining the weights for the variables. The fitness function used in the IABC model is listed in Eq. (11):

$$\min f(W) = \sum_{t=1}^{n} \left| \left(\sum_{d=1}^{D} w_d \times v_{t,d} \right) - R_{real,t} \right| \tag{11}$$

where $f(W)$ stands for the fitness value, $w = (w_1, w_2, \ldots, w_d)$ denotes the referenced days in the process of optimization, D means the total amount of referenced information, v indicates the variable information of reference, and R_{real} means the actual exchange rate value.

Step 5. To find out the forecasting results, take records of the optimized weighted ratio of every variable and calculate the outcome with the data in the next period of time.

$$R_{pd,t+1} = \sum_{d=1}^{D} w_d \times v_{t,d} \tag{12}$$

where R_{pd} stands for the forecasted exchange rate.

Step 6. Calculating the MAPE value by the forecasting result and the actual foreign exchange rate. The MAPE value is treated as the evaluation criterion of the models' forecasting abilities.

3 Experiments and Experimental Results

The data used in this research are the nine macro-economic variables and Consumer Confidence Index that Tsai et al. (2015) has applied, along with the new variable applied in this research – Monitoring Indicator. There are eleven variables in total, and the focused period of data is in the range of January 1st in 2006 to December 31st in 2010, which includes 1245 records in total. GARCH (1,1), EGARCH (2,2) and IABC models are applied to forecast exchange rate.

The comparison of the experimental results are made in two sets: the first set is for the monthly MAPE value between different models; and the second set is focused on the effect callused by adding the Monitoring Indicator in the reference variable.

3.1 Monthly MAPE of Every Model

In Fig. 1, there is no output for the first month in 2006 because the IABC developed in this research took the former 30 days for forecasting. Aside from January, the forecasting result obtained by the IABC model in the rest months all present better forecasting ability than the Time-series models.

In Fig. 2, the IABC model developed in this research showed better results in its January, March, April, June, August, September and October than Time Series.

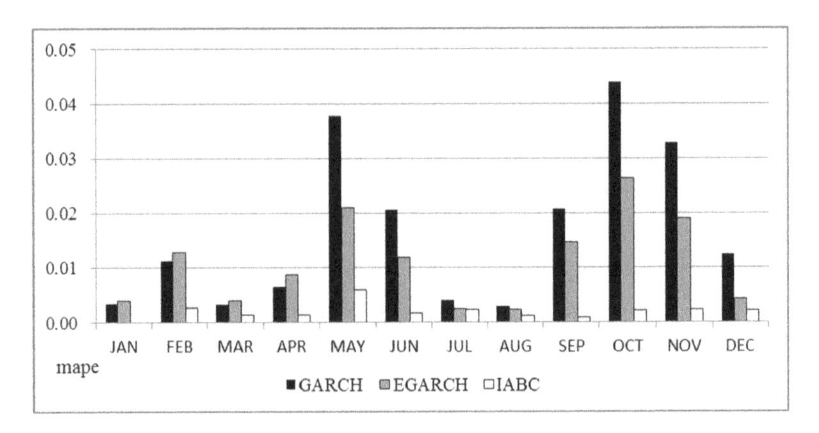

Fig. 1. Monthly MAPE for every model in 2006.

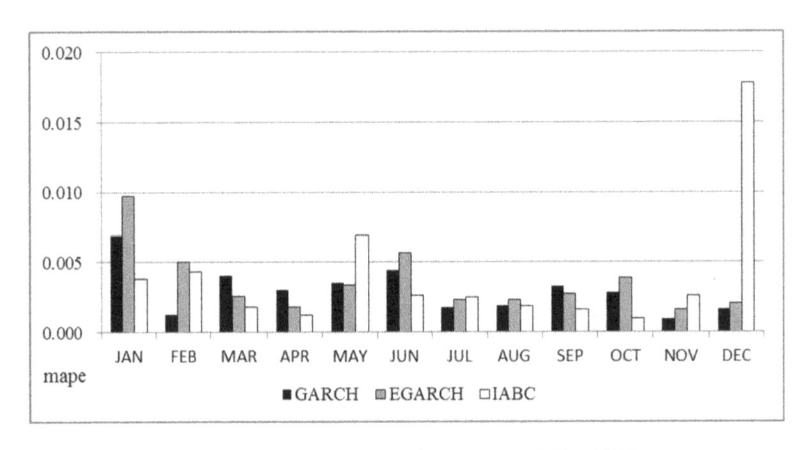

Fig. 2. Monthly MAPE for every model in 2007.

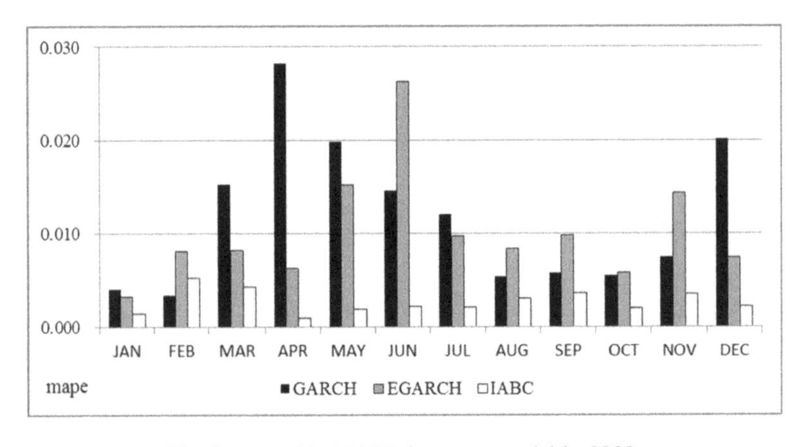

Fig. 3. Monthly MAPE for every model in 2008.

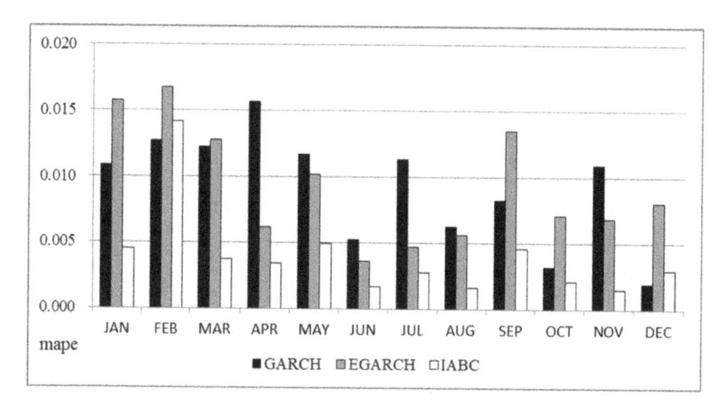

Fig. 4. Monthly MAPE for every model in 2009.

However, in December, IABC was stuck with the optimization problems. Thus, it performed worse forecasting ability.

Except for February, the IABC of the other months are superior to GARCH (1,1) and EGARCH (1,1) in Fig. 3. IABC model is comparatively stable than Time-series models in forecasting.

Except for February and December, the IABC model developed in this research performed better than GARCH (1,1) AND EGARCH (1,1) in Fig. 4.

Aside from February, the IABC model of the other eleven months are super to

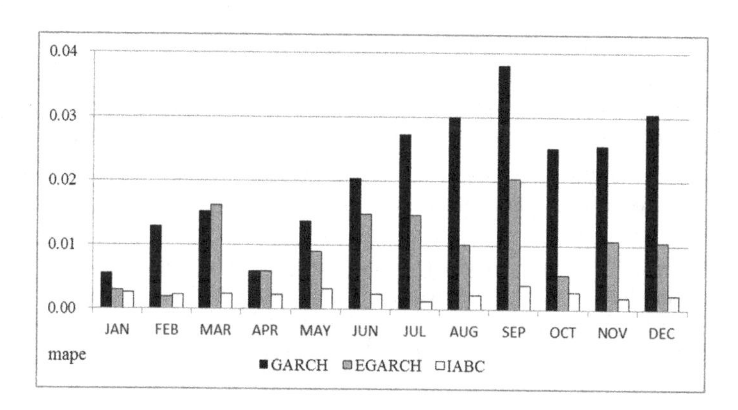

Fig. 5. Monthly MAPE for every model in 2010.

GARCH (1,1) and EGARCH (1,) in Fig. 5. IABC model is comparatively stable than Time-series models in forecasting.

Table 3. The before-and-after of adding monitoring indicator from 2006 to 2010

	Average MAPE value	Standard deviation
GARCH	0.6142	0.1440
EGARCH	0.6115	0.1284
IABC	0.2145	0.0725
MI-GARCH	0.0123	0.0069
MI-EGARCH	0.0089	0.0030
MI-IABC	0.0031	0.0008

3.2 Comparison Between the Before-and-After of Adding Monitoring Indicator

Table 3 shows the comparison of our experimental results with the results obtained by Tsai et al. (2015), which applied Consumer Confidence Index into Time-series models. The MI- prefix represents the models including the Monitoring Indicator as the considered variable. After adding the Monitoring Indicator, IABC shows the least deviation. The MAPE value of IABC and Time-series is lower when it is added the Monitoring Indicators as a new variable. It implies that adding the Monitoring Indicator in IABC can greatly lower the forecasting error and enhance the forecasting ability.

4 Conclusions and Future Works

This research firstly applied Time Series Analysis and IABC to get forecasted exchange rate, and further conducted MAPE. As the result showed, overall, IABC model was superior to Time Series Analysis GARCH (1,1) and EGARCH (1,1) model when forecasting. After adding the variable of Monitoring Indicator, Time Series Analysis models and IABC models are both effective in lowering deviation rates. Owing to time limited, the base period is set to conduct in only one year. Thus, for further suggestion, the base period could be longer to improve models' forecasting ability.

Acknowledgement. This work is partially supported by the Key Project of Fujian Education Department Funds (JA15323), Fujian Provincial Science and Technology Project (2014J01218), Fujian Provincial Science and Technology Key Project (2013H0002), and the Key Project of Fujian Education Department Funds (JA13211). We also acknowledge the treasurable comments from the reviewers.

References

1. Engle, Robert: Autoregressive conditional heteroscedasticity with estimates of the variance of united kingdom inflation. Econometrica **50**(4), 987–1007 (1982)
2. Bollerslev, Tim: Generalized autoregressive conditional heteroscedasticity. J. Econometrics **31**(3), 307–327 (1986)

3. Nelson, D.B.: Conditional heteroscedasticity in asset returns: a new approach. Econometrica **59**(5), 347–370 (1991)
4. Tsai, P.W., Muhammad, K.K., Pan, J.S., Liao, B.Y.: Interactive artificial bee colony supported passive continuous authentication system. IEEE Syst. J. **8**(2), 395–405 (2014)
5. Tsai, P.-W., Liu, C.-H., Liao, L.-C., Chang, J.-F.: Using consumer confidence index in the foreign exchange rate forecasting. In: Proceedings of 11th International Conference on Intelligent Information Hiding and Multimedia Signal Processing (IIH-MSP-2015), Adelaide, Australia, pp. 360–363, 23–25 September 2015

Advanced Multimedia Information and Image Processing

Myanmar-English Machine Translation Model

Khin Thandar Nwet(✉) and Khin Mar Soe

Natural Language Processing Lab, University of Computer Studies, Yangon, Myanmar
{khinthandarnwet,khinmarsoe}@ucsy.edu.mm

Abstract. Natural Language Processing (NLP) is the field that strives to fill the communication gap between the different sections of societies One of the NLP processes, Machine Translation (MT) is used to translate any language from native language by understanding and generating natural language. The research in Myanmar-English MT has started since 2010. However, the translation accuracy is not becoming raise since the complex syntactic structure of Myanmar Language and the scarceness of resources. We found that it takes a lot of time to collect language resources such as Myanmar-English aligned corpora and Treebank. This paper presents current work of Myanmar-English machine translation system based on statistical methods. The aim of this paper is to introduce the Myanmar-English translation model and the comparative study using Asian Language Tree-bank (ALT) data.

Keywords: Asian languages · Parallel corpus · Treebank · Machine translation

1 Introduction

There are many tasks needed to perform in machine translation. For Myanmar language, word segmentation is the early step to do since Myanmar language does not use spaces between words. After that, syntactic analysis, semantic analysis and synthesis analysis has to be done to complete the translation. There are three approaches used in MT: rule based, example based and statistical based. Our main motivation for this research is to investigate Myanmar-English MT based on statistical methods. We will evaluate the accuracy and also contribute the comparative study of the translation model using ALT data.

2 Related Work

Nowadays, the study of automatic translation of Myanmar to English is very few. In this section, previous works in machine translation on Myanmar language are reviewed. Recent Statistical machine translation systems based on phrase or word group and use probabilistic model by using source channel approach or direct probability model (log linear model).

Czajkowski and Wai [4, 7] studied Myanmar-English Bidirectional Machine Translation system by using transfer based approach. In the analysis stage, input source

J. Pan et al. (eds.), *Genetic and Evolutionary Computing*, Advances in Intelligent Systems and Computing 536, DOI 10.1007/978-3-319-48490-7_23

sentence is parsed using existing parsers. They used Stanford Parser for parsing English language [13] and Myanmar 3 parser for parsing Myanmar language [11]. They also used tree to tree transformation approach such as Synchronous Context Free Grammar (SCFG) rules to change source sentence structure to target sentence structure. The examples are shown in Fig. 1.

Input: သူမသည်လှပသောမိန်းကလေးဖြစ်သည်။

Output: She is a beautiful girl.

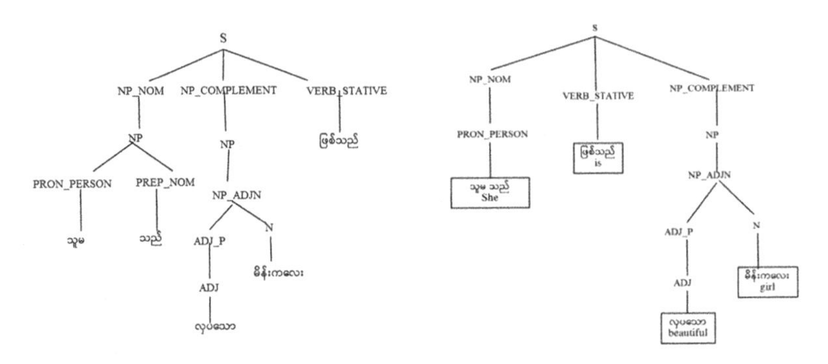

Fig. 1. Myanmar sentence structure before transformation and myanmar sentence after transformation

They translated English to Myanmar using tree to tree transformation approach using Synchronous Context Free Grammar (SCFG) rules. Morphological synthesis is also to improve smooth translation because Myanmar language is a morphologically rich language. It considers the articles (a, an, the) and cardinal number especially. They are translated as "တစ်ခု၊ တစ်ကောင်၊ တစ်ယောက်" and so on but in Myanmar Language this article is translated depend on noun. Therefore, to solve this problem, sense of noun is getting from Myanmar Word Net. They also used Myanmar-English bilingual lexicon is 13373 words.

Foster et al. [3] proposed string to tree and tree to string Statistical Machine translation for Myanmar language. They published evaluation of the quality of string-to-tree (S2T) and tree-to-string (T2S) statistical machine translation methods between Myanmar and Chinese, English, English, French, German in both direction. They used multilingual Basic Travel Expressions Corpus(BTEC), which is a collection of travel related expressions [12]. The BLEU score results for Myanmar to English is 44.53 and English to Myanmar is 42.83.

Thu et al. [15] studied Factored Machine Translation for Myanmar to English, Japanese and Vice Versa. Factored machine translation models extend traditional Phrase Based Statistical Machine translation (PB-SMT) by taking into account not only the surface form of the words, but also linguistic knowledge such as the dictionary form (lemma), part-of-speech (POS) and morphological tags. They also used Basic Travel Expressions Corpus (BTEC) [13]. The BLEU score results for Myanmar to English is 20.74. They assumed that due to the lack of training data for POS tag factor, the

Myanmar annotations for the factor intended to be incomplete, and potentially accurate. Most of All previous research is the use of small corpora. Most of the NLP works are based on Rules. In this work, we focused on machine translation with ALT data.

3 Overview of Myanmar-English Translation Model

Figure 2 shows overview of the machine translation for Myanmar to English. In the writing systems of many Asian languages, such as Myanmar, Chinese, Japanese and Thai, words are not delimited by spaces. There are no blanks in Myanmar text forward boundaries. In segmentation step, we used Myanmar Language Segmentator published by UCSY NLP Lab [14]. In Part-of-Speech tagging, the segmented sentence is tagged using bigram part-of-speech for Myanmar language [10]. They used 20 POS tags and 6 for finer tags. The category for a word, can be constructed from the features of that word. For instances of POS tag with category, " မိန်းကလေး " <girl> word must be tagged with NN.Person (Person category of Noun tag), "သို့ "<to> with PPM.Direction (Direction type of Postpositional Marker), "သူ" <he>with PRN.Person (Person type of Pronoun), "လှပသော" <beautiful> with JJ.Dem (Demonstrative sense of Adjective), "အလွန်" <very> with RB.State (State of Adverb) and so on. Our ALT data defined 14 POS tags to be used in the ALT corpus in order to get more detailed syntactic information of both source and target languages. They are Abbreviation (ABB), Adjective (ADJ), Adverb (ADV), Conjunction (CONJ), Foreign words (FOR), Interjection (INT), Noun (N), Number (NUM), Particle (PART), Post positional marker (PPM), Pronoun (PRON), Punctuation (PUNC), Symbol (SB), Verb (V). In translation phase, the tagged Myanmar sentence is translated to English sentence using Phrase Based Myanmar-English translation model proposed by Thet et al. [2] Myanmar language is inflected language and there are very few creations and researches of corpora in Myanmar, comparing to other language. Therefore, Myanmar phrases translation model is based on syntactic structure and morphology of Myanmar language (see Fig. 3).

Moreover, this translation model also interacts with Word Sense Disambiguation (WSD) [5] to solve ambiguities when a phrase has with more than one sense. For example, the polysemous Myanmar noun "တူ" would translate to three different English words in the following three sentences:

Myanmar-English bilingual corpus is proposed by [6] is used as a main knowledge source for this phrase translation and Word Sense Disambiguation. This Bitext corpora play an important role in the development of Machine Translation. Meta-data annotation includes: (i) Information about part-of-speech, (ii) Lemma information, (iii) Segmented words, (iv) Word/Phrase alignment, and (v) Locality information. The full format specification is available as a txt file. In total, the corpus consists of approximately 5000 parallel sentences for general domain (such as local newspaper, dictionaries, middle school text book, etc.) [16]. Moreover, Myanmar is a verb final language and reordering is needed when our language is translated from other languages with different word orders. This system used reordering rules by proposed [2], automatic reordering rule generation and application of generated reordering rules in stochastic reordering model (see Fig. 4).

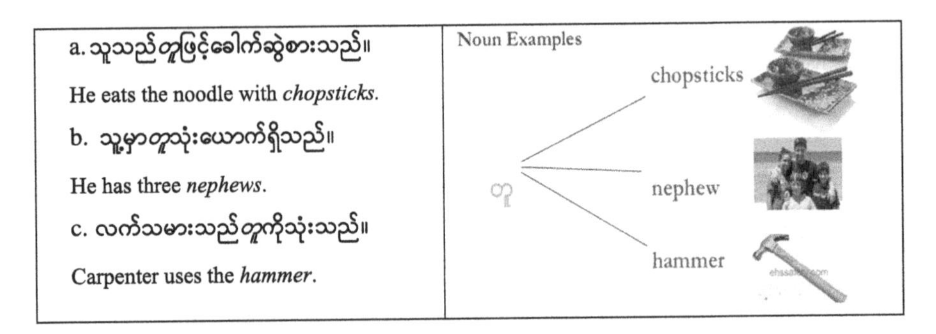

Fig. 2. One Myanmar word "တူ" (Tuu) has three English words sense example

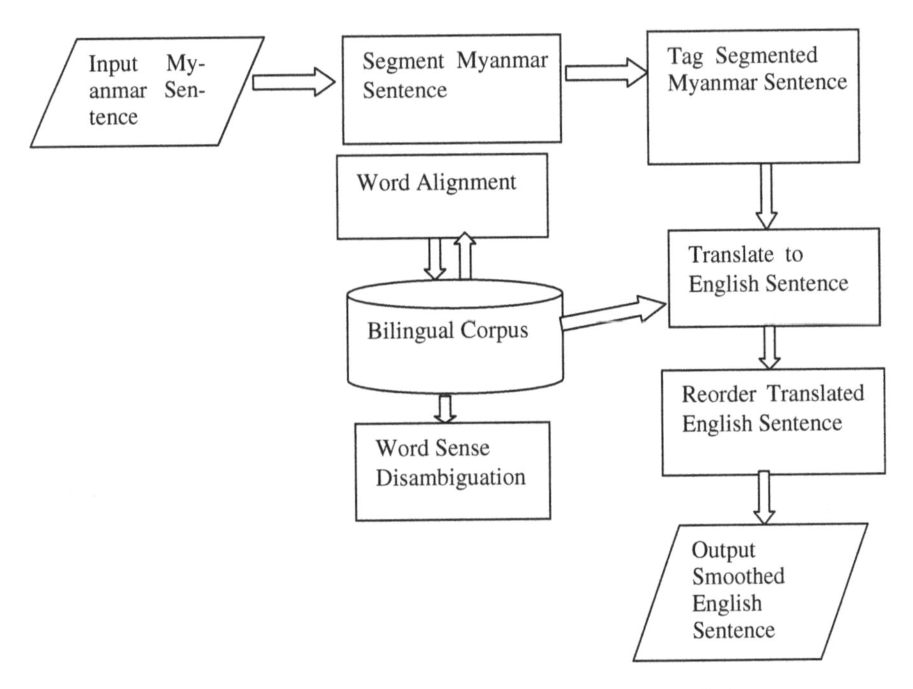

Fig. 3. Overview of Myanmar to English machine translation

The ALT project was first proposed by the National Institute of Information and Communications Technology, Japan (NICT) in 2014. NICT started to build Japanese and English ALT and worked with the University of Computer Studies, Yangon, Myanmar (UCSY) to build Myanmar ALT in 2015. ALT has about 20,000 sentences extracted from the English Wikinews. These were already translated into the six languages, in order to provide word segmentation, POS tagging, and syntax analysis annotations, in addition to the word alignment information. Figure 5 shows the word alignment annotation between an English sentence and the corresponding translated Myanmar sentence, and Fig. 6 shows the constituency tree building [9].

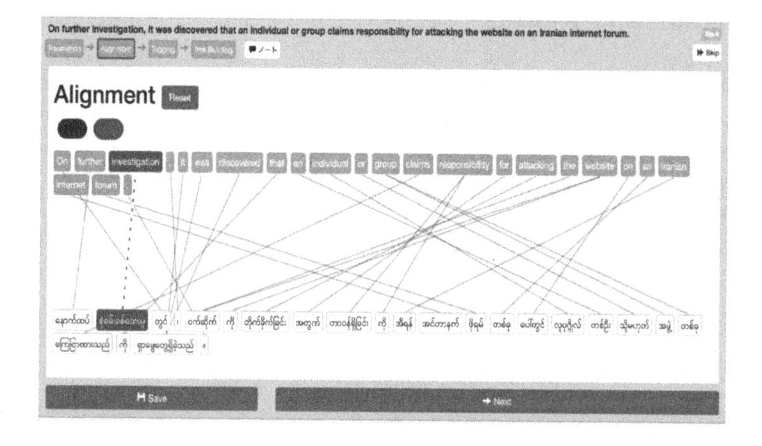

Fig. 5. Word alignment interface

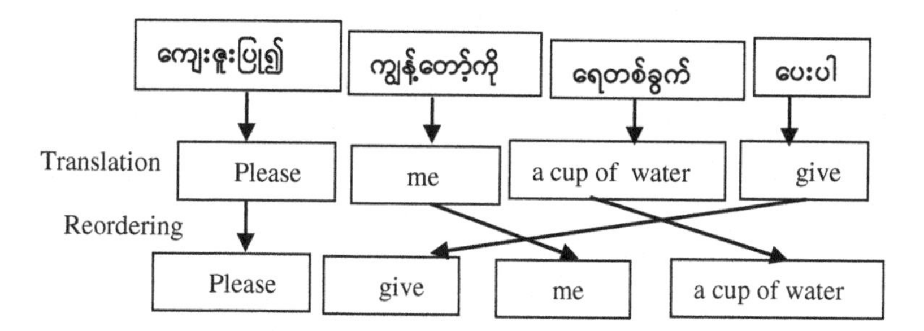

Fig. 4. Phrase based translation

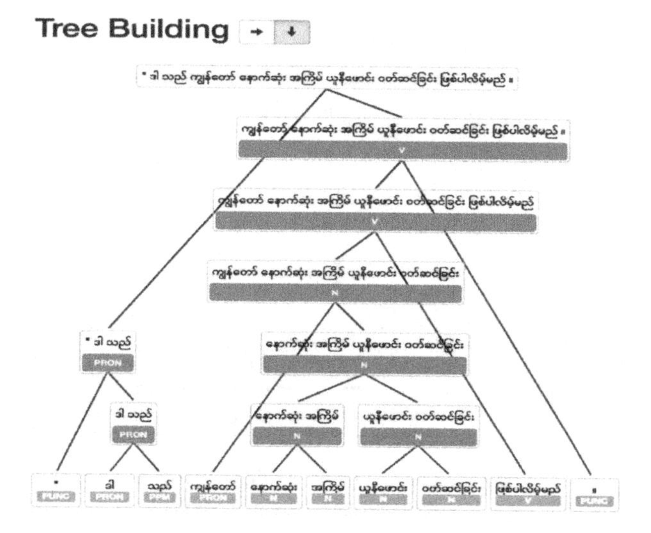

Fig. 6. Tree building interface

4 Evaluation and Results

BLEU is the best known and best adopted Machine Evaluation for (machine) translation [17]. BLEU is an automatic evaluation technique which is a geometric mean of n-gram matching. To compute the BLEU score, one has to count the number of n-grams in the test translation that have a match in the corresponding reference translations. The formula used to calculate the n-gram precision is simple. The words from a candidate translation that match with a word in the reference translation (human translation) are counted, and then divided by the number of words in the candidate translation. IBM's formula for calculating BLEU score is as follows [18]:

$$\text{BLEU} = BP \times exp\left(\sum\nolimits_{n=1}^{4} \frac{1}{n}\log(pn)\right). \tag{1}$$

where brevity penalty is calculated as:

$$\text{BP} = \min\left(1, \text{e}^{1-\text{r}/\text{c}}\right) \tag{2}$$

where c is the length of the corpus of hypothesis translations, and r is the effective reference (is calculated as the sum of the single reference translation from each set which is closest to the hypothesis translation) corpus length. The n-gram precision is calculated as:

$$P_n = \frac{\sum_{i=1}^{I} \sum_{ngram \in S_i} count(ngram)}{\sum_{i=1}^{I} \sum_{ngram \in S_i} count_{sys}(ngram)} \tag{3}$$

count(*ngram*) is the count of n-grams found both in s_i and r_i. $count_{sys}(ngram)$ is the count of n-grams found only in s_i.

According to the Fig. 7, current phrase translation BLEU score is 79.7 and they used 12817 sentences parallel corpus size [2]. The best results got by adding morphology and POS of Myanmar language to baseline system. Postpositional markers

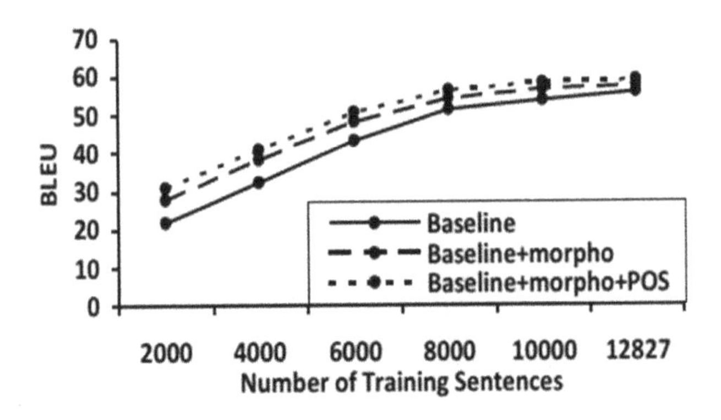

Fig. 7. BLEU scores for translating to English

have ambiguous meaning in translation. By using POS tags, the system reduced ambiguous in postpositional markers. Especially, ambiguous in Subject PPM မှာ (hmr) "has, have, had" and Place PPM မှာ (hmr) "at", Subject PPM က (ka;) "null" and Leave PPM က (ka) "from", Used PPM နှင့် (hnint) "with" and Compare PPM နှင့် (hnint) "and" Used PPM ဖြင့် (phyint) "with" and Cause PPM ဖြင့် (phyint) "because of" and Place PPM တွင် (twin) "at" and Extract PPM တွင် (twin) "among".

Myanmar word segmentation accuracy is 97 % [14] and Myanmar part of speech tagging accuracy is 597 % [10].

In [5], for evaluation purpose, the test sentences are grouped into two groups, 150 sentences for Type1 (Test sentences is taking in the training set) and Type2 (Test sentences that are composed of words in the training sentences, but not exactly the same sentences in the training set). The accuracy of type-I is 98 % and type-II is 90 %. [2, 5, 6] used the same training data.

The Word alignment for Myanmar to English translation accuracy is 89 % [6].

For reordering, the accuracy is 98.9 % in simple sentences, 95.4 in complex sentences and 93.6 in compound sentences [1].

Evaluation Result of Myanmar to English Machine Translation is 82.14 % and English to Myanmar Machine Translation is 80.45 [7].

In the future, more and more training data are going to be trained. The accuracy will be higher. We have to test ALT data for bidirectional Myanmar to English machine translation using human evaluation with bilingual judges.

5 Conclusion

In conclusion, we can say that the field of machine translation has been and continues a key focus of research on natural language processing and that extended to the development of many positive results. Moreover, perfection is still far away. Most of the previous works for Myanmar language machine translation used small corpora and rule based. We focused on construction of a statistical MT model at the end to increase the performance of the machine translation system.

This paper also discussed the ALT project. ALT is intended to accelerate NLP development in low resource Asian languages. The corpus consists of about 20,000 sentences from the news domain consisting of Asian language translations from a shared English source text together with accompanying word segmentation, word alignment, POS tagging, and syntax trees. ALT includes English, Indonesian, Japanese, Khmer, Malay, Myanmar and Vietnamese in the short term, and extend to other languages in the long term through collaboration with international research organizations.

Acknowledgment. This work is partly supported by the ASEAN IVO Project "Open Collaboration for Developing and Using Asian Language Treebank".

References

1. Smith, T.F., Waterman, M.S.: Identification of common molecular subsequences. J. Mol. Biol. **147**, 195–197 (1981)
2. May, P., Ehrlich, H.-C., Steinke, T.: ZIB structure prediction pipeline: composing a complex biological workflow through web services. In: Nagel, W.E., Walter, W.V., Lehner, W. (eds.) Euro-Par 2006. LNCS, vol. 4128, pp. 1148–1158. Springer, Heidelberg (2006). doi:10.1007/11823285_121
3. Foster, I., Kesselman, C.: The Grid: Blueprint for a New Computing Infrastructure. Morgan Kaufmann, San Francisco (1999)
4. Czajkowski, K., Fitzgerald, S., Foster, I., Kesselman, C.: Grid information services for distributed resource sharing. In: 10th IEEE International Symposium on High Performance Distributed Computing, pp. 181–184. IEEE Press, New York (2001)
5. Foster, I., Kesselman, C., Nick, J., Tuecke, S.: The physiology of the grid: an open grid services architecture for distributed systems integration. Technical report, Global Grid Forum (2002)
6. National Center for Biotechnology Information. http://www.ncbi.nlm.nih.gov
7. Wai, T.T., Htwe, T.M., Thein, N.L.: Automatic reordering rule generation and application of reordering rules in stochastic reordering model for English-Myanmar machine translation. Int. J. Comput. Appl. **27**(8), 19–25 (2011)
8. Zin, T.T., Soe, K.M., Thein, N.L.: Translation model of Myanmar phrases for statistical machine translation. In: Huang, D.-S., Gan, Y., Gupta, P., Gromiha, M. (eds.) ICIC 2011. LNCS, vol. 6839, pp. 235–242. Springer, Heidelberg (2012). doi:10.1007/978-3-642-25944-9_31. ISBN 978-3-642-25943-2
9. Thu, Y.K., Finch, A., Sumita, E., Pa, W.P, Htike, K.W.W.: String to tree and tree to string statistical machine translation for Myanmar language. In: 14th International Conference on Computer Applications, ICCA (2016)
10. Win, Y.Y., Nwe, T.H.: Myanmar-English bidirectional machine translation translation system by using transfer based approach. In: 13th International Conference on Computer Applications, ICCA (2015)
11. Aung, N.T.T., Thein, N.L.: Myanmar word disambiguation for Myanmar-English machine translation. Int. J. Comput. Appl. **27**(8) (2011)
12. Nwet, K.T., Thein, N.L., Soe, K.M.: Word alignment system based on hybrid approach for Myanmar-English machine translation. In: SICE Annual Conference, 13–18 September 2011. Waseda University, Tokyo, Japan (2011)
13. Win, Y.Y., Thida, A.: English to Myanmar translation system with numerical particle identification, Int. J. Inf. Technol. Comput. Sci. **6**, 37–43 (2016). Accessed June 2016 in MECS. http://www.mecs-press.org/
14. Win, A.T., Words to phrase reordering machine translation system in Myanmar-English using English grammar rules (2011)
15. Thu, Y.K., Finch, A., Sumita, E., Pa, W.P.: Introducing the Asian Language Treebank (ALT)
16. Htay, H.H., Murthy, K.N.: Myanmar word segmentation using syllable level longest matching. In: The 6th Workshop on Asian Language Resources (2008)
17. Myint, P.H., Htwe, T.M., Thein, N.L.: Bigram part-of-speech for Myanmar language. In: Proceedings of the 2011 International Conference on Information Communication and Management (ICICM 2011) (2011)
18. Phu, S.L.: Development of Lexico-conceptual knowledge resources and syntax analyzer for Myanmar language. Ph.D. thesis, University of Computer Studies, Mandalay

19. Kikui, G., Sumita, E., Takezawa, T., Yamamoto, S.: Creating corpora for speech to speech translation. In: Proceedings of EUROSOEECH 2003, pp, 381–384 (2003)
20. http://nlp.stanford.edu/software/lex-parser.shtml
21. http://www.nlpresearch-ucsy.edu.mm
22. Thu, Y.K., Finch, A., Sumita, E., Sagisaka, Y.: Factored machine translation for Myanmar to English, Japanese and Vice Versa, ICCA (2012)
23. Nwet, K.T.: Developing word to phrase alignment for Myanmar-English machine translation, ICCA (2016)

CRF-Based Named Entity Recognition for Myanmar Language

Hsu Myat Mo(✉), Khin Thandar Nwet, and Khin Mar Soe

Natural Language Processing Lab, University of Computer Studies, Yangon, Myanmar
{hsumyatmo,khinmarsoe}@ucsy.edu.mm, khin.thandarnwet@gmail.com

Abstract. Named Entity recognition (NER) is a subtask of information extraction and information retrieval that automatically identify proper nouns in texts and classify into predefined categories of name types. This paper introduces the effort on identification and classification of Named Entities in written Myanmar scripts in a statistical way. A statistical approach for NER of Myanmar Language using one of the supervised machine learning approaches called Conditional Random Fields (CRF) has been proposed for this task.

Keywords: Named entity recognition · Myanmar language · CRF

1 Introduction

Named Entity Recognition (NER) has been considered as an important task in most of the Natural Language Processing tasks and has been recognized as part of Information Extraction, Question and Answering, Text Summarization, Entity Relation Extraction and Machine Translation. The term NER was first used in the Message Understanding Conference-6 (MUC-6) in 1995. A well performing NER is important for further level of NLP techniques.

Named entity recognition (NER) is the process of automatically classifying different identifiers, named entities (NE), in accordance with a predefined set of types. Different types of name are usually person names, location names, organization names, date and time expression, etc. NER is a very difficult task due to the ambiguities nature of natural language. A word may belong to more than one name class and can be used in unbounded number of possible names.

There are efficient NER systems for languages like English and European Languages which give high f-score values and many attempts to some Asia Languages such as Japanese, Chinese, India, Thai, Malayalam Language, etc. have been applied with many techniques such as which make use of dictionary and patterns of NEs, Decision trees, Hidden Markov Model (HMM), Maximum Entropy Markov Model (MEMM), Conditional Random Fields (CRF), etc. by various researchers.

NER can be performed in various ways. NER systems have been created by using rule-based approach which applies linguistic grammar rules as well as by building

J. Pan et al. (eds.), *Genetic and Evolutionary Computing*, Advances in Intelligent Systems and Computing 536, DOI 10.1007/978-3-319-48490-7_24

statistical model, (i.e. machine learning approach) or hybrid approach that combines rule-based and machine learning based strategies.

Rule-based approaches are highly dependent on linguistic rules and thus it becomes language dependent and difficult to adapt into other languages. Moreover, resources specific to a language are used in Rule based approach which cause the difficulty that cannot be portable to other languages.

In machine learning based approach, statistical methods work by using annotated corpus as training data and builds a probabilistic model with the features of the data which are similar to the rules that are used in ruled-based approaches. The corpora with correctly labeled name entities are learned to produce the features of the data. The model then uses the features to calculate and identify the most probable NEs.

How to perform the task of identifying names in Myanmar text automatically is still challenging and complex compared to other languages for many reasons. One of the reasons is the lack of resources such as annotated corpus, name lists, name dictionaries, etc. which means that Myanmar is resource-constrained language. Moreover, there is no concept of capitalization which is the main indicator of proper names for some other languages like English.

This paper introduces the work for developing CRF based NER for Myanmar Language in a statistical way with the intension of automatically inducing name entities in scripts and to develop a based line NER that could be used for further development work to improve NER for Myanmar Language. Moreover, this work is also aimed to support in the works of statistical machine translation project, information extraction, and summarization for Myanmar Language.

The structure of this paper is as follows: Sect. 2 describes the related work in the same area. Conditional Random Fields (CRF) is briefly explained in Sect. 3. Section 4 gives brief introduction to Myanmar Language and discusses the nature of NE in Myanmar language and then CRF based NER with data preparation followed by describing named tag set used in Sect. 5. The experiment followed by evaluation is shown in Sect. 6. Finally, conclusion is described in Sect. 7.

2 Related Works

Thi Thi Swe and Hla Hla Htay presented a method for Myanmar Named Entity Identification using a hybrid method. This method is a combination of ruled based and statistical N-grams based method which use name database. They classified Myanmar NEs into three classes, namely person name (PER), organization name (ORG) and location name (LOC) [1].

Thida Myint and Aye Thida proposed Myanmar Named Identification algorithm. In the algorithm, the system defines the names by using some of the POS information, Name entity identification rules and clue words in the left and/or the right contexts of NEs carry information for NE identification [2].

Sudha Morwal and Nusrat Jahan performed Named Entity Recognition using Hidden Markov Model (HMM) and discussed some experimental results on Indian Languages like Hindi, Urdu and Marathi [3].

Vijayakrishna R and Sobha L proposed a domain focused Tamil Named Entity Recognizer of tourism domain. It handles nested tagging of named entities with a hierarchical tagset containing 106 tags. They have experimented building Conditional Random Fields (CRF) models by training the noun phrases of the training data [4]. Different approaches for NER are proposed in [6, 7].

3 Conditional Random Fields

Conditional models are used to label the observation sequence x_* by selecting the label sequence y_* that maximizes the conditional probability $p(y_*|x_*)$. The conditional nature of such models means that no effort is wasted on modeling the observation and one is free from having to make unwanted independence assumptions about these sequences; arbitrary attributes of the observation data may be captured by the model, without the modeler having to worry about how these attributes are related [11]. Additionally, CRFs avoid the label bias problem, a weakness exhibited by maximum entropy Markov models (MEMMs) and other conditional Markov models based on directed graphical models [7, 11]. Conditional Random Field is an example of discriminative models.

Conditional Random Fields [7] (CRFs) are a probabilistic framework for labeling and segmenting sequential data, based on the conditional approach. A CRF is a form of undirected graphical model that defines a single log-linear distribution over label sequences given a particular observation sequence.

Lafferty [7] defined the probability of a particular label sequence y given observation sequence x to be a normalized product of potential functions, each of the form as:

$$p(y|x,\lambda) = \frac{1}{Z(x)} \exp(\sum_{j} \lambda j F j(y,x)) \tag{1}$$

where $F_j(y, x)$ is either a state function $s(y_{i-1}, y_i, x, i)$ or transition function $t_j(y_{i-1}, y_i, x, i)$, λ_j is the weight of indicating the precision of feature f_j, Z(x) is a normalization factor.

A set of real-valued features b(x, i) of the observation that should hold the model distribution is constructed to expresses some characteristic of the empirical distribution of the training data.

ဖြူ သည် ရန်ကုန် တွင် နေသည် ။

For example, in the above sentence which means "Phyu lives in Yangon", the feature observation of word ရန်ကုန် at position 3 is constructed as follow:

b(ရန်ကုန်,3)= $\begin{cases} 1 \; \textit{if the observation at position 3 is the word} \text{ "ရန်ကုန်"}, \\ 0 \; \textit{otherwise.} \end{cases}$

Each feature function takes on the value of one of these real-valued observation features b(x, i) if the current state (in the case of a state function) or previous and current states (in the case of a transition function) take on particular values.

4 Myanmar Language

4.1 Nature of Myanmar Language

Myanmar language is the official language of the Republic of the Union of Myanmar. Myanmar Language Commission (MLC) standardized that it is composed of nine parts of speech in Myanmar grammar such as noun, pronoun, adjective, verb, adverb, post-positional marker, particle, conjunction and interjection. It is written from left to right and usually with no space between words. Myanmar language is mainly characterized as a SOV (subject, object and verb) language; would probably defined as postpositional language and it is also regarded as a free order of word language which means that the part of speech of the word in the text can vary in accordance with its position in the sentence.

Like other languages, Myanmar Language also has ambiguity problem in both syntactic and sematic meanings. Statistical ways to solve most of the important issues in Myanmar Natural Language Processing have been applied but effective statistical approaches to Myanmar NER have not been tried yet.

4.2 Nature of NE in Myanmar Language

As described, how to perform the task of identifying names in Myanmar text is still challenging because of the nature of language mentioned above and also the ambiguity nature of EN types. The ambiguity of NE types may lead to problem in classifying named entities into predefined types.

For example, the word စံပယ် which means the flower jasmine can be the name of person or the name of road or the name of companies because Myanmar NE are given with no definite rule; any word can be NE, resulting problems in the task of NER.

In this paper, one of the machine learning approaches, CRF is applied to identify NE in Myanmar text.

5 CRF-Based NER

5.1 Data Preparation

The work flows of training and testing process for CRF based NER are shown in Fig. 1.

Firstly, data are collected and prepared for training and testing. Sentences written in Myanmar scripts are collected from Myanmar news articles websites. Data for training and testing are prepared through the process of segmentation, manually label the POS tags and NE tags. Then, template file for the training is created. In order to infer

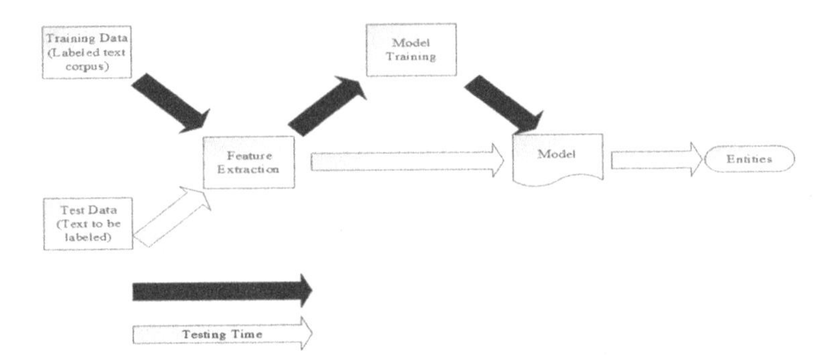

Fig. 1. NER work flow

whether a phrase or word is NE of not, it is dependent to attributes. Feature templates help CRF engine to form features from the attributes of the training data. Different kinds of features such as POS information, word-level features, sentence-level features and list look-up features, etc. can be used to train the model. In this work, word token and POS are only used as features. Feature template used in shown in Fig. 2.

မောင်မြနှင့်မောင်ကောင်းတို့သည်သူငယ်ချင်းများဖြစ်ကြသည်။

For the above sentence, which means that Mg Mya and Mg Kaung are friends, the training data format and the template features are defined as shown in Figs. 3 and 4 respectively.

By using the model file produced from the training process, test data is given as input to get the NE tagging. Test data is also preprocessed with the similar form of training data. Testing sentences are also segmented into tokens and POS tagged. Finally, the CRF NER model produced the result word tokens with NE tags labeled.

5.2 Named Entity Tagset Used

The list of NE types employed in this work is described in Table 1. 15 types of NEs are used as label.

```
#template
U00:%x[-2,0]        /* unigram feature templates */
U01:%x[-1,0]
U02:%x[0,0]
U03:%x[1,0]
U04:%x[2,0]
U05:%x[-2,0]/%x[-1,0]/%x[0,0]
U06:%x[-1,0]/%x[0,0]/%x[1,0]
U07:%x[0,0]/%x[1,0]/%x[2,0]
U08:%x[-1,0]/%x[0,0]
U09:%x[0,0]/%x[1,0]
U10:%x[-2,1]
U11:%x[-1,1]
U12:%x[0,1]
U13:%x[1,1]
U14:%x[2,1]
U15:%x[-2,1]/%x[-1,1]
U16:%x[-1,1]/%x[0,1]
U17:%x[0,1]/%x[1,1]
U18:%x[1,1]/%x[2,1]
B                  /* bigram feature template */
```

Fig. 2. Feature template file

မောင်မြ	PRP	PERSON
နှင့်	CONJ	O
မောင်ကောင်း:PRP		PERSON << current token
တို့	PART	O
သည်	PPM	O

Fig. 3. Training data

Template	Expanded feature
%x[0,0]	မောင်ကောင်း
%x[0,1]	PRP
%x[-1,0]	နှင့်
%x[-2,1]	PRP
%x[0,0]%x[0,1]	မောင်ကောင်း/PRP

Fig. 4. Template feature

Table 1. NE types.

NE tags	Meaning	Example
PNAME	Name of Person	မိုးမိုးအေး
LOCATION	Name of Location	ရန်ကုန်မြို့
ORGANIZATION	Name of Organization	ဘီဘီစီ
GOVERNMENT	Name of Government	နိုင်ငံခြားရေးဝန်ကြီးဌာန
MATERIAL	Name of Material Things	၆ဂကေဗ္ဗိအေမီးစက်
TIME	Hour, Minutes and Seconds	၆နာရီ၊၇မိနစ်၊၁၀စက္ကန့်
DATE	Date	စနေနေ့
MONTH	Names of months	ဧပြီလ
YEAR	Year format	၂၀၁၆ခုနှစ်
SEASON	Seasons of the Year	မိုးရာသီ
DISTANCE	Distance of places	၅မိုင်
MEASURE	Measurement	၃၀ ဒီဂရီ
MONEY	Money	၁၀၀၀၀၀ကျပ်
NUMBER	Numerical values	၅၀
PERCENT	Percentage values	၅%၊၅ရာခိုင်နှုန်း

6 Experiments and Evaluation

The total of 500 sentences including about 16000 words has been used as training set for the CRF-based NER system. 100 test sentences including 2500 words of test data have been tested with the CRF model. Different standard measures such as Precision, Recall and F-measure are used for evaluation parameters.

Recall is the ratio of number of NE words retrieved to the total number of NE words actually present in the file.

Precision is the ratio of number of correctly retrieved NE words to the total number of NE words retrieved by the system.

The F-measure is computed by the weighted harmonic mean of precision and recall.

မောင်အေးချမ်းတို့ သူငယ်ချင်းတစ်စုသည် ရန်ကုန်ကွန်ပျူတာတက္ကသိုလ်တွင်၂၀၀၅ဒီဇင်ဘာမှစ၍အတူတကွပညာသင်ကြားနေကြသည်။

For the above input sentence, which means that Mg Aye Chan and his friends are studying together at the University of Computer Studies, Yangon from December, 2005, the NER result is shown in below. The result will show that Mg Aye Chan is the name of person, University of Computer Studies, Yangon is the name of organization, 2015 is year and December is the name of month.

မောင်အေးချမ်းတို့သူငယ်ချင်းတစ်စုသည်ရန်ကုန်ကွန်ပျူတာတက္ကသိုလ်တွင် ၂၀၀၅ဒီဇင်ဘာမှစ၍အတူတကွ
ပညာသင်ကြားနေကြသည်။

The precision of 50 %, recall of 57.14 % and F-measure got 53.33 for the given data. It is because of the very little amount of training data and test data. In the future, more and more training data are going to be trained with more NE types.

7 Conclusions

Named Entity Recognition is the key sub task of most of the current hot researches in NLP. Current NER for Myanmar Language has been performed in rule-based approach and cannot be used effectively. This paper is intended to approach NER for Myanmar in a statistical way by using CRF. A baseline NER system that could be used for further development work to improve NER for Myanmar Language is going to be modeled. This work is proposed to be a statistical NER and also intended to use in statistical Myanmar-English machine translation system.

Acknowledgment. This work is partly supported by the ASEAN IVO Project "Open Collaboration for Developing and Using Asian Language Treebank".

References

1. Swe, T.T., Htay, H.H.: A hybrid method for Myanmar named entity identification and transliteration into English. Master thesis, University of Computer Studies, Yangon, October 2009
2. Myint, T., Thida, A.: Name entity recognition and transliteration in myanmar text. Ph.D. Research, University of Computer Studies, Mandalay, May 2014
3. Morwal, S., Jahan, N.: Named entity recognition using hidden markov model (HMM): an experimental result on Hindi, Urdu and Marathi languages. Int. J. Adv. Res. Comput. Sci. Softw. Eng. **3**(4) (2013)
4. Vijayakrishna, R., Sobha, L.: Domain focused named entity recognizer for tamil using conditional random fields. In: Proceeding of the IJCNLP-08 Workshop on NER for South East Asian Languages, January 2008

5. Sobhana, N.V., Mitra, P., Ghosh, S.K.: Conditional random field based named entity recognition in geological text. Int. J. Comput. Appl. (0975-8887) **1**(3), 119–125 (2010)
6. Wibawa, A., Purwarianti, A.: Indonesian named-entity recognition for 15 classes using ensemnle supervised learning. In: 5th Workshop on Spoken Language Technology for Under-resourced Languages, SLTU 2016, 9–12 May 2016
7. Lafferty, J., McCallum, A., Pereira, F.: Conditional random fields: probabilistic models for segmenting and labeling sequence data. In: ICML 2001 Proceedings of the Eighteenth International Conference on Machine Learning, pp. 282–289 (2001)
8. Sutton, C., McCallum, A.L.: An introduction to conditional random fields. Found. Trends Mach. Learn. **4**(4), 267–373 (2012)
9. Hiremath, P., Shambhavi, B.R.: Approaches to named entity recognition in Indian languages: a study. Int. J. Eng. Adv. Technol. (IJEAT) **3**(6), August 2014. ISSN: 2249–8958
10. Jayan, J.P., Rajeev, R.R., Sherly, E.: A hybrid statistical approach for named entity recognition for Malayalam language. In: International Joint Conference on Natural Language Processing, Nagoya, Japan, 14–18 October 2013, pp. 58–63 (2013)
11. Wallach, H.M.: Conditional random fields: an introduction. University of Pennsylvania CIS Technical Report MS-CIS-04-21, 24 February 2004
12. Saha, S.K., Ghosh, P.S., Sarkar, S., Mitra, P.: Named entity recognition in Hindi using maximum entropy and transliteration. Polibits **38**, 33–42 (2008)
13. Jeyashenbagavalli, N., Srinivasagan, K.G., Suganthi, S.: An automated system for Tamil named entity recognition using hybrid approach. In: International Conference on Intelligent Applications (2014)

Back-Propagation Neural Network Approach to Myanmar Part-of-Speech Tagging

Hay Mar Hnin[1], Win Pa Pa[1(✉)], and Ye Kyaw Thu[2]

[1] Natural Language Processing Laboratory,
University of Computer Studies, Yangon, Myanmar
haymarhnin123@gmail.com, winpapa@ucsy.edu.mm
[2] Language and Speech Science Research Laboratory,
Waseda University, Tokyo, Japan
wasedakuma@gmail.com

Abstract. Part-of-Speech (POS) tagging is the process of assigning a POS label to each of a sequence of words. It is also a lowest level of syntactic analysis and useful for many natural language processing (NLP) tasks such as subsequent syntactic parsing and word sense disambiguation. We developed an annotated corpus and POS tagger for Myanmar language based on back-propagation neural network (BPNN) model. In our experiments, BPNN model is trained with 3gram, 4gram and 5gram. The results show that the BPNN model with 4 g is able to achieve considerable higher F-scores on the POS tagging task than 3 g and 5 g models for both close and open test sets. Moreover, BPNN POS tagging approach performed better than proposed HMM with rule based.

Keywords: Part-of-Speech (POS) Tagging · Back-propagation Neural Network (BPNN) · Hidden Markov Model (HMM) · Myanmar language

1 Introduction

Many words in natural languages are ambiguous and have more than one POS tag. POS tagging or learning syntactic categories of words, which disambiguates ambiguity of words in the context of the sentence. It is an important preprocessing step for natural language processing such as speech recognition, test to speech synthesis, parsing, information retrieval and machine translation. Although there are numerous machine learning approaches already proposed for POS tagging, Myanmar POS tagging is still in its early stages. Myanmar language is one of the under-resourced languages and in particular POS tagged corpora are scare. As far as the authors are aware there have been only three published methodologies (two supervised and one unsupervised) for Myanmar language POS tagging and all of them are HMM based techniques. Khine Zin (2009) [1] proposed two approaches; one is rule based and another is HMM with rule based. 1 million words corpus manually tagged with their defined 36 POS tags is used for their study. 2gram HMM with rule based approach achieved 97.56 % accuracy and

J. Pan et al. (eds.), *Genetic and Evolutionary Computing*, Advances in Intelligent Systems and Computing 536, DOI 10.1007/978-3-319-48490-7_25

significantly higher than accuracy 89.56 % of rule based approach. Phyu Hninn Myint et al. (2011) [2] proposed similar 2 g HMM for training and used Viterbi algorithm for decoding. Post editing process with lexical rules were done on output of HMM for getting better performance. They prepared 1,000 sentences of news domain corpus that was manually tagged with their defined 20 POS tags for training and used 3,000 words lexicon for post editing. 3 test sets that contained 3,000 untagged sentences for each test set were used for evaluation. Their experimental results achieved over 90 % accuracy for all 3 test sets. Ye Kyaw Thu et al. (2014) [3] attempted to increase statistical machine translation (SMT) performance for Myanmar, by applying POS tags induced with a unsupervised novel bilingual infinite HMM (B-iHMM) approach. They extended iHMM approach for bilingual POS tags based on Tamura et al. (2013) [4] non-parametric Bayesian method for inducing POS tags from dependency trees. Their experimental results show that phrase based SMT with words that were tagged with induced POS tags gains over 2 points in BLEU for Myanmar to English translation.

The main contribution of this paper, is the first study of BPNN POS tagging for Myanmar language. Moreover we developed the manually POS tagged Myanmar corpus under general domain for POS tagging experiments. We did evaluation with one closed test set and two opend test sets on trained 3-gram, 4-gram and 5-gram BPNN models and achieved highest F-scores 99 with 4-gram and 5-gram closed test sets and 80 with 3-gram open test set1. Futhermore, the experimental comparison results with two open test sets proved BPNN outperforms proposed HMM with rule.

2 Back-Propagation Neural Network (BPNN)

Back-propagation (BP) is used to calculate gradients of a loss function with respect to given parameters such as weights. Computing procedure is based on the application of the chain rule and calculation proceeds backwards through the network with respect to the computations performed forward to compute the loss itself. The goal of BP is to assign correct weights and minimizes a continuous loss function or objective function. This algorithm of explicit, efficient error BP in arbitrary, discrete, possible sparsely connected, NN-like networks was originally introduced in the 1970 master's thesis of Linnainmaa [8,9]. David et al. [10] proposed a new learning procedure of BP that proved the ability to create useful new features distinguishes BP from earlier, simpler methods such as the perceptron-convergence procedure. The proposal of David et al. [10] made BP algorithm become the workhorse of learning in neural networks. The algorithm can be decomposed into four steps (1) Feed-forward computatin (2) Backpropagation to the output layer (3) Backpropagation to the hidden layer and (4) Weight updates. It will keep running until the value of the error function has become sufficiently small [11].

As BPNN is a supervised learning technique, input and output classes are known. For that, output class label are manually predefined before training is started that are actual output class label result. Output classes of output layer for network are networks output classes. If actual and network output class are not match, this network is backpropagated from output layer to closed hidden layer to next hidden layer to input layer. Backpropagated from output layer to hidden layer is second stop backpropagation to output layer. In this step, calculate for all output nodes of output layer. By using these error values for output nodes of output layers, calculate error values for all nodes of hidden layer, this is third step backpropagation to hidden layer. Finally weight are updated. Initial weights and bias values are randomly initialized from 0 to 1 or –0.5 to 0.5 or –1 to 1. In this step, weight and bias values are updated until error from output layer and hidden layer are low enough threshold values of predefined minimum error values. It iteratively processed for network until error is small enough or count of iterative reached predefined maximum epoch for network. To construct the network, need to encode input and output for input data. By using represented input and output values, randomly initialized weight and bias values, predefined maximum epoch, minimum error, learning rate and momentum are used to train network. We used the maximum epoch = 5000, learning rate = 0.5 and momentum = 0.01 since they are the best for this BPNN trainings. An architecture of 3gram BPNN for Myanmar language POS tagging is shown in Fig. 1.

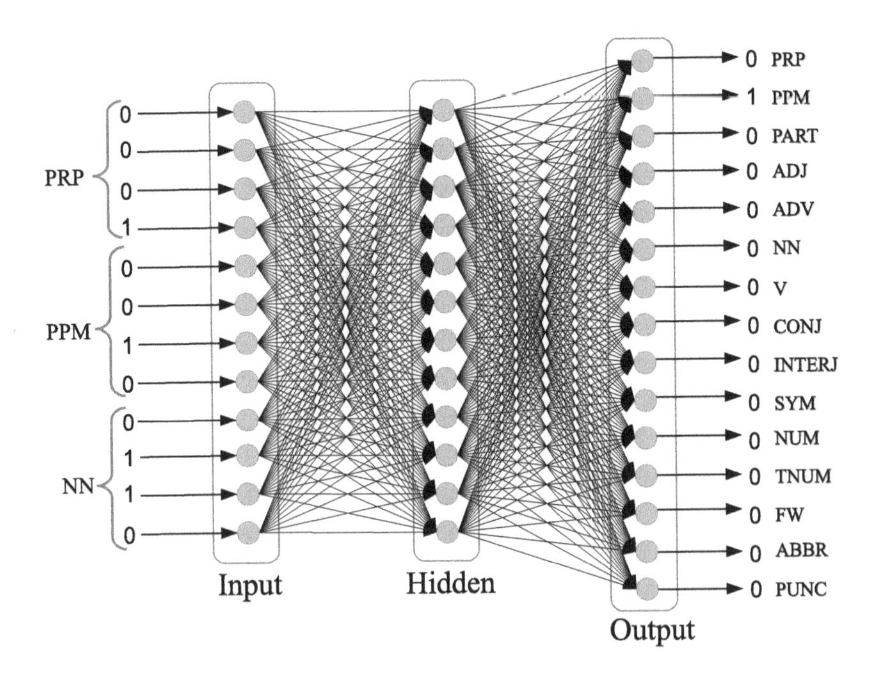

Fig. 1. Back-propagation neural network for POS tagging

3 Data Preparation

3.1 Corpus Building

The corpus is manually collected 5,000 sentences (82,892 words) from different Myanmar language resources such as newspapers (including market economy news, international news, instructional program news), books, journal and Wikipedia (Myanmar) [12–15,17]. Currently most of the Myanmar users are using partial-unicode encoding font (i.e. Zawgyi font) and we have to converted into Unicode encoding [16]. Word segmentation and tagging with defined POS tags for each word was done manually.

3.2 Closed and Open Test Set

There are three sets of test data with equal size, 250 sentences each: one close data set (3,060 words) and two open data sets. Open test set 1 (4,362 words) is the same domain with the training data. Open test set 2 (6,944 words) is selected from the Asian Language TreeBank Corpus [7] which was built from Wiki-news in English to Asian languages treebank including Myanmar language and they are quite different from training data in length, domain, style and POS patterns.

3.3 Data Representation of Input and Output

In this POS tagging, 3-gram, 4-gram and 5-gram representation format are used to encode input sentence as binary format. Each node in input layer is own each bit of input words representation. 12 input nodes for 3-gram (refer Fig. 1), 16 input nodes for 4-gram and 20 input nodes for 5-gram are defined from input words for training network. For example, 4 bit representation of 3 g POS tag input for the word သွား (go in English) of သူ/PRP သည်/PPM ကျောင်း/NN သို့/PPM သွား/V သည်/PPM ။/PUNC sentences is 0010 0111 0010. Here, the middle binary number 0111 is the POS tag of သွား the first binary number 0010 is the POS tag of previous word သို့ and the last binary number 0010 is the following word of သွား, သည် (refer Table 1).

3.4 Word Segmentation

In Myanmar texts, words composed of single or multiple syllables are usually not separated by white space. Spaces are used for easier reading and generally put between phrases, but there are no clear rules for using spaces in Myanmar language. Therefore, word segmentation is a necessary prerequisite for POS tagging. Myanmar word can usually be identified by the combination of root word, prefix and suffix. For example, a Myanmar word 'သွားသည်' (go) can be segmented into two units: one is root verb, 'သွား' and the other unit is postpositional marker 'သည်' and forms a complete verb. Conditional Random Fields Myanmar word

Table 1. An example of four bit binary representation for input 3 g word and output POS tag.

Input Word(Binary Data)		Output POS-tag (Binary Data)	
သူ	(0000 0001 0010)	PRP	(100000000000000)
သည်	(0001 0010 0110)	PPM	(010000000000000)
ကျောင်း	(0010 0110 0010)	NN	(000001000000000)
သို့	(0110 0010 0111)	PPM	(010000000000000)
သွား	(0010 0111 0010)	V	(000000100000000)
သည်	(0111 0010 1111)	PPM	(010000000000000)
။	(0010 1111 0000)	PUNC	(000000000000001)

segmentation [5] is used for this POS tagging. The segmented example Myanmar sentence, (Every person is responsible to maintain the environment.) is shown as follow:

Unsegmented sentence: လူတိုင်းသဘာဝပတ်ဝန်းကျင်ကိုထိန်းသိမ်းသင့်ပါသည်။

Word segmented sentence:
လူတိုင်း_သဘာဝ_ပတ်ဝန်းကျင်_ကို_ထိန်းသိမ်း_သင့်_ပါ_သည်_။

Here, ထိန်းသိမ်းသင့်ပါသည် is complete verb form that means "should maintain", but we segmented this into four parts ထိန်းသိမ်း_သင့်_ပါ_သည် that will be easy to extract the important information.

3.5 Part-of-Speech Tag-Set

There are 10 POS tags for Myanmar language that is generally defined by Myanmar Language Commission [6]. They are Noun, Pronoun, Adjective, Adverb, Verb, Post-positional marker, Particles, Conjunction, Interjection and Punctuation. There is no preposition in Myanmar language and the similar words are appear after Nouns and they are called postpositional marker. In this paper, 5 more POS tags are defined, to assist more accurate data information for other NLP applications, such as Machine Translation, Information Retrieval, Text summarization. The new tags are symbol, number, text number, foreign word and abbreviation. An example of POS tagged sentence is shown as follows:

လူ|NN တိုင်း|PART သဘာဝ|NN ပတ်ဝန်းကျင်|NN ကို|PPM ထိန်းသိမ်း|V သင့်|PART ပါ|PART သည်|PPM ။|PUNC

The detail information of defined fifteen POS tags are shown in Table 2

Table 2. Part-of-Speech Tag-set for Myanmar.

POS Tag	Brief Definition	Examples
Pronoun (PRP)	takes the place of a noun	သူ(he), ကျွန်တော်(I), ငါ(I), နင် (you), မင်း (you), ထို (that)
Noun (NN)	names, activities, events, persons, objects, abstract ideas	ခဲတံ (pencil), ကျန်းမာရေး (health), မြန်မာ (Myanmar)
Adjective (ADJ)	Can be before or after a noun and usually ends with သော သည့် မည့်	လှပ_သော (beautiful), ထင်ရှား_သည့် (obvious)
Adverb (ADV)	Always before a verb, sometimes in repeated syllables patterns and usually ends with စွာ	လှပ_စွာ(beautifully), လှလှပပ (beautifully), အလွန် (very)
Verb (V)	always suffixed with one or more particles to show the tenses, politeness etc.	ရှိ_သည် (has), သွား_မည် (will go), လုပ်_ပြီး_ပြီ (has done)
Conjunction (CONJ)	joins words, phrases or sentences	သောကြောင့် (therefore), ၍ (and)
Postpositional Marker (PPM)	after pronoun or noun, similar to preposition in English but no exact words to express English	သည်, က, မှာ, သို့, ကို, အား, ဖြင့်
Particle (PART)	suffixed or prefixed to nouns, verbs, adjectives and adverbs	များ (s/es), ခဲ့ (ed), နိုင် (can)
Interjection (INTERJ)	is expressing of emotion	အို (Oh!), ဟေး (Hey!)
Symbol (SYM)	braces, currencies, special symbols and mathematical operators	&, $, %, {, }, ¥, <, \, &, *etc.
Number (NUM)	numeric word	၀၁၂၃၄၅၆၇ (01234567), ၀.၂၃ (0.23)
Text Number (TNUM)	words for number like one for 1, two for 2 in English	တစ်ထောင် (one thousand), ငါးဆယ့်သုံး (fifty three)
Foreign Word (FW)	Transliteration of foreign words	ဝင်းဒိုး (Windows), အက်တမ် (atom), နိုက်ထရိုဂျင်ပါအောက်ဆိုက် (nitrogen peroxide)
Abbreviation (ABBR)	generally initial of words in Myanmar languages	အထက (abbreviation of basic education high school)
Punctuation (PUNC)	divide words, phrases or sentences	။ (as a full stop in English), ၊ (as a comma in English)

4 Evaluation Criteria

The POS tagging performance of BPNN was measured using the commonly used precision (Eq. 2), recall (Eq. 3), and F-score (Eq. 1) defined as follows.

$$\text{F-score} = \frac{2 \times Recall \times Precision}{Recall + Precision} \tag{1}$$

$$Precision = \frac{\#of\ correct\ tokens}{\#of\ tokens\ in\ test\ corpus} \tag{2}$$

$$Recall = \frac{\#of\ correct\ tokens}{\#of\ tokens\ in\ system\ output} \tag{3}$$

5 Experimental Results

The performance of BPNN POS tagging is evaluated on different input 3-gram, 4-gram and 5-gram. HMM with rule approach [1] is used as baseline and BPNN results are compared with the baseline results. The detail results are shown in Table 3 and it is clear that almost all of BPNN results outperforms HMM with rule. The computational process of HMM with rule is first detecting input sentence with their defined POS tagging rules and if the system cannot handle with rules they applied HMM approach [1]. And thus, we will use "rule+HMM" annotation for [1]. rule+HMM approach has almost the same performance with 5gram BPNN for Open test set 1. The results of Open test set 1 show better than Open test set 2. Among all BPNN experiment, 4 g has the highest training accuracy and the highest test result for Open test set 2.

Table 3. Part of speech tagging performance using 3 different n-grams with BPNN models.

	N-grams	Closed test set	Open test set 1	Open test set 2
Precision	3-gram	0.9767	0.7397	0.6172
	4-gram	0.9912	0.8051	0.7725
	5-gram	0.9863	0.6872	0.7236
	rule+HMM	–	0.6876	0.6116
Recall	3-gram	0.9732	0.7283	0.6061
	4-gram	0.9912	0.8029	0.7734
	5-gram	0.9863	0.6767	0.7245
	rule+HMM	–	0.6546	0.6067
F-Score	3-gram	0.9700	0.7300	0.6100
	4-gram	0.9900	0.8000	0.7700
	5-gram	0.9900	0.6800	0.7200
	rule+HMM	–	0.6700	0.6100

6 Discussion

According to Table 4, the number of new POS patterns of open test data 2 are more than open test data 1. As we mentioned in Sect. 3.2, domain of open test data is pure news domain [7] (i.e. translated from English wikinews to Myanmar) and generally, sentences contained foreign words and longer than both training and open test data 1. We assume these factors effects the experimental results between open test data 1 and 2 (refer Table 3). The experimental results proved that our BPNN POS tagging with 4-gram outperforms proposed rule+HMM approach for both open test 1 and open test 2. Interesting point is F-Score of

BPNN POS tagging with 3-gram and 5-gram are comparable with rule+HMM (refer F-score of open test set 1 and 2 in Table 3). From our studies, we found that demerit point of rule+HMM was applying the small number POS tagging rules at first on unambiguous words and follow by POS tagging with HMM modeling. The performance might be increase if we changed the processing order and using rules as a post-editing for errors.

We did POS tagging error analysis based on the experimental results and we found out-of-vocubalary (OOV) words are tagged as foreign word (FW POS tag), for example တောင်ငူနေပြည်တော်|FW (Taungoo palace in English). Another prominent errors are tagging Myanmar text number words with noun tag, for example ခုနစ်သိန်းကိုးသောင်း|NN (seven hundred ninety thousand in English). From our studies, we found that these errors are relating to frequency of POS tag patterns in the training data of 3-gram, 4-gram and 5-gram.

Table 4. Number of POS pattern comparison.

	Training data	Open test 1	Open test 2
	Total pattern	New pattern (%)	New pattern (%)
3-gram	1,192	10.85	13.04
4-gram	4,278	16.26	22.30
5-gram	10,652	29.61	37.56

7 Conclusion

This paper aims to show that BPNN can work well for Myanmar POS Tagging. As shown in the experimental results, BPNN outperforms the existing Myanmar POS Tagging approach. This paper presents the first result of BPNN POS tagging for Myanmar language. The experiments are done on both close and two open test sets. Although open Test set 2 contains some spoken style sentences, many OOV words and new POS tag patterns, BPNN models are able to provide around 80 % accuracy and higher than proposed HMM with rule. We plan to make detail analysis on disambiguation errors of ambiguous Myanmar words based on morphological information in the near future.

References

1. Zin, K.K.: Hidden markov model with rule based approach for part-of-speech tagging of Myanmar language. In: Proceedings of the 3rd International Conference on Communications and Information Technology, Vouliagmeni, Athens, Greece, pp. 123–128 (2009)
2. Myint, P.H., Htwe, T.M., Thein, N.L.: Bigram part-of-speech tagger for Myanmar language. In: Proceedings of 2011 International Conference on Information Communication and Management, Singapore, pp. 147–152 (2011)
3. Thu, Y.K., Tamura, A., Finch, A., Sumita, E., Sagisaka, Y.: Unsupervised POS tagging of low resource language for machine translation. In: Proceedings NLP2014, Hokkaido, Sapporo, Japan, pp. 590–593 (2014)
4. Tamura, A., Watanabe, T., Sumita, E.: Part-of-speech induction in dependency trees for statistical machine translation. In: Proceedings of ACL 2013, Sofia, Bulgaria, pp. 841–851 (2014)
5. Pa, W.P., Thu, Y.K., Finch, A., Sumita, E., Genetic, E.C.: Proceedings of the Ninth International Conference on Genetic and Evolutionary Computing, ICGEC2015, Yangon, Myanmar, 26–28 August 2015, vol. II, pp. 447–456 (2015)
6. Thadda, M.: Myanmar Language Commission. Ministry of Education, Myanmar (2005)
7. Thu, Y.K., Pa, W.P., Utiyama, M., Finch, A., Sumita, E.: Introducing the Asian Language Treebank (ALT). In: Proceedings of the Tenth International Conference on Language Resources and Evaluation (LREC 2016), pp. 1574–1578, May 2016
8. Seppo, L.: The representation of the cumulative rounding error of an algorithm as a Taylor expansion of the local rounding errors. Master thesis, University of Helsinki (1970)
9. Seppo, L.: Taylor expansion of the accumulated rounding error. BIT Numer. Math. **16**(2), 146–160. Kluwer Academic Publishers (1976)
10. Rumelhart, D.E., Hinton, G.E., Williams, R.J.: Learning representations by back-propagating errors. In: Neurocomputing: Foundations of Research, pp. 696–699. MIT Press (1988)
11. Rojas, R., Networks, N.: A Systematic Introduction. Springer, New York (1996)
12. The commerce journal (**စီးပွားရေးသတင်းစာဂျာနယ်**), http://www.commercejournal.com.mm
13. The wikipedia Myanmar. http://my.m.wikipedia.org
14. Burmese classic. http://www.burmeseclassic.com/
15. Online Burmar/Myanmar library. http://www.burmalibrary.org/show.php?cat=1449&lo=&sl=
16. The unicode standard 9.0 of Myanmar. http://www.unicode.org/charts/PDF/U1000.pdf
17. Yu, U.T.: **အမှတ် (၁) အခြေခံပညာအထက်တန်းကျောင်း၊ သံဖြူဇရပ်မြို့ နှစ် (၅၀) ပြည့် ရွှေရတုအထိမ်းအမှတ်မဂ္ဂဇင်း , မဂ္ဂဇင်း ထုတ်ဝေရေး ဆပ်ကော်မတီ** September 2012

Efficient Algorithm for Finding Aggregate Nearest Place Between Two Users

Su Nandar Aung(✉) and Myint Myint Sein

University of Computer Studies, Yangon, Myanmar
{sunandaraung, myint}@ucsy.edu.mm

Abstract. Keyword search over a large amount of data is an important operation in a wide range of domains. Spatial keyword search on spatial database has been well studied for years due to its importance to commercial search engines. Specially, a spatial keyword query takes a user location and user-supplied keywords as arguments and returns object that is nearest k objects from user current location and textually relevant to the user required keyword. Geo-textual index structure plays an important role in spatial keyword querying. This paper proposes the geo-textual index structure that intends to reduce unnecessary cost in processing spatial keyword queries and searching time for required results. The proposed index is used for searching most relevance results between two users that is based on the most spatial and textual relevance to query point and required keyword within given range. It can search the required result point with minimum IO costs and CPU costs. In this system, we also discuss how to answer inconsistencies and errors in the user's typed queries.

Keywords: Spatial keyword queries combined index · Proposed index construction · Problem statement · ANN keyword search algorithm

1 Introduction

How to efficiently index and search location-specific information is being a key problem for location based search engines. A straightforward approach is to treat geographical words which represent location information as common keyword, and to retrieve web pages with specified location names in the same way to keyword matching. However, simple keyword matching neglects underlying spatial relationships, therefore, does not support advanced spatial queries. To solve the problem, it is necessary to design an efficient index structure that considers both spatial and textual textures of web pages. This development calls for techniques that enable the indexing of data that contains both text descriptions and geo-locations in order to support the efficient processing of ANN keyword query that take a geo-location and a set of keywords as arguments and return relevant content that matches the arguments. The spatial relevance is measured by the distance between the location associated with the candidate document to the query location, and the textual relevance is computed in the same way as in traditional search engines. ANN keyword query is being supported in real-life applications. In these systems, inconsistencies and errors can exist in user queries and data. For example, a user might be looking for a restaurant close to their

J. Pan et al. (eds.), *Genetic and Evolutionary Computing*, Advances in Intelligent Systems and Computing 536, DOI 10.1007/978-3-319-48490-7_26

location. The corresponding query is "Finding nearest restaurant to the location (16.774301791/96.159639358)". This query returns the lists of restaurant that are the nearest neighbors of that location.

Sometime, users may not know the exact spelling of the keyword that they want to look for. As query keyword exist spelling errors, the query returns the incorrect answers to the users. Finding relevant answers to such queries is also important. Various types of geo-textual indices have been proposed in recent years.

A scalable index structure solution should satisfy three requirements:

(i) The index requires low maintenance cost and should be efficient for highly frequent insert operations,
(ii) The index shows effective storage utilization.
(iii) The index answers queries efficiently.

In the sate-of-the-art solutions, the most popular one is to embed the inverted index into the R-tree and its variants, named IR tree. The IR-tree can take advantage of spatial and textual information simultaneously in the pruning stage. When handling a large dataset, the IR-tree suffers from two main drawbacks. First, as the data objects in the R-tree can be overlapping and covering each other, the search process in the R-tree might suffer from unnecessary node visits and higher IO cost. Second, the IR-trees suffer from high update cost. Each node has to maintain an inverted index for all the keywords of documents associated with this node's MBR. When a node is full and split into two new nodes, all the textual information in the node has to be re-organized. As the R-tree need to reorganized, it suffers from higher CPU costs. This system intends to reduce IO costs, CPU costs and searching time for ANN keyword search. Aggregate nearest neighbor keyword search algorithm is developed using proposed index structure. Moreover, this system also takes into account the inconsistencies and errors of the user's typed keyword using approximate string matching method. Then it evaluate the searching time of the proposed system based on the user's input for all query types. In this experiment, it can be seen that using proposed index structure is faster and less IO and CPU costs than the index structure that combines R-tree or its variants and inverted index.

2 Related Works

In [4], they proposed and solved aggregate nearest neighbor (ANN) queries in spatial databases. In this article, they proposed the novel problem of aggregate nearest neighbor retrieval, a generalized form of NN search, where there are multiple query points and the optimization goal depends on an input function. In [12], they study an interesting generalization of nearest neighbor search and the processing of such queries for the case where the position and accessibility of spatial objects are constrained by spatial (e.g., road) networks. They consider alternative aggregate functions and techniques that utilize Euclidean distance bounds, spatial access methods, and/or network distance materialization structures. H. Htoo [11] proposed a powerful method that can be adapted to the verification phase, named a single-source multi-target A* (SSMTA*) algorithm which simultaneously searches the shortest paths from a query point to multiple target points. In [14], they propose an added flexibility to the query definition, where the similarity is an aggregation over the distances between p and any subset of objects in Q. Next, they present algorithms for answering Flexible aggregate similarity search (FANN) queries exactly and approximately. It also returns near optimal answers with guaranteed constant factor approximations in any dimensions. Aggregate k-Nearest Neighbor (k-ANN) queries are required to develop a new promising Location-Based Service (LBS) which supports a group of mobile users in spatial decision making.

To overcome the problem, [18] presented a procedure for computing approximate results of k-ANN queries. Various types of spatial keyword queries have been proposed. For spatial keyword search, the index structure is created for both spatial and textual relevance. Most index structures [10, 13, 17] use R-tree and its variants as spatial index and inverted file for text index. They all combine both indices depending on the combination schemes [17]. Among them [8] integrates signature file instead of inverted file into each node of the R-tree. Inverted file-R*tree (IF-R*) and R*-tree-inverted file (R*-IF) [7] are two geo-textual indices that loosely combine the R*-tree and inverted file. S. N. Aung [1–3] uses R*-tree for spatial index and inverted file for text index. In [5] the posting list of term contains all its term bitmaps rather than documents. The IR tree [6] creates each nodes of the R-tree with a summary of the text content of the objects in the corresponding sub tree. Li et al. [16] proposed an index structure, which is also called IR tree that stores one integrated inverted file for all the nodes.

3 Geo-Textual Index Construction

The geo-textual is built for the whole dataset for generating 2-dimensional points for data. Geo-textual index tree can be constructed with the procedure shown in BUILDGEOTEXTUAL algorithm. The proposed index structure is shown in Fig. 1.

Algorithm. BUILDGEOTEXTUAL (*P*, *depth*)

Input: a set of points that can handle coordinate points and corresponding value set *P* and the current depth .
Output: The root of a index storing *P*.

1. **if** *P* contains only one point
2. **then return** a leaf storing this point
3. **else if** *depth is even*
4. **then** Split *P* into two subsets with a vertical line *l* through the median latitude-coordinate of the points in *P*. Let P_1 be the set of points to the left of *l* or on *l*, and let P_2 be the set of points to the right of *l*.
   else Split *P* into two subsets with a horizontal line *l* through the median longitude-coordinate of the points in *P*. Let P_1 be the set of points below *l* or on *l*, and let P_2 be the set of points above *l*.
5. v_{left} ← BUILDGEOTEXTUAL (P_1, *depth*+1)
6. v_{right} ← BUILDGEOTEXTUAL (P_2, *depth*+1)
7. Create a node *v* storing *l*, make v_{left} the left child of *v*, and make v_{right} the right child of *v*.
   return *v*.

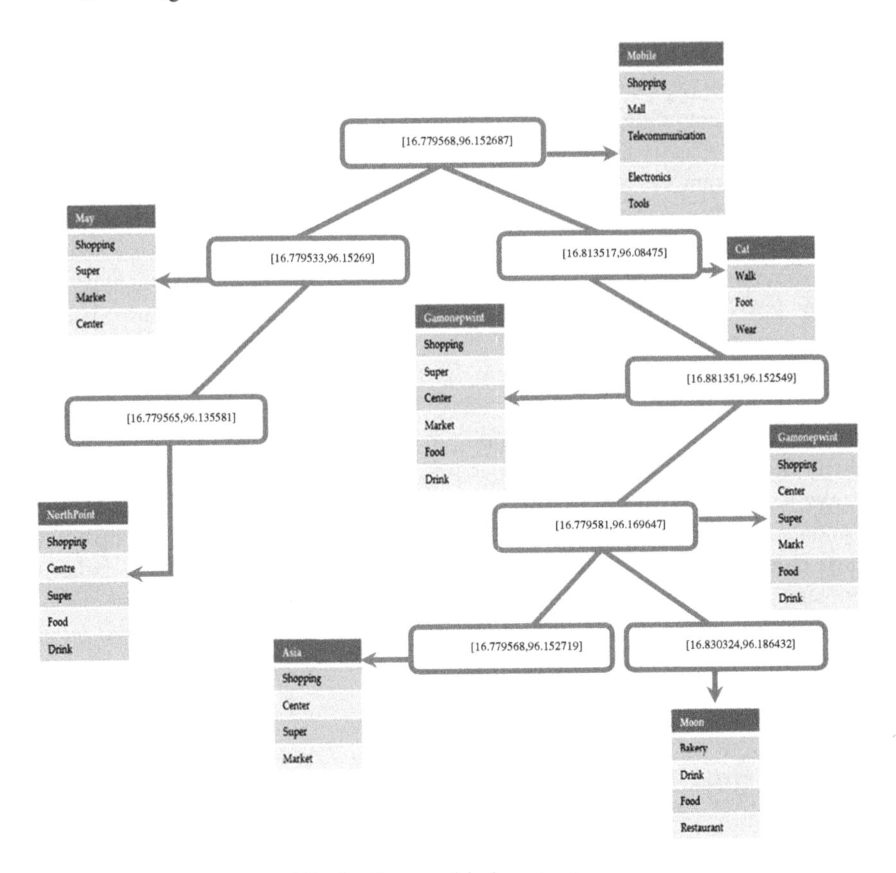

Fig. 1. Proposed index structure

4 Problem Statement and Algorithm for ANN Keyword Query

Aggregate Nearest Neighbor Keyword (ANN) query is defined by $q = (q1_{loc}, q2_{loc}, k)$ that take three parameters, where $q1_{loc}$ is a first user's location, $q2_{loc}$ is a second user's location and k is their point of interest name (keyword). According to these three parameters, the query returns one POI that is the minimum total distance from each query point to the POI. Distance from each user to each object in D is calculated by great circle distance and POI searching or keyword searching is using the same approach as KNN keyword query. Algorithm-2 shows the process of aggregate nearest neighbor keyword search between two users. It returns the minimum distance point between two users that contain their required keyword.

```
Algorithm-2. Aggregate Nearest Neighbor Keyword Search
Input: Query A;
Output: Minimum total distance point between two users that contains required keyword;
ANNKeywordSearch(Q)
Q: Query that contains Q.loc1, Q.loc2, Q.keyword;
pq: Priority Queue;
SEARCH: tuple kd-tree, bounding_points end tuple;
earth_radius <– 6371;
threshold <–(Q.keyword/2)-1;
points_distance <–GreatCircleDistance(Q.loc1,Q.loc2)/2;
midpoint<–MIDPOINT(Q.loc1,Q.loc2);
Bound_points<–BoundPoints(midpoint,points_distance,earth_radius);
pq.add(SEARCH(Bound_points));
while pq.size>0 do
        point<–pqremove();
        edit_distance<–HAMMING(point.keyword,Q.keyword);
        if (edit_distance ≤ threshold) then
                distance1<–GreatCircleDistance(point.loc,Q.loc1);
                distance2<–GreatCircleDistance(point.loc,Q.loc2);
                sum_distance<–distance1,distance2; pqResult<–(point,sum_distance);
        end if
return (pqResult.remove());
```

5 Approximate Keyword Matching

This system uses modified Hamming Distance to handle such errors even two strings are not the same lengths. The modified Hamming Distance is commonly used in three available query types in this system. The ***modified Hamming distance*** is modified the original Hamming distance to compare two strings that have arbitrary length i.e. have different string length. The modified Hamming distance calculates not only the difference between lengths of two strings but also the different between the numbers of position at which the corresponding characters are different to answer approximate string matching properly. Time for string matching is the same with the original Hamming distance but it is more useful for approximate keyword searching. The following algorithm shows the process of modified Hamming Distance.

```
Algorithm. Modified Hamming Distance
Input: An array s1[0....n-1] of n characters representing stored text and
        An array s2[0....m-1] of m characters representing query string
Output: distance between two strings
HAMMING (s1[0..,n-1], s2[0...,m-1])
distance <– 0;
if s1.length>s2.length then
        distance<–s1.length-s2.length;
        strlength<–s2.length;
else
        distance<–s2.length-s1.length;
        strlength<–s1.length;
end if
        j<–0;
while j<strlength do
        if s1[j]!=s2[j] then
                distance<–distance+1;
        end if
        j<–j+1;
return distance;
```

6 Architecture of Proposed System

The architecture of the proposed system is illustrated in Fig. 2. To provide the best users' requirements, it employs a client-based approach to retrieve the current location of the user, while we use the conventional client/server approach to process spatial keyword queries. The main objective of this system is to give most relevance answers to the users based on their current location and required keywords.

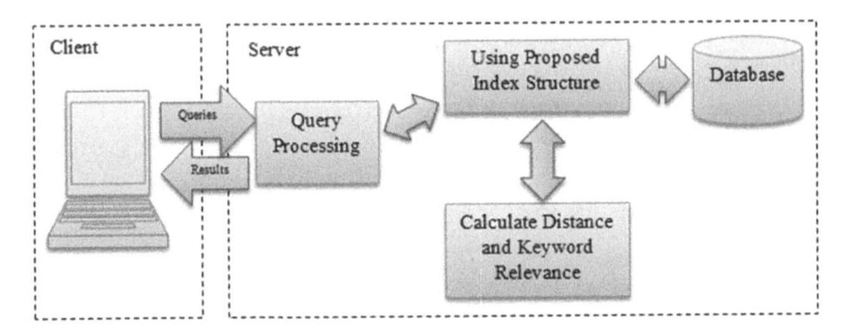

Fig. 2. System architecture

The system does not consider the services that are not relevance to the distance from the user's location and are not satisfy approximate keyword searching threshold for keyword search. The Google Maps infrastructure is applied to get the geo-location of the user by clicking on the Google Map interface. This approach allows us to get the user's current location without require any GPS devices that can provide the latitude and longitude position of the user. And then user need to specify required keyword, and desired query type and send that query to the server for processing by the HTTP post operation. On the server side, the spatial objects with text are index by the proposed geo-textual index structure. Depend on the user's querying data, the system calculate the distance and keywords relevance using the proposed index structure and the results are sent back to the client browser and displayed detail information for the results using Google Maps in the users' browser or client. The users can click the relevant objects shown on the map for more detailed information. The browser side in computers provides interfaces to users for generating required queries and viewing the returned objects on the web browser. This component presents a map and provides interactions with the map using the Google Maps API.

7 System Performance and Evaluation

This system is implemented and tested with about 5000 objects including 80 services. This system consists of client side and server side. Both are implemented with Java servlet on an Intel(R) Core(TM) i5 processor 2430 M (2.40 GHz) machine with 4 GB memory running on window-7 32-bits operating system. The available query was run 20 times and the average time was recorded as evaluation results.

This system evaluated the index construction depend on the number of available objects. The numbers of objects range from 1000, 2000, 3000, 4000 and 5000. This evaluation time tests to determine construction time of both spatial location and textual data pair.

$$T_{construct} = T_{dp} - T_{ic} \tag{1}$$

Where $T_{construct}$ = Total Index Construction Time
T_{dp} = Time of Spatial and Textual Pair
T_{ic} = Time of Index Construct
The index construction depends on objects count is shown in Fig. 3.

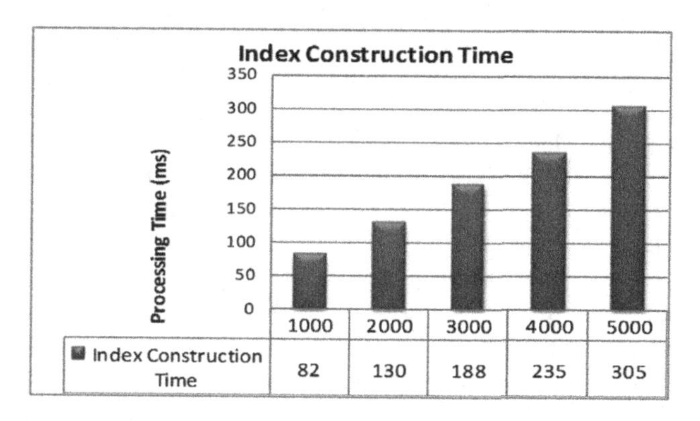

Fig. 3. Index construction time

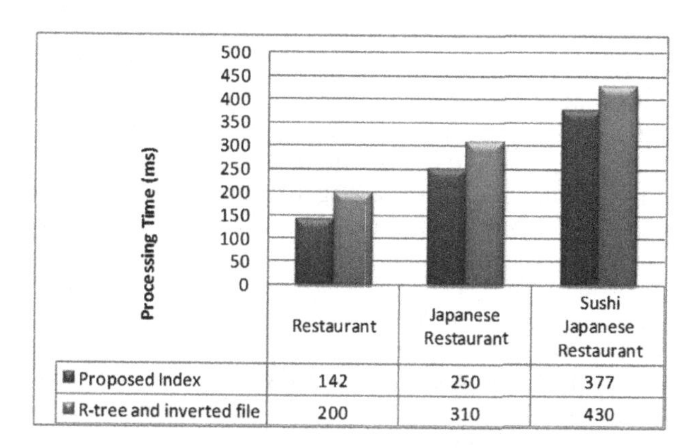

Fig. 4. ANN keyword search algorithm

The processing time for available query is recorded as start and end time of each step. Among total processing time, the time for searching and processing in proposed time is fewer than the time for selecting final result objects from database. So, this experimental results show that using proposed index structure is time efficient. The processing time varied depends on the query type. For each processing time is calculated as:

$$T_p = T_{index} + T_{db} \tag{2}$$

Where T_p is the total processing time and T_{index} is the computation time in proposed index. Figure 4 is described the searching times by using Eq. (2).

8 Conclusions and Further Extensions

The hybrid index structure is presented for aggregate nearest neighbor keyword query searching with minimum IO costs and CPU costs. This index structure can avoid searching in overlapping area. So it can reduce searching time in overlap area. Moreover, it can't cause node overflow, so it doesn't need to re-organize the textual data and spatial data. Many Further extensions can be considered for efficient hybrid index structure for spatial database. In addition, this system will consider the static user location, so the user location will be considered as a continuous moving object using the proposed index structure as a further work. Moreover, user's typed keyword language is allowed with Myanmar language, which is one of ongoing research works for spatial keyword search. Like keyword search in this system, type-ahead search can be considered that will return spatial objects whose names or descriptions are valid completions of the query string typed so far and which rank highest in terms of proximity to the user's location.

References

1. Aung, S.N., Sein, M.M.: Hybrid geo-textual index structure for spatial range keyword search. Proc. Comput. Sci. Eng. Int. J. (CSEIJ) **4**(5/6), 21–28 (2014)
2. Aung, S.N., Sein, M.M.: Geo-textual index structure for approximate keyword search within given range on spatial database. In: Proceedings of 7th International Conference on Science, Technology, Engineering and Management (ICSTEM 2015), Singapore, pp. 49–54, January 2015
3. Aung, S.N., Sein, M.M.: Modify compact R-tree dynamic index structure for myanmar GIS database. In: ICCA2014, Yangon, Myanmar, pp. 201–204, February 2014
4. Aung, S.N., Sein, M.M.: Efficient combined index structure for K-nearest neighbors keyword search on spatial database. In: Proceedings of the 13th International Conference on Computer Applications, Yangon, Myanmar, pp. 324–328, February 2015
5. Aung, S.N., Sein, M.M.: K-nearest neighbours approximate keyword search for spatial database. Proc. Int. J. Adv. Electron. Comput. Sci. (IJAECS) 2(4), April 2015. ISSN:2393-2835
6. Aung, S.N., Sein, M.M.: Index structure for nearest neighbors search with required keywords on spatial database. In: Zin, T.T., Lin, J.-W., Pan, J.-S., Tin, P., Yokota, M. (eds.) Genetic and Evolutionary Computing. AISC, vol. 388, pp. 457–467. Springer, Heidelberg (2015)

7. Aung, S.N.: Nearest neighbor public services search system for myanmar land. In: ICT Virtual Organization of ASEAN Institutes and NICT Forum 2015, Kuala Lumpur, Malaysia, 26 November 2015
8. Aung, S.N.: Finding nearest services for emergency cases. In: Workshop on ICT Application for Responding Natural Disasters and Environmental Changes, Yangon, Myanmar, 22 January 2015
9. Ohsawa, Y., Htoo, H., Nyunt, N.J., Sein, M.M.: Generalized bichromatic homogeneous vicinity query algorithm in road network distance. In: Morzy, T., Valduriez, P., Bellatreche, L. (eds.) ADBIS 2015. CCIS, vol. 539, pp. 60–67. Springer, Heidelberg (2015). doi:10.1007/978-3-319-23201-0
10. Htoo, H., Ohsawa, Y., Sonehara, N., Sakauchi, M.: Aggregate nearest neighbor search methods using SSMTA* algorithm on road-network. In: Morzy, T., Härder, T., Wrembel, R. (eds.) ADBIS 2012. LNCS, vol. 7503, pp. 181–194. Springer, Heidelberg (2012)
11. Zhang, D., Tan, K.L., Tung, A.K.H.: Scalable Top-K spatial keyword search. In: EDBT/ICDT 2013, 18–22 March 2013
12. Chen, L., Cong, G., Jensen, C.S., Wu, D.: Spatial keyword query processing: an experimental evaluation. In: Proceedings of the VLDB Endowment, vol.6, no.3 (2013)
13. Cao, X., Chen, L., Cong, G., Jensen, C.S., Qu, Q., Skovsgaard, A., Wu, D., Yiu, M.L.: Spatial keyword querying. In: Atzeni, P., Cheung, D., Ram, S. (eds.) ER 2012 Main Conference 2012. LNCS, vol. 7532, pp. 16–29. Springer, Heidelberg (2012)
14. Rocha-Junior, J.B., Gkorgkas, O., Jonassen, S., Nørvåg, K.: Efficient processing of top-k spatial keyword queries. In: Pfoser, D., Tao, Y., Mouratidis, K., Nascimento, M.A., Mokbel, M., Shekhar, S., Huang, Y. (eds.) SSTD 2011. LNCS, vol. 6849, pp. 205–222. Springer, Heidelberg (2011)
15. Li, Z., Lee, K.C.K., Zheng, B., Lee, W.-C., Lee, D.L., Wang, X.: Ir-tree: An efficient index for geographic document search. IEEE TKDE **23**(4), 585–599 (2011)
16. Zhou, Y., Xie, X., Wang, C., Gong, Y., Ma, W.-Y.: Hybrid index structures for location-based web search. In: CIKM, pp. 155–162 (2005)
17. Göbel, R., Henrich, A., Niemann, R., Blank, D.: A hybrid index structure for geo-textual searches. In: CIKM, pp. 1625–1628 (2009)
18. Guttman, A.: R-trees: A dynamic index structure for spatial searching. In: SIGMOD, pp. 47–57 (1984)

Optimal Route Finding for Weak Infrastructure Road Network

K-zin Phyo(✉) and Myint Myint Sein

University of Computer Studies, Yangon, Myanmar
kzinphyo@ucsy.edu.mm, myintucsy@gmail.com

Abstract. As large road network and housing areas increased, the roles of emergency services are more demanding. But the infrastructures of road networks are weak in many developing countries. During the emergency cases (e.g. Accident or Fire), there is a great need to find out the closest emergency services and the optimal route for rapidly available facilities. According to the weak road network infrastructure, there are some difficulties for Emergency Vehicles. In some townships, there are narrow and close roads. If the drivers mistakenly choose these roads, it can cause problems and delays. The best route between the location of emergency services and accident is very useful and provide to minimize the loss of life and property in the event of fire. In the proposed system, modified Dijkstra's algorithm is applied to find optimal route by avoiding by close and narrow roads for emergency vehicles.

Keywords: Emergency vehicles · Modified dijkstra's algorithm · Optimal route finding · Closed and narrowed roads · Infrastructures of road network

1 Introduction

In Myanmar, fire is the most frequent disaster as on average approximately 900 cases are reported every year and large amount of property and lives are unfortunately lost through fire. Most of the fire cases were due to negligence, while others were caused by electric short-circuit, arson and forest fire. In 2011, total of 107 fires broke out in Myanmar. The February fire in this year destroyed 592 houses and buildings, leaved 2, 343 people homeless, killed four people and injured six others. Total of 648 fires broke out across Myanmar in the first nine months (Jan.-Sept.) of 2012, suffering a loss of 210,000 U.S. dollars. Over the period, 103 fire cases occurred in August and September, killed six people, injured 19 others and destroyed 615 houses. In 2013, 2014 and 2015 the number of fire incident cases is 1673, 1629 and 1708 respectively and we had lost several millions kyats and valuable lives per year [1]. To save the human life and property, the fire and emergency vehicles are essentially required to reach the incident location in a short time. Due to the increase of buildings, population and social development, the road network is become weak and complex in many developing country. In fire emergency situation, finding the suitable route to reach the incident location in time is critical concern.

It is possible to reduce the fired area and prevent the remaining unfired building when the fire vehicles reach as fast as they can. There are some difficulties in verification

J. Pan et al. (eds.), *Genetic and Evolutionary Computing*, Advances in Intelligent Systems and Computing 536, DOI 10.1007/978-3-319-48490-7_27

of incident cite based on the received information. The buildings, landmarks, street networks incorporated into the developed digital map can provide to identify and confirm the location of the incident site from receiving emergency call. If the optimal route is unknown, it is quite difficult to reach the incident site and can cause delays. So, the development of well-organized optimal route finding for urgent need of emergency services is vital to save the valuable lives and property. Based on our previous works [1–5], the proposed optimal route finding system is designed to solve the problems caused by close and narrow roads for emergency vehicles.

2 Related Works

Recently, much work was carried in the application of exiting studies for emergency response system to consider the shortest path analysis. N. Kumar et al., [6] develop the GIS based transport system which assist fastest, shortest and safest route to reach hospitals within Allahabad city. Although it is possible to determine the fastest and shortest route using GIS based network analysis but it not always work as link on a real road network in a city tends to possess different levels of congestion during different time periods of a day. Geographic Information System (GIS) based healthcare emergency response systemhad developed to identify the optimal route from the location of incident to any healthcare service providers, and the optimal route was modeled based on the distance (the shortest path) to the closest healthcare service providers. It focuses on finding a way to quickly locate an incident or case [7].

Route Analysis for Decision Support System is proposed to find shortest route between one facility to another at the time of disaster situation [8]. The research part of this work will comprise of Geographic Information Systems (GIS) technologies, GIS Web services and how these interact with each other. R. Fadlallaet al., [9] proposed the system to produce digital route guided maps and to improve services in case of emergencies such as accidents. This had been done by utilizing the capabilities of GIS in network analysis and visualization to enhance decision making in route selection to the nearest hospital by mapping the services area based on travel time. Public bus transportation system is developed to calculate the shortest route by using A* algorithm [2, 4]. The problem of identifying the shortest path along a road network is a basic problem in network analysis. In the case of any incident, it is important to respond the risk and to reach the incident site as soon as possible. The system is proposed the optimal part to reach the incident location within short time. This proposed work can provide the effectiveness and efficiency of fire emergency service in Myanmar and can reduce delays caused by closed and narrowed roads.

3 System Design

This proposed system finds the optimal routes between the emergency services and the incident site by using Dijkstra's Algorithm. The overview of the system is shown in Fig. 1. When an emergency call is coming to the system, it takes the information of the incident site such as street name and nearest landmark or residential address. By using these address information the system identify the location of the incident site. And then it displays the closest emergency services from incident site. Finally, the optimal route for fire station to incident site is calculated and the related route information will be display on the road map. Once knowing the optimal route to reach incident location, the fire trucks can reach this location in short time and it will provide to reduce the fired area, preventing the remaining area and save valuable human lives.

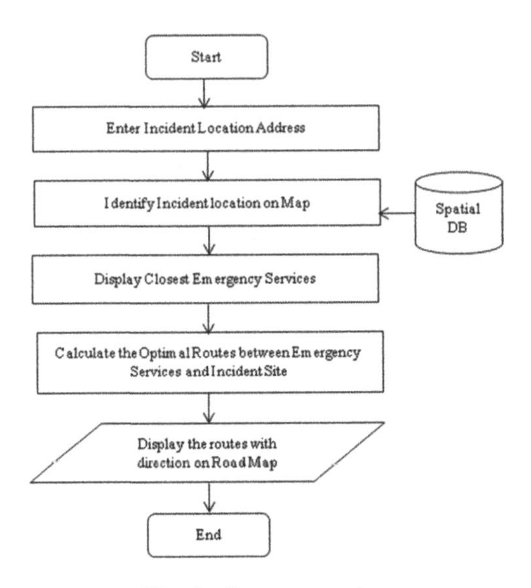

Fig. 1. System overview

3.1 Data Creation of Spatial Database

Recently, the optimal route computing system has been developed for Yangon Division. This system includes a spatial-database of the emergency services and the incident site locations. The emergency services locations (e.g. Fire Stations) of Yangon Region are collected from Myanmar Fire Service Department, Google Earth and GPS GARMIN etrex-10 device. The collected data are stored in Spatial-Database and used to develop this system. Seven attributes such as object ID, name, address are included in created special database and shape are presented in gometrically. In Yangon Division there are 41 Fire Emergency Services (fire stations). Table 2 shows the locations of some fire stations are described with related latitude and longitude (Table 1).

Table 1. Attributes for spatial database

File name	Data type
FID	Object ID
Shape	Geometry
Name	Text
Address	Text
Facility	Text
Township	Text
Contact Info	Text

Table 2. Locations of the fire stations in Yangon division

No	Name of fire station	Latitude	Longitude
1	Hlaing	16° 51' 01 .55"	96° 07' 28 .53"
2	Hmawbi	17° 05' 46.42"	96° 03' 25.24"
3	South Dagon	16° 51' 15 .94"	96° 13' 19 .80"
4	Mingalardon	17° 02' 31.54"	96° 08' 40. 07"
5	Shwepyithar	16° 58' 16.33"	96° 04' 36. 09"
6	Shwepyithar B	16° 57' 29.89"	96° 04' 36. 12"
7	Shwepaukkan	16° 55' 37.99"	96° 11' 04. 76"
8	North Okkalapa	16° 55' 00. 85"	96° 09' 29. 96"
9	Wabargi	16° 54' 57. 99"	96° 08' 56 .36"
10	Tarmway	16° 48' 12 .35"	96° 10' 26 .48"
11	North Dagon	16° 57' 30. 85"	96° 17' 45. 33"
12	North Dagon B	16° 52' 40 .44"	96° 12' 25 .82"
13	Sauchaung	16° 48' 14 .62"	96° 07' 58. 93"
14	Insein	16° 53' 10 .27"	96° 06' 04. 45"
15	Hlaing Thayar B	16° 52' 30. 02"	96° 04' 05. 47"

3.2 Optimal Route Analysis

When the incident occurred, it is very important to reach the incident site in time. In this paper, the optimal route between the fire station and incident sites is calculated by using Modified Dijkstra's Algorithm. It is used to find the shortest path from one node to another node in a graph. It is applied only on positive weights [7]. The procedure of Modified Dijkstra's algorithm is as follows:

```
for each vertex v in G
        dist[v] = infinity ;
        previous[v] = undefined ;
end for
       dist[source] = 0 ;
       Q = the set of all nodes in Graph ;
while Q is not empty:
 u = vertex in Q with smallest distance in dist[] ;
        remove u from Q ;
        if dist[u] = infinity:
               break ;
        end if
for each neighbor v of u:
        tdist= dist[u] + dist_between(u,v) ;
               if  tdist < dist[v]:
                 dist[v] = tdist ;
                 previous[v] = u ;
                decrease-key v in Q;
               end if
 end for
end while
        return dist[destination];
```

4 Experimental Result

The proposed system is tested on the Road Map of Yangon Region for emergency case (Fire). Figure 2 shows the locations of fire stations in Yangon Region. When a fire incident occurs, the location information of emergency call could not determine accurately the incident location. This system will help to determine the place of fire incident location by using the residential address or street name. After identifying the incident site, the system displays the emergency services near the incident location and then calculates the optimal routes between each emergency services and the incident site. Figures 3, 4 and 5 show the optimal route between the incident location and each fire stations with related route direction information, respectively.

The performance of optimal route finding is achieved by computing the Yangon region road network with the number of edges (streets) 87038 and the number nodes (junctions) 27852. The efficiency of the modified Dijkstra's algorithm is tested with in terms nodes and time. Figure 6 shows the performance evaluation of the proposed work.

Fig. 2. Fire stations in Yangon division

Fig. 3. Optimal route from station 1 to incident site

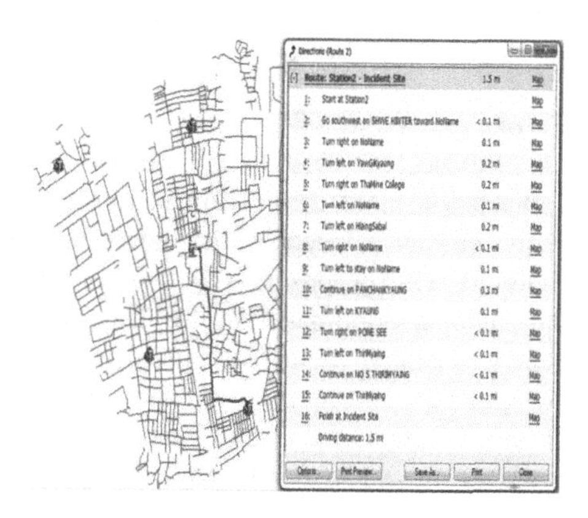

Fig. 4. Optimal route from station 2 to incident site

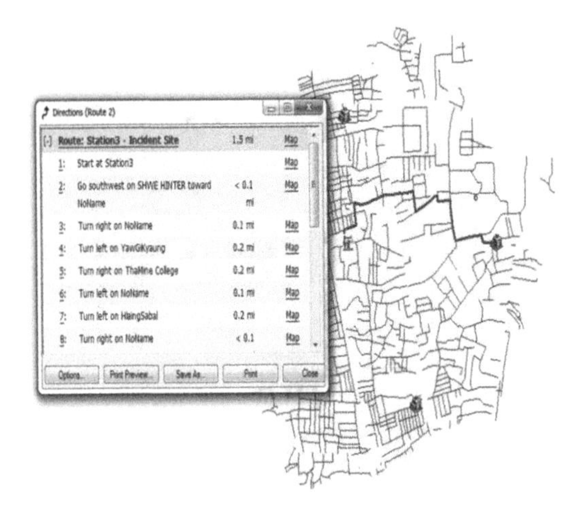

Fig. 5. Optimal route from station 3 to incident site

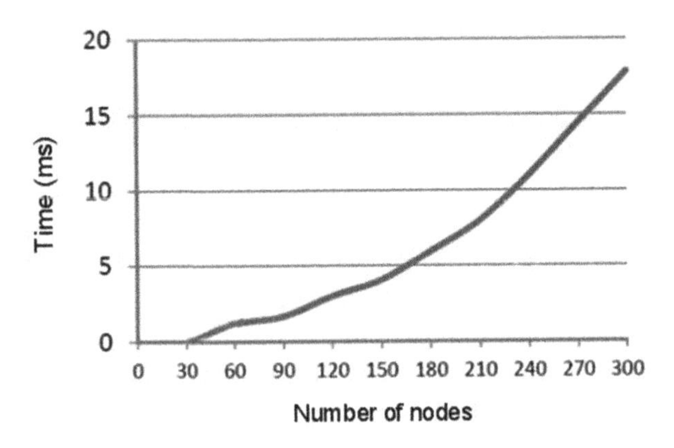

Fig. 6. Performance of the algorithm

5 Conclusions

The increasing numbers of population and the vehicles of the same time cause congestion on road networks. Besides, the closed and narrowed roads can cause delay for emergency vehicles to reach the incident location in time. The proposed system is especially designed for the problems faced by the emergency service vehicles while travelling on the road network. Our proposed system significantly solves the problem like location finding and calculates optimal route for fire emergency vehicles.

The proposed system was aimed to increase the effectiveness and efficiency of emergency services in Myanmar. The main advantages of this system is that it will make the work of the drivers much easier and make sure they get to the accident location within short time and we can save valuable human lives and properties. It can be used in the emergency services (e.g. ambulance and hospital) by changing the related databases. As future work, we will advance and develop the propose work on the mobile devices.

References

1. Phyo, K., Sein, M.M.: Optimal route finding to support fire emergency service. In: 13th International Conference on Computer Applications, ICCA (2015)
2. Zar, M.T., Sein, M.M.: Public transportation system for yangon region. In: Proceedings of the 13th International Conference on Computer Applications (ICCA 2015), Yangon, Myanmar, February 2015
3. Zar, M.T., Sein, M.M.: Finding shortest path and transit nodes in public transportation system. In: The 9th International Conference on Genetic and Evolutionary Computing (ICGEC 2015), Yangon, Myanmar, 26–28 August 2015
4. Zar, M.T., Sein, M.M.: Using A* algorithm for public transportation system in yangon region. In: Proceedings of International Conference on Science, Technology, Engineering and Management (ICSTEM, 2015), Singapore, February 2015
5. Ohsawa, Y., Htoo, H., Nyunt, N.J., Sein, M.M.: Generalized bichromatic homogeneous vicinity query algorithm in road network distance. In: Morzy, T., Valduriez, P., Bellatreche, L. (eds.) ADBIS 2015. CCIS, vol. 539, pp. 60–67. Springer, Heidelberg (2015). doi:10.1007/978-3-319-23210-0-8
6. Kumar, N., Kumar, M., Kumarsrivastva, S.: Geospatial path optimization for hospital: a case study of Allahabad city, Uttar Pradesh. Int. J. Mod. Eng. Res. **4**(10), 9–14 (2014)
7. Gubara, A., Ahmed, Z., Amasha, A., El Ghazali, S.: Decision support system network analysis for emergency applications. In: The 9th International Conference on Informatics and Systems (INFOS), pp. 40–46 (2014)
8. Bhanumurthy, V., Bothale, V.M., Kumar, B., Urkude, N., Shukla, R.: Route analysis for decision support system in emergency management through GIS technologies. Int. J. Adv. Eng. Glob. Technol. **3**(2), 345–350 (2015)
9. Elsheikh, R.F.A., Elhag, A., Sideeg, S.E.K., Mohammed, A.E.: Route network analysis in Khartoum city. SUST J. Eng. Comput. Sci. (JECS) **17**(1), 50–57 (2016)

Analysing the Effect of Disaster

Thida Aung(✉) and Myint Myint Sein

University of Computer Studies, Yangon, Myanmar
aung.demom85@gmail.com, myintucsy@gmail.com

Abstract. Analysing the damage area is the critical task for recovery and reconstruction for the urban area after the disaster. The purposed method is developed to detect the damage areas after the disaster using the satellite images. Most countries are exposed to a number of natural hazards such as Tsunami, Cyclone and landslide. It needs to estimate the destroying areas using the change detection techniques. In this approach, the pre and post satellite images are used to detect the damage areas. The main focus of the paper is to develop an approach that estimates the destroying areas combining the Morphological Building Index (MBI) and Slow Feature Analysis (SFA). He system output the Tchange map for the damage area. The results indicate that the proposed approach is encouraging for automatic detection of damaged buildings and it is a time saving method for monitoring buildings after the disaster happened.

Keywords: Morphological building index · Change detection · Disaster effect · Slow feature analysis

1 Introduction

Natural disasters such as earthquakes, land-slides, floods, fires and storms have increased in frequency and intensity over recent years. Observation of damaged buildings is vital for emergency management professionals, helping them for directing the rescue teams in short time to right location. Remote Sensing (RS) and Geographic Information Technologies (GIS) are an efficient tool for rapid monitoring of damaged buildings in urban regions [6, 9]. Satellite data has been used since 1970's. Today Climate changes and urban growth effect increasing pressure around the world wide. The results are: urban growth, intensified agriculture, decreases of forested areas, loss of biodiversity accelerated land degradation and soil erosion. The great demands are introduced on land use planning. Remote sensing data and techniques, and geographic information systems (GIS), provide efficient methods for analysis of land covers/uses. [1, 7] Image registration is required in remote sensing (multispectral classification, environmental monitoring, change detection, image mosaicing, weather forecasting, creating super-resolution images, integrating information into geographic information systems (GIS)), in medicine (combining computer tomography (CT) and NMR data to obtain more complete information about the patient, monitoring tumor growth, treatment verification, comparison of the patient's data with anatomical atlases), in cartography (map updating), and in computer vision (target localization, automatic quality control), to name a few.

J. Pan et al. (eds.), *Genetic and Evolutionary Computing*, Advances in Intelligent Systems and Computing 536, DOI 10.1007/978-3-319-48490-7_28

Object detection in satellite images has been an important research topic in computer vision for many years. Some useful applications of this subject are; updating of geographic information system (GIS) databases, urban city planning and land use analysis. In order to solve this complex problem, integrating the power of multiple algorithms, cues, and available data sources is also implemented recently to improve the reliability and robustness of the extraction results [11]. Recent researches in this area focus on automatic and unsupervised extraction of buildings. Lizy Abraham and Dr. M. Sasikumar [10] introduced an approach to the problem of automatic and unsupervised extraction of building features irrespective of rooftop structures in multispectral satellite images. The algorithm instead of detecting the region of interest, eliminates areas other than the region of the interest which extract the rooftops completely irrespective of their shapes. Mahak Khurana and Baishali Wadhwa proposed modified grab cut partitioning algorithm that detect the buildings in image which will take input from the previous objective and rather than min-max evaluation used in grab cut. They use bio inspired optimization which will find a global optimal solution for maximum energy better than min max algorithm [12].

For analyzing the effect of disaster, the damage area detection system is developed based on the simultaneously taking the advantage of MBI (Morphology Building Index) and SFA (Slow Feature Analysis). In this paper, This purposes a set of novel building change indices (BCI) for the automatic building change detection. This paper is organized as follows: The system overview is discussed in Sect. 2 and Methodology is expressed in Sect. 3. In Sect. 4, experiments can be seen and Sect. 5 gives the discussion and conclusion.

2 Change Detection of Damage Areas

In recent year, most countries are exposed to flooding and landslides during raining season and drought during dry season. It needs to estimate changing the building areas after the disasters and to know how much damaged building areas are reconstructed in the progress of natural hazard relieving cycle.

The purpose system is divided into six parts. They are

(1) Input the two successive images
(2) Preprocessing such as image registration
(3) Morphological Building Index (MBI)
(4) Slow Feature Analysis (SFA)
(5) Output the change result.

The two years images are applied in the proposed system, The flow chart for the purposed system is shown in Fig. 1.

2.1 Image Registration

Image registration is a crucial step in most image processing tasks for which the final result is achieved from the combination of various resources. The ground control point

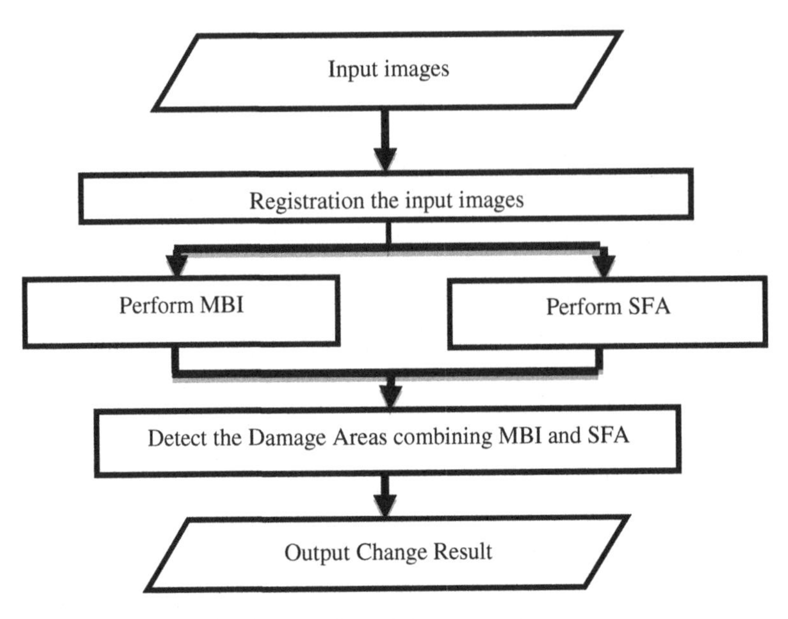

Fig. 1. Overview of the proposed system

is used to register the input successive images using image pixel addresses in terms of a map coordinate base. The most commonly used registration transformation is the affine transformation which is sufficient to match two images of a scene taken from the same viewing angle but from different position. It can tolerate more complex distortions. Affine transform can be categorized based on the geometric transforms for a planar surface element as translation, rotation, scaling, stretching, and shearing. The general 2D affine transformation can be expressed as shown in the following equation.

$$\begin{bmatrix} x_2 \\ y_2 \end{bmatrix} = \begin{bmatrix} t_x \\ t_y \end{bmatrix} + \begin{bmatrix} a_{11} & a_{12} \\ a_{21} & a_{22} \end{bmatrix} \begin{bmatrix} x_1 \\ y_1 \end{bmatrix} \tag{1}$$

where (x_2, y_2) is the new transformed coordinate of (x_1, y_1). The matrix $\begin{bmatrix} a_{11} & a_{12} \\ a_{21} & a_{22} \end{bmatrix}$ can be rotation, scale or shear. The scale of both x and y axes can be expressed as

$$Scale = \begin{bmatrix} S_x & 0 \\ 0 & S_y \end{bmatrix} \tag{2}$$

The two shear factors are represented on a and b are the shear factor along the x axis and y axis, respectively.

$$Shear = \begin{bmatrix} 1 & a \\ 0 & 1 \end{bmatrix}, \; Shear = \begin{bmatrix} 1 & 0 \\ b & 1 \end{bmatrix} \tag{3}$$

2.2 Morphological Building Index

The basic idea of MBI is to build the relationship between the spectral-structural characteristics of buildings and the morphological operator, which are summarized as follows.

- Brightness
- Local contrast
- Size and Directionality
- Shape

This method uses multispectral bands for high resolution images. Now we use low resolution image of three band color. The modified MBI is defined by describing the characteristic of building feature especially color of building roof and image intensity value. In order to solve this problem, modified MBI is proposed as the following steps:

Step 1: Enhancement of Image

The input low resolution registered image is transformed to high contrast image by applying with only red intensity value and stored as the brightness value which is computed by Eq. 4.

$$g = T(f_R(x, y)) \tag{4}$$

where $f_R\ (x,\ y)$ is the intensity transformation of red color-space image, g is the result of enhanced red band image using histogram adjust In [6], Original MBI is applied in multispectral band images of high resolution satellite images.

Step 2: Construction of MBI

The spectral-structural characteristics of buildings (e.g., contrast, size and directionality) are represented using the Differential Morphological Profile (DMP). The construction of MBI contains three steps.

(i) *White top-hat by Reconstruction* can be computed by Eq. 5.

$$W\_TH(d, s) = g - \gamma_b^{re}(d, s) \tag{5}$$

where γ_b^{re} represents the opening-by-reconstruction of the brightness image, and s indicates a flat and disk-shaped linear structuring element (SE), respectively.

(ii) *Morphological Profiles (MP)* of the white top-hat is defined as Eqs. 6 and 7.

$$MP_{W\_TH}(s) = W\_TH(s) \tag{6}$$

$$MP_{W\_TH}(s) = 0 \tag{7}$$

(iii) *Differential Morphological Profiles (DMP)* of the white top-hat is calculated as Eq. 8.

$$DMP_{W\_TH}(s) = |MP_{W\_TH}(s + \Delta s) - MP_{W\_TH}(s)| \quad (8)$$

where Δs is the interval of the profiles and $s_{min} \leq s \leq s_{max}$..

MBI is defined as the average of the DMPs of the white top-hat profiles defined in Eqs. 9 and 10 since buildings have large local contrast within the range of the chosen scales. Thus

$$MBI = \frac{\sum_s DMP_{W\_TH}(s)}{D \times S} \quad (9)$$

$$S = \left(\frac{S_{max} - S_{min}}{\Delta S}\right) + 1 \quad (10)$$

where D and S denote the numbers of disk and scale of the profiles, respectively [5, 6].

Step 3: Building extraction

The final building extraction step is decided by using predefined threshold value in order to classify these *MBI (x)* pixels because of different resolutions and image capturing time.

$$\begin{aligned} IF\ MBI(x) &\geq t_1, \\ THEN\ map_1(x) &= 1 \\ ELSE\ map_1(x) &= 0 \end{aligned}$$

where $MBI(x)$ and $map_1(x)$ indicate the value of MBI and the initial label for pixel x. t_1 is threshold value and set $t_1 = 5$ for the best result for the system [1, 2].

2.3 Slow Feature Analysis

Given a bitemporal spectral vector pair $x^i = [x_1^i, \ldots, x_N^i]$ and $y^i = [y_1^i, \ldots, y_N^i]$, where i indicates the pixel number and N is the dimension of the band, the input is normalized with zero mean and unit variance, expressed as Eqs. (11) and (12);

$$\hat{x}_j^i = \frac{x_j^i - \mu_{x_j}}{\sigma_{x_j}} \quad (11)$$

$$\hat{y}_j^i = \frac{y_j^i - \mu_{y_j}}{\sigma_{y_j}} \quad (12)$$

where μ_{x_j} is the mean and σ_{x_j} is the variance for band j of image X.

The SFA algorithm is reformulated with the normalized multi temporal pairs and rewrite in Eq. (13) as

$$\frac{1}{P}\sum\nolimits_{i=1}^{P}\left(g_j(\hat{x}^i) - g_j(\hat{y}^i)\right)^2 \text{ is minimal} \tag{13}$$

where P is the number of bitemporal spectral vector pairs in the input data set.

Finally, the SFA variable can be represented as

$$SFA_j = w_j\hat{x} - w_j\hat{y} \tag{14}$$

Where

$$\frac{w_j^T A w_j}{w_j^T B w_j} = \frac{w_j^T \sum_{\Delta} w_j}{w_j^T \frac{1}{2}(\sum_x + \sum_y) w_j}. \tag{15}$$

3 Detect the Change of Damage Area Combining MBI and SFA

In Morphological Building Index (MBI), it leads to a number of false alarms involving non-building urban structures such as soil and roads. Its accuracy is relates to the radiometric conditions of the image. In Slow Feature Analysis (SFA), it alone is not suitable for building change detection since it provides high commission error. It is not only related to the change of buildings, but also to other urban structures.

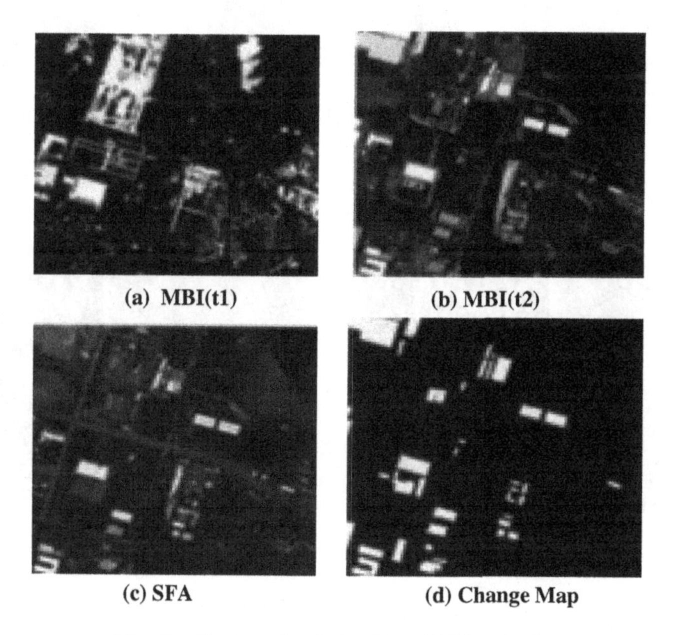

(a) MBI(t1) **(b) MBI(t2)**

(c) SFA **(d) Change Map**

Fig. 2. The Result of Combing MBI and SFA

The result of MBI and SFA are compared to detect the change of the damage areas. Decreasing damage building Area are estimated by comparing with $MBI(t_2)$ and SFA. $MBI(t_1)$ and $MBI(t_2)$ represents building components extracted by MBI for time t_1 and t_2, respectively as shown in Fig. 2.

4 Experimental Results

In this paper the study area was located in coastal city of Tacloban, Philippines. The aerial images are taken after the Typhoon Haiyan. The purposed method is applied to detect the damage building area. The Fig. 3 shows the reference map of Tacloban City. In this figure the color rectangle can be seen to classify the settlements types, transportation and hydrology. The blue rectangle represents water regions; brown, pink, light green, grey shows residential settlement, educational settlement, and recreational settlement, respectively. Then lines of blue and yellow presents the costal and roads of Tacloban city. The damage area can be calculated comparing the result of MBI(t_1), MBI(t_2) and SFA. The damage area of the specified types of settlement is obtained as Fig. 5 by comparing the reference map of Tacloban City

Fig. 3. Reference Map

Fig. 4. Before Haiyan Cyclone

Fig. 5. Damage Area After Cyclone

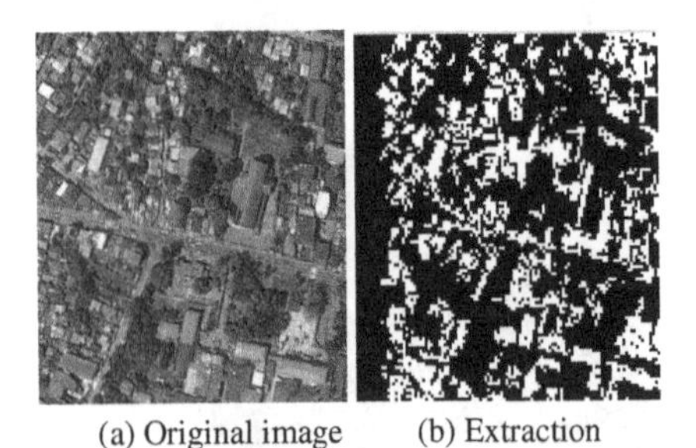

(a) Original image (b) Extraction

Fig. 6. Tacloban City (before Haiyan Typhon)

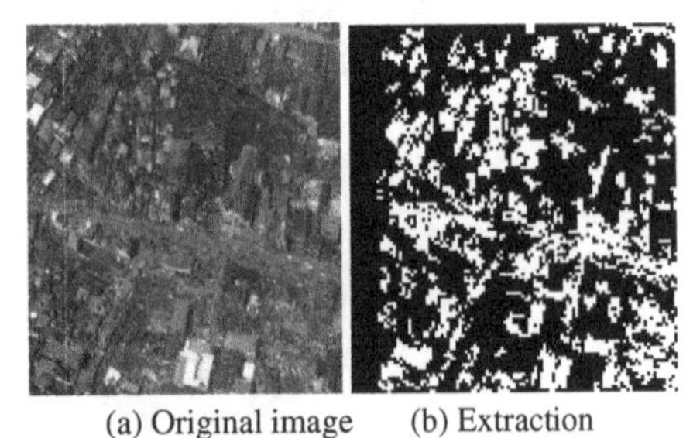

(a) Original image (b) Extraction

Fig. 7. Tacloban City (after Typhon)

The effectiveness of the SFA and MBI is assessed on the Google Earth images of the Tacloban City, Philippine. The following Fig. 7 shows the damage area of Tacloban City after Haiyan Cyclone in 8 November, 2013. To test the performance of the proposed system, we use these evaluation measures (completeness, correctness, quality) in Fig. 8.

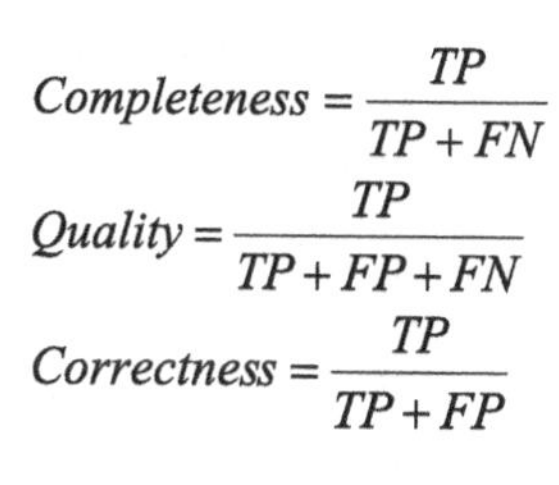

$$Completeness = \frac{TP}{TP + FN}$$

$$Quality = \frac{TP}{TP + FP + FN}$$

$$Correctness = \frac{TP}{TP + FP}$$

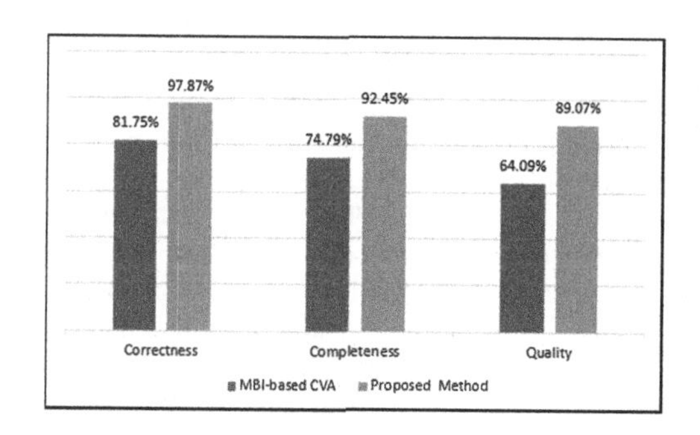

Fig. 8. Accuracy assessment of the proposed method

5 Conclusions

In this study, the combination of MBI and SFA method is purposed for automatically extracting the damage areas of the urban region. This system can solve the various satellite images only with three spectral colors without using multispectral band images. Since the purposed method uses unsupervised technique, the training data is not required.

So it can save the time for the training samples than other supervised method. But the errors may sometimes occur when many crowed cars on the road lead to miss building extraction because of the urban downtown area is our research area. Besides the purposed method is used manual threshold value in extracting the building area so it costs time to set the manual threshold value. In the future we will purpose the change detection system to know the increasing and decreasing rate of the building areas of the urban regions.

References

1. Nwe, N.M.M., Sein, M.M.: Detecting the environmental changes from satellite image. In: The Proceedings of the third Malaysian Software Engineering Conference, ACRS 2007, November 21–24 (2007)
2. Nwe, N.M.M., Sein, M.M.: Detecting the environmental changes area of satellite image. In: Proceedings of the Sixth International Conference on Computer Application, Yangon Myanmar, pp. 232–236 (2008)

3. Moe, K.C., Sein, M.M.: Detection the changes of urban area in yangon within ten years. In: Proceedings of the 1st International Conference on Energy, Environment and Human Engineering (ICEEHE), Yangon, Myanmar (2013)
4. Moe, K.C., Sein, M.M.: Automatic building change detection and open space area extraction in urban areas. In: Proceedings of the 12th International Conference on Computer Applications (ICCA2014), Yangon, Myanmar, pp. 291–296 (2014)
5. Moe, K.C., Sein, M.M.: An unsupervised technique for building change detection in urban areas. In: International Journal of Computer Application(IJCA), November on (106_18), New York, pp. 31–35, November 2014
6. Moe, K.C., Sein, M.M.: Building area extraction of urban region based on GIS. In: Proceedings of the 13th International Conference on Computer Applications (ICCA2015), Yangon, Myanmar, pp. 329–334 (2015)
7. Moe, K.C., Sein, M.M.: Urban growth detection using morphology of satellite image. In: Proceedings of International Conference on Science, Technology, Engineering and Management (ICSTEM,2015), Singapore, pp. 43–47 (2015). ISBN: 978-93-84209-89-6
8. Moe, K.C., Sein, M.M.: Urban growth detection using morphology of satellite image. Int. J. Adv. Electornics Comput. Sci. (IJAECS) **2**(3), March 2015
9. Moe, K.C., Sein, M.M.: Urban build -up building change detection using morphology based on GIS. In: International Journal of Genetic and Evolutionary Computing (ICGEC), Myanmar, 26–28 August 2015
10. Abraham, L., Sasikumar, N.: Unsupervised building extraction from high resolution satellite images irrespective of rooftop structures. Int. J. Image Process. (IJIP), **6**(4) (2012)
11. Bhadauria, A.S., Bhadauria, H.S., Kumar, A.: Building Extraction from Satellite Images **12**(2), 76–81 (2013). e-ISSN: 2278-0661, P-ISSN: 2278-8727
12. Doxani, G., Karantzalos, K. Tsakiri, M.: Automatic change detection in urban areas under a scale-space, object-oriented classification framework (2009)

Deep Learning Model for Integration of Clustering with Ranking in Social Networks

Thi Thi Zin[1(✉)], Pyke Tin[1], and Hiromitsu Hama[2]

[1] Faculty of Engineering, University of Miyazaki, Miyazaki, Japan
thithi@cc.miyazaki-u.ac.jp, pyketin11@gmail.com
[2] Research Center for Industry Innovation, Osaka City University, Osaka, Japan
hama@ado.osaka-cu.ac.jp

Abstract. Now a day Deep Learning has become a promising and challenging research topic adaptable to almost all applications. On the other hand Social Media Networks such as Facebook, Twitter, Flickr and etc. become ubiquitous so that extracting knowledge from social networks has also become an important task. Since both ranking and clustering can provide overall views on social network data, and each has been a hot topic by itself. In this paper we explore some applications of deep learning in social networks for integration of clustering and ranking. It has been well recognized that ranking systems without taking cluster effects into account leads to dumb outcomes. For example ranking a database and deep learning papers together may not be useful. Similarly, clustering a large number of things for example thousands of users in social networks, in one large cluster without ranking is dull as well. Thus, in this paper, based on initial N clusters, ranking is applied separately. Then by using a deep learning model each object will be decomposed into K-dimensional vector. In which each component belongs to a cluster which is measured by Markov Chain Stationary Distribution. We then reassign the objects to the nearest cluster in order to improve the clustering process for better clusters and wiser ranking. Finally, some experimental results will be shown to confirm that the proposed new mutual enforcement deep learning model of clustering and ranking in social networks, which we now name DeepLCRank (Deep Learning Cluster Rank) can provide more informative views of data compared with traditional clustering.

Keywords: Deep learning · Social ranking · Markov chain · Clustering

1 Introduction

Deep Learning has been a popular and challenging research topic among the researchers in machine learning and artificial intelligence. Due to its versatile applications and enforcement by neural networks power, deep learning gains an added value in many aspects [1–3]. Many algorithms, theories, and large-scale training systems towards deep learning have been developed and successfully adopted in real tasks, such as speech recognition, image classification, and natural language processing [4–7]. Some examples are shown in Fig. 1. However, the adoption of deep learning in clustering has not been adequately investigated particularly with respect to social networks. Moreover only a few works have been done deep learning models for

J. Pan et al. (eds.), *Genetic and Evolutionary Computing*, Advances in Intelligent Systems and Computing 536, DOI 10.1007/978-3-319-48490-7_29

clustering. To the best of our knowledge, we have not seen the integration of deep learning models with clustering and ranking concepts in social networks.

Clustering based on similarity measures has long been recognized as a classical machine learning problem. It has never been integrated with the architecture of deep learning framework. But there have been some approaches using two stages to learn similar to deep learning concept by embedding features into the stages for making clustering objects [8–11]. In addition, the techniques employed have used a metric such as Euclidean or cosine distance, in the second stage. Since these metrics are induced by human nature, the chosen metrics may not be appropriate for the embedded feature space. Therefore the existing methods may not handle the problems occurred in clustering and ranking simultaneously.

In our daily life applications, things are interconnected so that they behave individually as well as in groups. Especially, the occurrence of social and information networks and Internet of Things (IoT) are growing on the brinks of emerging technologies. Generally speaking, these networks contain objects of multiple types. So, a tremendous variety of techniques have been proposed to make better understanding of these networks and their potential appearances such as clustering similar objects and ranking their properties. On one hand, ranking evaluates objects in social networks and the Internet of Things by defining a ranking function. The ranking function in here means a mathematical function that describes the characteristics of an object. Such ranking functions can make two objects able to be compared whether the objects are similar or not in quantitatively and qualitatively. Some most popular ranking systems are PageRank, HITS and hybrid ranking to name a few [12–14]. On the other hand, clustering objects into groups based on a certain proximity measure are also a challenging task. This will lead to the concept which states that similar objects are in the same cluster, whereas dissimilar ones are in different clusters. After all, as two fundamental analytical tools, ranking and clustering demonstrate overall views of social networks, and hence be widely applied in different social network settings.

Recently a new concept of deep learning has been widely used for image classification and object recognition systems. Therefore the concept of deep learning to be

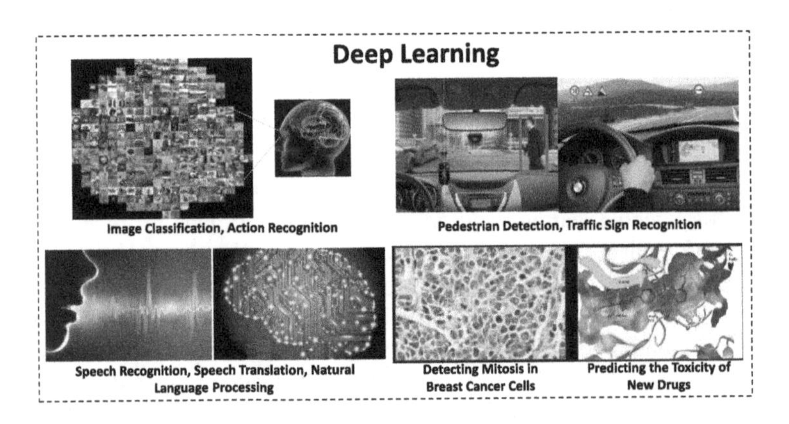

Fig. 1. Examples of deep learning

integrated with the clustering and ranking concepts may lead to a new paradigm research area having rich potentialities in techniques. Usually, we can consider clustering and ranking as a mutual enforcement for making understand social networks and IoT. Most of research works have treated the two concepts separately. But the correct use of the two concepts together will lead to a fruitful outcome. For example, ranking objects in social network based only on global scene without taking which clusters they belong to often leads to dumb results, e.g., ranking database and deep learning works for analysis together may not make much sense. Again, clustering a large number of objects (e.g., thousands of images) in one cluster without distinction is not a correct answer. Therefore, we propose DeepLCRank, an integrated model for clustering and ranking in social network by using a deep learning model. The experimental results show that DeepLCRank can generate more accurate clusters than the state-of-the-art link-based clustering method in a more effective and comprehensive way. Moreover, the combined outcomes of clustering with ranks make the network information more accurate and more significant.

In the following, we organize the paper starting with Sect. 2 on related work. In Sect. 3, we propose the DeepLCRank algorithm, taking bi-type social network as an example. Section 4 is a systematic experimental analysis on both synthetic and real datasets. We discuss and conclude the paper in Sect. 5.

2 Some Related Works

Deep learning modeling is a technique of modeling nonlinear complex phenomena based on convolutional neural network concepts. Deep learning models have also variety of successful applications in diversity of research fields. To name a few, it includes speech recognition, computer vision and deep neural networks [15–17]. It is also regarded as an effective method for learning high level concepts from low level features. This is a merit point of deep learning models. Exploiting the merit points of deep learning, we conducted experiments on utilizing the learned representations for a spectrum of different tasks, On the other hand in social network platforms, there are three popular systems of ranking such as Page Rank, HITS and Hybrid Rank which are successfully applied to the Internet of Things. The first one is concerned with link analysis the second one introduces the concept of authority scores and the third one employed the combination of two concepts, Link Analysis and Authority Values.

However, both PageRank and HITS are designed on the network of web pages, which is a directed homogeneous network, and the weight of the edge is binary. But DeepLCRank aims at ranking popularity of web objects by using deep learning models. We have considered the rank scores as a good measure for clustering. Instead of proposing a totally new strategy for ranking, we aim at finding empirical rules in the specific area of social networks data set, and providing ranking function based on these rules, which works well for the specific case. The real novelty lies in our framework is that it tightly integrates ranking and clustering through deep learning process.

We realize that clustering is another way to summarize a network and discover the underlying structures, which partitions the objects of a network into subsets (clusters) so that objects in each subset share some common trait. In clustering, proximity between objects is often defined for the purpose of grouping "similar" objects into one cluster, while partitioning dissimilar ones far apart. Therefore, similarity extraction methods should be applied first, which is an iterative PageRank-like method for computing structural similarity between objects. Without calculating the pairwise similarity between two objects of the same type, DeepLCRank uses conditional ranking as the measure of clusters, and only needs to calculate the distances between each object and the cluster center.

3 Integration of Clustering with Ranking via Deep Learning

In this section, we shall describe the proposed system architecture for integrating the two concepts clustering and ranking in social networks by using deep learning model. In doing so, we divide the section into two parts. One is concerned with definitions and problem formulation. The second one is the detail analysis of the step by step procedure for clustering and ranking through the employment of deep learning.

3.1 Definitions and Problem Formulation

Definition 1: A Social Community is a group of cohesive users and their interest images (nodes) such that there is much more links inside that cohesive group than links connecting that group to nodes outside it.

Problem Formulation: First, the definition of a social community can be illustrated by the graph G = (V = (U, I), E) in which V = (U, I) stands for a set of vertices (or nodes) represented by the pairs of users U and their interested images I and E represent a set of edges (or links). Figure 2(a) is an illustration of graph G.

Then the problem of interest will be finding one or more communities S = (V_s, E_s) within given social network as a subset of G. Suppose $d(v_i)$ and $d(v_e)$ denote the internal and external degrees of a node v respectively. According to the previous intuition, a good community would be a sub graph, such that for each node we maximize $d(v_i)$ and minimize $d(v_e)$. This can be seen in Fig. 2(b), where edges linking nodes inside are drawn in black and edges linking nodes in to nodes outside are drawn in red. In the context of Deep Learning a neural network shown in Fig. 3 will be

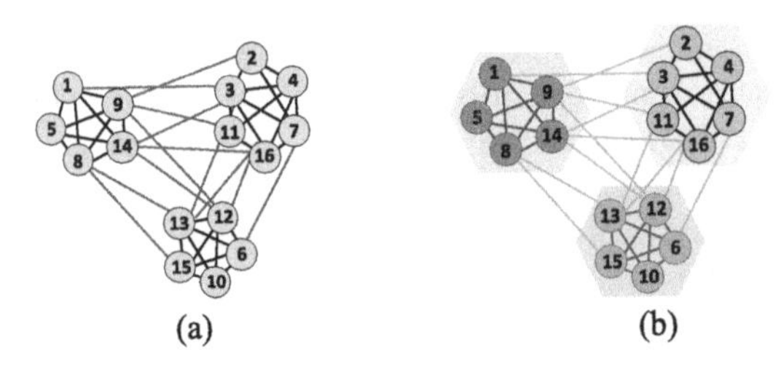

Fig. 2. (a) A graph G with 16Nodes and 45Edges and (b) Three communities (color-shadowed in red, green, and blue) defined for the graph G (Color figure online)

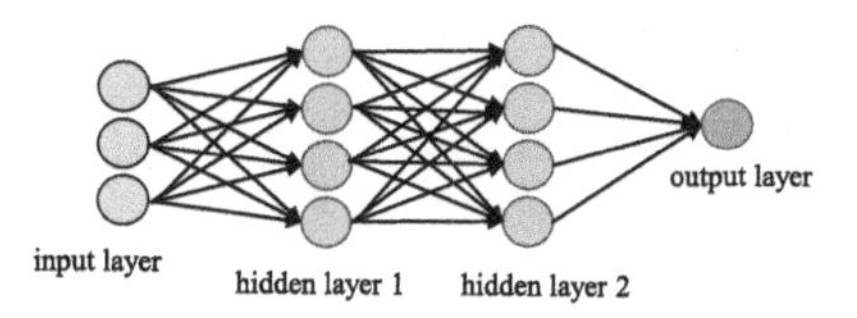

Fig. 3. Two layer deep learning model

responsible for the extraction and embedding of features in an (usually) unsupervised fashion to perform the final learning task, for example a classification or clustering with ranking.

3.2 Proposed Implementation Scheme for Clustering with Ranking

- First let us consider the overview of the proposed system which is described in Fig. 4. For implementation we use the step by step procedure as follows,
- For a given graph G = (V = (U, I), E), compute a pairwise similarity matrix from G, using a Markov similarity function P = (p_{ij}) for v_i and v_j for all nodes i, j which belong to V.
- In this concern, Random walks have been used as a similarity measure for a variety of problems in content recommendation and community detection. They are also the foundation of a class of output sensitive algorithms which use them to compute local community structure information graph. It is this connection to local structure that motivates us to use a stream of short random walks as our basic tool for extracting information from a network. In addition to capturing community information, using random walks as the basis for our algorithm gives us two other desirable properties. First, local exploration is easy to parallelize. Several random walkers (in different threads, processes, or machines) can simultaneously explore different parts of the same graph. Secondly, relying on information obtained from short random walks make it possible to accommodate small changes in the graph structure without the need for global computation. We can iteratively update the learned model with new random walks from the changed region in time sub-linear to the entire graph.
- Compute a diagonal matrix D, whose elements d(i) for i = 1,2,… $|V|$ are the node degrees for nodes vi ϵ V for i = 1,2,… $|V|$.
- Instead of clustering data directly we apply a non-linear transformation to the dataset. Suppose the non-linear transformation be $f(\alpha, x) : X \rightarrow Z$ where α is a learnable parameter and Z is a feature space.
- Employ Markov Chain based clustering algorithm MC by computing the powers of the pairwise similarity matrix calculated in step 1 for the feature space Z. Let the computed clusters be denoted by $C_1 \ldots C_l$ having cluster μ_i for i = 1, 1,…, l. We then define the encoding hidden layer function $f(x)$ = μ_i for x belongs to C_i and zero otherwise.

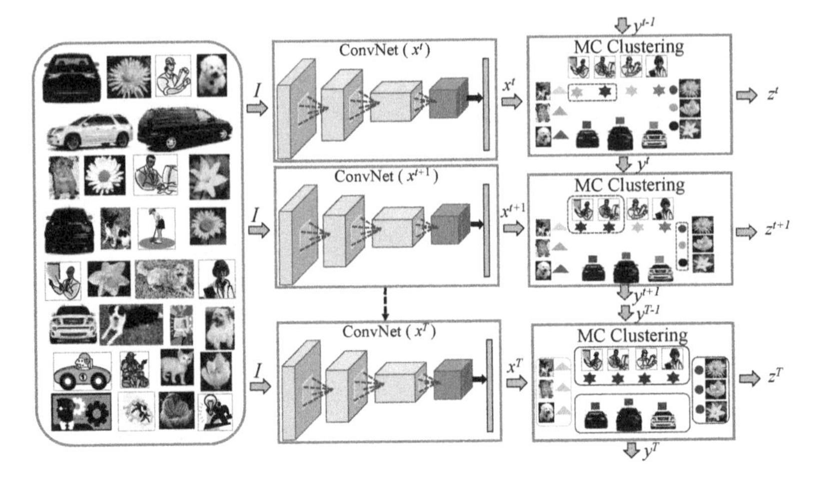

Fig. 4. Overview of proposed system

- Assume the initial input data as the matrix X = D-1P, the desired number of layers of MC as l, the desired number of clusters for each layer of as C, and the final desired number of communities as k.
- For each layer l_1, l_2, …, l_m, compute pooling data $X_l = MC(X_l\text{-}1, C_l)$ where C_l = (number of columns of X_l-1)/2.
- As the last step, we apply standard Markov Chain based clustering to X_l to the desired k communities of G with ranking procedure.

4 Illustrative Simulation Results

In order to evaluate the proposed method of clustering with ranking we use two data sets. First we consider a set of images from two concepts of holiday photos of Flick.com and occurrences of natural disasters of YouTube. Each set has sub concepts as shown in Table 1. Then we choose 100 users in each concept. The Markov chain transition matrix 10 for mixed images is derived as shown in Table 2. With the ranking function based on Markov Chain stationary distribution, the rank algorithm gives the results are rather dumb (because of the mixture of the concepts) and are biased towards (i.e., ranked higher for) the disaster concept. What is more, such dull or biased ranking result is caused not by the specific ranking function we chose, but by the inherent incomparability between the two concepts. However, by using deep learning concept, we have computed for 3 layers and present the result of the third layer of deep learning model as in Table 3.

Table 1. Set of images from Flickr and YouTube

Holiday	$I_1, I_2, \ldots, I_{10}$
Disaster	$I_{11}, I_{12}, \ldots, I_{20}$

Table 2. Transition matrix for 10 mixed images

	I_1	I_2	I_3	I_4	I_5	I_{12}	I_{15}	I_{17}	I_{18}	I_{20}
I_1	0.4	0.2	0.149	0	0	0.0005	0	0.0005	0.1	0.15
I_2	0.2	0.35	0.2	0.0245	0.0009	0.0005	0.0246	0.0005	0.15	0.049
I_3	0.149	0.2	0.0996	0.0003	0.2	0.201	0.0001	0	0.1	0.05
I_4	0	0.0245	0.0003	0.65	0.2652	0.0004	0.0001	0	0.05	0.0095
I_5	0.2	0.0009	0.2	0.2652	0.3148	0.0001	0	0	0.01	0.009
I_{12}	0.201	0.0005	0.201	0.0004	0.0001	0.288	0.1	0	0.2	0.009
I_{15}	0.0001	0.0246	0.0001	0.0001	0	0.1	0.8	0.0749	0.0002	0
I_{17}	0	0.0005	0	0	0	0	0.0749	0.4	0.27	0.2546
I_{18}	0.1	0.15	0.1	0.05	0.01	0.2	0.0002	0.27	0.1198	0
I_{20}	0.15	0.049	0.05	0.0095	0.009	0.009	0	0.2546	0	0.4689

Table 3. Results of third layer of deep learning model

	I_1	I_2	I_3	I_4	I_5	I_{12}	I_{15}	I_{17}	I_{18}	I_{20}
I_1	0.1552	0.1176	0.0993	0.0720	0.060	0.0717	0.0913	0.1090	0.1035	0.1198
I_2	0.1552	0.1176	0.0993	0.0720	0.060	0.0717	0.0913	0.1090	0.1035	0.1198
I_3	0.1552	0.1176	0.0993	0.0720	0.060	0.0717	0.0913	0.1090	0.1035	0.1198
I_4	0.1552	0.1176	0.0993	0.0720	0.060	0.0717	0.0913	0.1090	0.1035	0.1198
I_5	0.1552	0.1176	0.0993	0.0720	0.060	0.0717	0.0913	0.1090	0.1035	0.1198
I_{12}	0.1552	0.1176	0.0993	0.0720	0.060	0.0717	0.0913	0.1090	0.1035	0.1198
I_{15}	0.1552	0.1176	0.0993	0.0720	0.060	0.0717	0.0913	0.1090	0.1035	0.1198
I_{17}	0.1552	0.1176	0.0993	0.0720	0.060	0.0717	0.0913	0.1090	0.1035	0.1198
I_{18}	0.1552	0.1176	0.0993	0.0720	0.060	0.0717	0.0913	0.1090	0.1035	0.1198
I_{20}	0.1552	0.1176	0.0993	0.0720	0.060	0.0717	0.0913	0.1090	0.1035	0.1198

Cluster results can be seen from the outcomes of Layer 3 as follows.

Class Rank 1 = $\{I_1\}$
Class Rank 2 = $\{I_2, I_{17}, I_{18}, I_{20}\}$
Class Rank 3 = $\{I_3, I_4, I_5, I_{12}, I_{15}\}$

5 Conclusion

We had proposed a new method for a network community by integration of clustering with ranking in the framework of deep learning concept. We only illustrated how social network data can be used by the aids of deep learning and Markov Chain models. For simplicity, we demonstrated by using only synthetic data. In future we will explore the proposed method on real world datasets.

Acknowledgment. This work is partially supported by KAKENHI 25330133 Grant-in-Aid for Scientific Research (C).

References

1. Goroshin, R., et al.: Unsupervised feature learning from temporal data. arXiv preprint arXiv: 1504.02518 (2015)
2. Hadsell, R., et al.: Dimensionality reduction by learning an invariant mapping. In: Proceedings of 2006 IEEE Conference on CVPR, New York, vol. 2, pp. 1735–1742 (2006)
3. Han, X., et al.: MatchNet: unifying feature and metric learning for patch-based matching. In: Proceedings of 2015 IEEE Conference on Computer Vision and Pattern Recognition (CVPR), Boston, pp. 3279–3286 (2015)
4. Huang, P., et al.: Deep embedding network for clustering. In: Proceedings of 22nd International Conference on Pattern Recognition (ICPR), Sweden, pp. 1532–1537 (2014)
5. Khashabi, D., et al.: Clustering with side information: a probabilistic model to a deterministic algorithm. arXiv preprint arXiv:1508.06235 (2015)
6. Krizhevsky, A., et al.: ImageNet classification with deep convolutional neural networks. In: Proceedings of 25th ANIPS, California, pp. 1097–1105 (2012)
7. Mnih, V., et al.: Human-level control through deep reinforcement learning. Nature **518** (7540), 529–533 (2015)
8. Rippel, O., et al.: Metric learning with adaptive density discrimination. arXiv preprint arXiv: 1511.05939 (2015)
9. Schroff, F., et al.: A unified embedding for face recognition and clustering. In: Proceedings of 2015 IEEE Conference on CVPR, Boston, pp. 815–823 (2015)
10. Shao, M., et al.: Deep linear coding for fast graph clustering. In: Proceedings of 24th International Conference on Artificial Intelligence, Argentina, pp. 3798–3804 (2015)
11. Simo-Serra, E., et al.: Discriminative learning of deep convolutional feature point descriptors. In: Proceedings of 2015 International Conference on Computer Vision (ICCV), Santiago, pp. 118–126 (2015)
12. Bilmes, J.A.: A gentle tutorial on the EM algorithm and its application to parameter estimation for gaussian mixture and hidden Markov models. Int. CS Inst. **4**(510), 126 (1998). California
13. Brin, S., et al.: The anatomy of a large-scale hyper textual web search engine. Comput. Netw. ISDN Syst. **30**(1–7), 107–117 (1998). Netherlands
14. Tin, P., et al.: A novel hybrid approach to image ranking system. ICIC Express Lett. Part B Appl. Int. J. Res. Surv. **6**(3), 743–748 (2015). Japan
15. Tian, F., et al.: Learning deep representations for graph clustering. In: Proceedings of 28th AAAI Conference, Canada, pp. 1293–1299 (2014)
16. Tang, L., et al.: Leveraging social media networks for classification. Data Min. Knowl. Disc. **23**(3), 447–478 (2011)
17. Perozzi, B., et al.: DeepWalk: online learning of social representations. In: Proceedings of 20th ACM SIGKDD International Conference on Knowledge Discovery and Data Mining, New York, pp. 701–710 (2014)

An Automatic Target Tracking System Based on Local and Global Features

Thi Thi Zin(✉) and Kenshiro Yamada

Faculty of Engineering, University of Miyazaki, Miyazaki, Japan
thithi@cc.miyazaki-u.ac.jp

Abstract. Understanding human and object behaviors in video surveillance systems is an important factor and particular interest as far as public security is concerned. In this paper, we propose a novel method for object classification and tracking processes. It has been well recognized that the two processes play important roles in video surveillance system to make them intelligent. The proposed system is able to detect and classify human and non-human in different weather conditions. The system is capable of correctly tracking multiple objects despite occlusions and object interactions. Some experimental results are presented by using self-collected data.

Keywords: Tracking · Human · Non-human · Object classification · Aspect ratio · Motion feature

1 Introduction

Target tracking and classification of objects in a video sequence have been played important roles in many video surveillance applications. In today modern society, due to terrorist attacks and suspicious people and objects occurrences in public places, more and more video surveillance cameras are demanding for their intelligent works. These systems have been used for security monitoring [1, 2], as well as for traffic flow measuring [3], accident detection on highways, and routine maintenance in nuclear facilities. In this aspect, there have been many studies to develop tracking algorithms that are robust to partial occlusions [4–6] and that can cope with a short-term loss of observations [7, 8]. A complementary line of research has focused on learning static occlusion maps using large sets of observations accumulated over time [9].

In this paper we consider the problem of target tracking and object classification in video sequences. Our aim is to detect suspicious things and people then track targets which can take motion in non-rigid deformations, rotations or partial occlusions.

2 Some Related Works

We describe here only some tracking methods related to our system. The first method is based on their moving regions. This method identifies and tracks a blob or a Bounding Box (BB) moving objects in 2D space [10, 11]. A benefit of this method is that it is time efficient, and works well for small numbers of moving objects. The second one is

J. Pan et al. (eds.), *Genetic and Evolutionary Computing*, Advances in Intelligent Systems and Computing 536, DOI 10.1007/978-3-319-48490-7_30

a contour based object tracking by using color for tracking in both static and dynamic objects [12, 13]. The method mostly depends on the boundary curves of the moving object. The third category is a method of tracking based on a 3D moving object model. It can solve the partially occlusion problem, but it can only ensure high accuracy for a small number of moving objects [14]. The last category is feature based tracking which is to select common features of moving objects and tracking these features continuously. Even if partial occlusion occurs, a fraction of these features is still visible, so it may overcome the partial occlusion problems.

3 Proposed Object Segmentation and Tracking Method

The proposed object segmentation and tracking method is composed of two components; namely segmentation and parameter estimation of extracted object regions and the task of target tracking.

3.1 Extraction of Target Object

For extraction of the target objects, we perform Intra-Frame difference. The pixels are considered foreground if the difference is over a predefined threshold. The basic scheme of intra-frame segmentation can be written as $|I_t - I_{t-1}| > r$. Then we apply morphological connectivity operator to obtain constant gray-level regions with sharp contours corresponding precisely to those of the original signal. After this the image is partitioned into many regions with different labels. In this stage we also derive a binary image from the partitioned regions based on motion criterion. Furthermore, the BB parameters are re-estimated. Some sample results after applying morphological operations are shown in Fig. 1. In the background subtraction results may include the shadow of the object. For shadow, we derive the maximum amount of white pixels within the height of between 0 and 15 % of BB and the minimum value in the interval of 15 to 30 % of BB. If the minimum amount is less than 80 % of the maximum amount we regard that as the shadow is present. So the shadows which are less the 15 % of BB are removed. The shadow removal results are shown in Fig. 2.

3.2 Tracking of Target Objects

In the process of tracking target objects, we first observe the characteristics of the pixel relationships between those in the front and those in the rear of current image frame. Then the occurrences of changes are used to determine whether the object is the same object or not. For this, we calculate the centroid of the target object in the current frame. Then comparison is made with those in the previous frame. If the object distance within the threshold at that time, it is determined as the same object otherwise it will be recognized as different object.

For multiple target objects, each candidate will be determined individually. The conceptual flow diagram of the target tracking is described in Fig. 3.

3.2.1 Tracking Procedures

Color Histogram Procedure: In order to realize target tracking algorithm, we first transform the input images from RGB color space to HSV color space. Then the pixels in each BB will be distributed into 72 bins. In continuations, the distributions of objects in current frame and previous frame are compared by using Euclidean distance among them.

(a) Input image

(b) Foreground image before operation

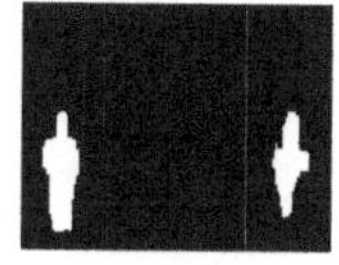

(c) Foreground image after operation

Fig. 1. Results of morphological operations

(a)Before Shadow Removal

(b)After Shadow Removal

Fig. 2. Shadow removal results

Centroids of Objects Procedure: As far as centroids are concerned, there can arise three cases.

(i) Two centroids of exists in the movable range of one target object. In this case the discrimination of the target object can be performed by using a color histogram.
(ii) A centroid can present in two regions in the same movable range. Again there can be two possibilities. Possibility of one of the targets disappeared and possibility of two-subjects had coalesced into one. In the second case, we must verify that coalesce in advance by using the size of BB. Inevitably it can be seen that BB is increased from the fact that the two objects overlap. Some examples are shown in Fig. 4.
(iii) Two centroids present in the overlapping movable range of 2 objects. Again two situations can happen. They are two targets of the previous frame are not coalesce and two targets in previous frame have been coalesced.

Bounding Box and Motion Features: To make an accurate tracking process, we use here some BB and motion features. The particular features employed in for tracking are:

(i) **Aspect ratio**: $A = H/W$, where H and W are the height and width of BB.
(ii) **Center of gravity of moment**:$M_g = \frac{m_x}{m_y} = \frac{m_{20}}{m_{02}}$, where $m_{pq} = \sum\sum_{(x,y)\in G} x^p y^q$.
(iii) **Dispersion measure**: This is defined as the ratio of the length of the target region area (the white pixel part) and the area around derived from equation $D = \frac{L^2}{G}$, where D, L, G represent dispersion measure, perimeter and area of the target, respectively.
(iv) **Amount of white pixels**: This is the ratio of the quantity of white pixels relative to the size of BB and use to judge sitting in a smaller aspect ratio for humans or animals.

(v) **Split measure**: BB is divided into four at the middle point. High ratio of upper body is expected than those of dogs or cats. In each region the extreme numerical difference is used to differentiate animal and human beings.

(vi) **Speed:** This is to be used to determine the exact value has moved forward who has moved back in the picture. Scale correction values are set manually scale settings for video images taken in advance.

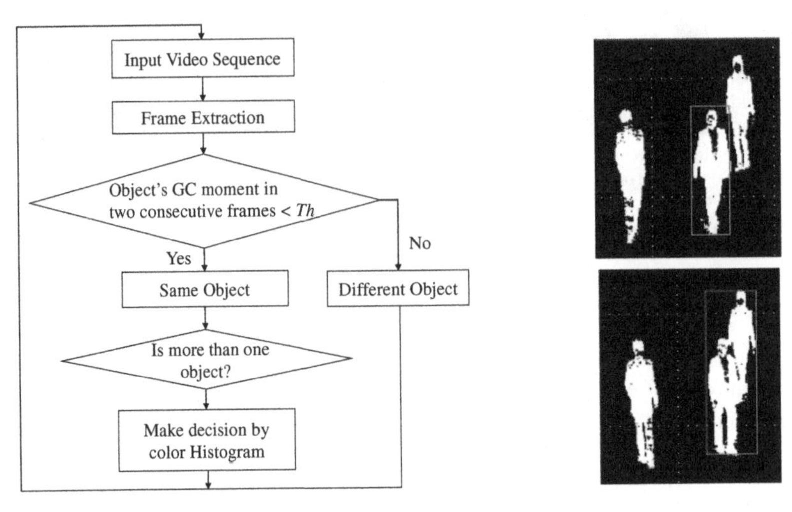

Fig. 3. Tracking block diagram

Fig. 4. Change of BB by coalescence

Updating Scheme: Since the features are about to change one frame to another, we use the following updating scheme to derive corresponding results as accurate as possible.

$T_n = F_n + \frac{T_{n-1}}{2}$, where T_n is nth frame and F_n is the corresponding feature vector.

Rules for Tracking and Classification: In order to a make a decision for tracking and classification we use the rules by estimating the features defined in the above. The threshold values for tracking and detection are described in Tables 1 and 2 with associated assigned points (p).

Table 1. Set scores

	Threshold values			
Aspect ratio	2.5 ~ 3.5(4p)	2 ~ 2.5(3p)	1 ~ 2(2p)	0 ~ 1(0p)
		3.5 ~ 4(3p)	4 ~ 5(2p)	7 ~ (0p)
Center of gravity	0 ~ 0.2(4p)	0.2 ~ 0.4(3p)	0.4 ~ 1(2p)	2.5 ~ 3.5(4p)

Table 2. Evaluation scores

	Threshold values			
Predicted Object	Human upright walking [7–8 p]	Crouching human or child [6 p]	Human sitting or animal [2 ~ 5 p]	Animal
Dispersion M	20 ~ 30 (2 p)	20 ~ 35 (3 p)	~20 (4 p)	
White pixels		0.4 ~ 0.6 (4 p)	0.5 ~ 0.75 (1 p)	
4 Split	0.5 ~ 1.5 (1 p)			
Speed	3 ~ 4 (1 p)			
Optional scores	All scored in previous frame			

4 Some Experimental Results and Discussion

In this section, we shall present some experimental results by using self- collected video sequences taken in University of Miyazaki, Japan. We have used DG-SC385 Panasonic network cameras save video: 1280 × 780, 3frame/s under various weather conditions such as cloudy occasionally snow, strong winds. Experimental results in Table 3 show that the decision for "human" by 1 and the decision "not human" is 0. In this case the higher aspect ratio and center of gravity moment, identified as human. Also, kept out the above two values are consistently high scores in Total 24 to 26 points. The results are shown in Fig. 5. The result of pedestrians with an umbrella is shown in Fig. 6. In this case, BB is vertically large and the aspect ratio is 2 or 3 with lower values. But this fact covered by the moment, distributed, all recognized as human beings. The performance evaluation results are described in Table 4. Figure 7 illustrates results where a dog is walking. But without addition of balancing basically has a complex shape of dog can be judged accurately. The performance evaluation results are described in Table 5. The results show that all conditions are generally better results. Compared to others, the case of umbrella carrying case only is lower. The process being heavily characterized by the shape of the aspect ratio and the weighting of scores might be the reason for lower accuracy. More experimental results and their performance evaluation are shown in Fig. 8 and Table 6.

Fig. 5. Sample result for tracking

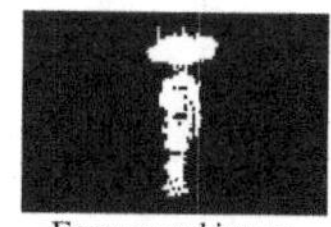

Fig. 6. Experimental result of pedestrian with an umbrella

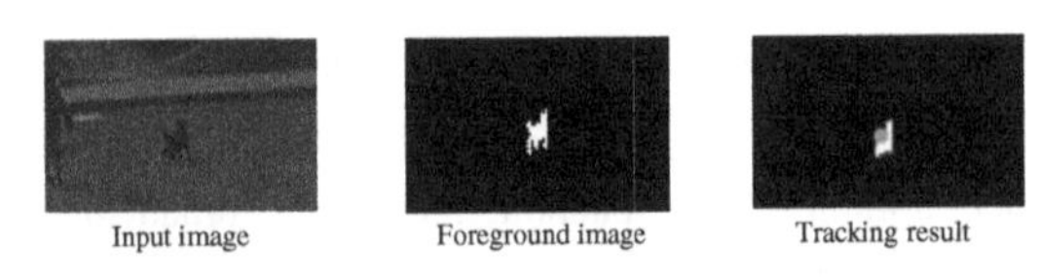

Fig. 7. Experimental result of dog walking

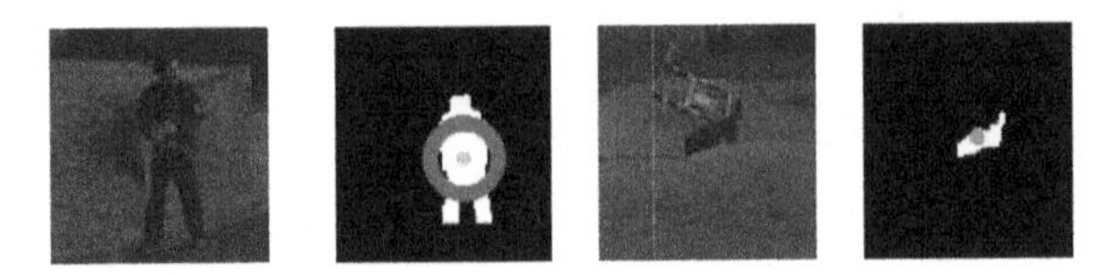

Fig. 8. More experimental results

Table 3. Results by aspect ratio and moment

Frame	Total	Previous frame	Aspect	Moment	Dispersion	White region	Velocity quarterly	Divided	Result
1	24	12.039	3	4	0	0	1	0	1
2	24	12.0195	3	4	0	0	1	0	1
3	24	12.0097	3	4	0	0	1	0	1
4	26	12.0048	2	4	3	0	1	0	1
5	26	13.0024	4	4	0	0	1	0	1
6	25	13.0012	3	4	0	0	1	0	1
7	25.5	12.5006	4	4	0	0	1	0	1
8	26.8	12.7503	4	4	0	0	1	1	1
9	26.4	13.3751	4	4	0	0	1	0	1
10	26.2	13.1875	4	4	0	0	1	0	1

Table 4. Results by pedestrains with an umbrella

Frame	Total	Previous frame	Aspect	Moment	Dispersion	White region	Velocity	Quarterly divided	Result
1	28	13.9541	2	4	3	0	0	0	1
2	28	13.97705	2	4	3	0	0	0	1
3	28	13.98853	2	4	3	0	0	0	1
4	28	13.99426	2	4	3	1	0	0	1
5	28	13.99713	2	4	3	0	0	0	1
6	27	13.99857	2	4	3	0	0	0	1
7	24.5	13.49928	3	4	0	0	0	0	1
8	23.3	12.24964	3	4	0	0	0	0	1
9	2.6	11.62482	2	4	3	0	0	0	1
10	25.3	12.31241	2	4	3	0	0	0	1

Table 5. Accuracy of pedestrains with an umbrella result

	Pedestrian	Umbrella	Carrier case	Bending, sitting	Child	Dog
No. of frames	10	10	30	35	10	25
Correct frames	10	10	21	31	10	20
Accuracy rate (%)	100	100	70	89	100	80

Table 6. Accuracy of more experimental results

Frame number	1 ~ 100	101 ~ 200	201 ~ 300	1 ~ 300
Correct frames	73	74	51	198
Accuracy rate (%)	73	74	51	66

5 Conclusion

In this paper we had proposed a target tracking and classification by using various shape and motion features. The experimentation results show that the proposed method is easy to handle and gives some promising results. In future we would do more experiments for multi-targets associations.

Acknowledgment. This work was supported by JSPS KAKENHI Grant Number JP15K14844.

References

1. Stringa, E., Regazzoni, C.S.: Real-time video-shot detection for scene surveillance applications. IEEE Trans. Image Process. **9**, 69–79 (2000)
2. Donato, D.P., et al.: An autonomous mobile robotic system for surveillance of indoor environments. Intl. J. Adv. Rob. Syst. **7**(1), 019–026 (2010)
3. Abdul, M.S.R., Abeer, S.J.: The proposed design of the monitoring system for security breaches of buildings based on behavioral tracking. Int. J. Comput. Sci. Netw. Secur. **14**(7), 37–40 (2014)
4. Wu, B., Nevatia, R.: Detection and segmentation of multiple, partially occluded objects by grouping, merging, assigning part detection responses. Int. J. Comput. Vision **82**(2), 185–204 (2009)
5. Xing, J., Ai, H., Lao, S.: Multi-object tracking through occlusions by local track lets filtering and global track lets association with detection responses. In: Proceedings of IEEE Conference on Computer Vision and Pattern Recognition, pp. 1200–1207, June 2009
6. Andriluka, M., Roth, S., Schiele, B.: People-tracking-by-detection and people-detection-by-tracking. In: Proceedings of IEEE Conference on Computer Vision and Pattern Recognition, Anchorage, Alaska, June 2008
7. Li, Y., Gu, L., Kanade, T.: A robust shape model for multi-view car alignment. In: Proceedings of IEEE Conference on Computer Vision and Pattern Recognition, Miami, Florida, pp. 2466–2473, June 2009
8. Renno, J., Greenhill, D., Orwell, J., Jones, G.: Occlusion analysis: learning and utilising depth maps in object tracking. In: Image and Vision Computing, pp. 430–441, March 2008

9. Ryoo, M.S., Aggarwal, J.K.: Observe-and-explain: a new approach for multiple hypotheses tracking of humans and objects. In: Proceedings of IEEE Conference on Computer Vision and Pattern Recognition, Anchorage, Alaska, June 2008
10. Haritaoglu, I., Harwood, D., Davis, L.: W4: real-time surveillance of people and their activities. IEEE Trans. PAMI **22**(8), 809–830 (2000)
11. Smith, K., et al.: Evaluating multi-object tracking. In: IEEE Computer Society Conference on Computer Vision and Pattern Recognition, San Diego, June 2005
12. Jaya, P., Balakrishnan, G.: Contour based object tracking. Int. J. Comput. Sci. Inf. Technol. **5**(3), 4128–4130 (2014)
13. Patel, C.I., Ripal, P.: Contour based object tracking. Int. J. Comput. Electr. Eng. **4**(4), 525–528 (2012)
14. Fakhreddine, A.: A new 3D model - based tracking technique for robust camera pose estimation. Int. J. Adv. Comput. Sci. Appl. **3**(4), 31–37 (2012)
15. Patel, M.J., Bhumika, B.: A comparative study of object tracking. Tech. Int. J. Innovative Res. Sci. Eng. Technol. **4**(3), 1361–1364 (2015)

Hybrid Intelligent Systems

Antlion Optimization Based Segmentation for MRI Liver Images

Abdalla Mostafa[1(✉)], Mohamed Houseni[2], Naglaa Allam[2], Aboul Ella Hassanien[3,5], Hesham Hefny[1], and Pei-Wei Tsai[4]

[1] Institute of Statistical Studies and Research, Cairo University, Cairo, Egypt
abdalla_mosta75@yahoo.com
[2] Radiology Department, National Liver Institute, Menofia University, Shibin Al Kawm, Al Minufiyah, Egypt
[3] Faculty of Computers and Information, Cairo University, Cairo, Egypt
[4] College of Information Science and Engineering, Fujian University of Technology, Fuzhou, China
[5] Scientific Research Group in Egypt (SRGE), Cairo, Egypt

Abstract. This paper proposes an approach for liver segmentation, depending on Antlion optimization algorithm. It is used as a clustering technique to accomplish the segmentation process in MRI images. Antlion optimization algorithm is combined with a statistical image of liver to segment the whole liver. The segmented region of liver is improved using some morphological operations. Then, mean shift clustering technique divides the segmented liver into a number of regions of interest (ROIs). Starting with Antlion algorithm, it calculates the values of different clusters in the image. A statistical image of liver is used to get the potential region that liver might exist in. Some pixels representing the required clusters are picked up to get the initial segmented liver. Then the segmented liver is enhanced using morphological operations. Finally, mean shift clustering technique divides the liver into different regions of interest. A set of 70 MRI images, was used to segment the liver and test the proposed approach. Structural Similarity index (SSIM) validates the success of the approach. The experimental results showed that the overall accuracy of the proposed approach, results in 94.49 % accuracy.

Keywords: Antlion optimization · Mean shift clustering · Segmentation

1 Introduction

The automated diagnosis systems that use medical imaging, e.g. CT liver [13], can be handled using current bio-inspired algorithms. These algorithms come out as a result of the observation of nature. They try to mimic the behaviour of animals, insects and birds. Recently, a lot of researchers are trying to understand the behaviour of the swarms to create algorithms, to solve the sophisticated optimization problems in various fields. Mostafa et al. in [8] used the Artificial Bee Colony optimization algorithm to segment the liver in CT image.

J. Pan et al. (eds.), *Genetic and Evolutionary Computing*, Advances in Intelligent Systems and Computing 536, DOI 10.1007/978-3-319-48490-7_31

The segmented clusters are improved using simple region growing technique. Linag et al. in [2] picked up a multilevel threshold using combined Ant Colony Optimization (ACO) algorithm with Expectation and Maximization algorithm. Then they used this threshold to segment the complex structured objects in an image. ABC is also used by Cuevas et al. [1] to compute the image threshold for image segmentation. Sivaramakrishnan et al. [10] used a combination of ABC algorithm and Fish Swarm Algorithm in mamograph images to detect breast tumors. Sankari [9] improved the Expectation-Maximization (EM) algorithm by using Glowworm Swarm Optimization algorithm. GSO extracted the image clusters as initial seed points, and passed to EM algorithm for segmentation. In this paper, Antlion optimization algorithm (ALO) is combined with a statistical image of liver to segment the whole liver. The segmented region of liver is improved using some morphological operations. Then, mean shift clustering technique divides the segmented liver into a number of regions of interest.

The remainder of this paper is ordered as follows. Section 2 gives an overview about ALO algorithm and Mean Shift clustering technique. Section 3 describes in detail the steps of the proposed approach for segmenting liver and getting ROIs. It includes preparing a statistical image, using ALO algorithm for clustering and using mean shift to extract ROIs. Section 4 shows the experimental results of the proposed approach. Finally, conclusions and future work are discussed in Sect. 5.

2 Methods

2.1 Antlion Optimization Algorithm

Antlion optimization (ALO) algorithm is one of the newest bio-inspired optimization algorithms proposed by Mirjalili. It mimics the hunting mechanism of an insect called Antlion, that digs a trap for catching insects, specially ants. It digs a cone-shaped hole in the sands and hides at the bottom, waiting for an insect to be trapped [3]. The algorithm depends on three main parameters, including the size of search space (antlions and ants), the number of clusters and the number of iterations.

Operators of ALO Algorithm: The ALO algorithm mimics the movement of ants in nature, and the process of building traps and hunting by antlions. This section describes the process as follows [3].

- **Random walks of the ants:** The ant walks randomly inside the search space according to the following equation:

$$X_i^t = \frac{(X_i^t - a_i) \times (d_i - C_i^t)}{(d_i^t - a_i)} + C_i \tag{1}$$

  where a_i is the minimum value of random walk of ith variable, C_i^t is the minimum value of ith variable at t^{th} iteration, and d_i^t represents the maximum of ith variable at t^{th} iteration. If ant random walk is outside the search space, it is normalized to minimum or maximum value of the search space.

- **Trapping in antlions pits:** The random walks of the ants are affected by the antlion traps. In Eq. (2), it is shown that the random walks of the ants are determined by a range of values (c,d). Where c is the minimum of all variables for i^{th} ant and d is the maximum of all variables for i^{th} ant.

$$c_i^t = Antlion_j^t + C^t, \qquad d_i^t = Antlion_j^t + d^t \tag{2}$$

- **Building trap:** A roulette wheel is used to model the antlion capability of hunting. Since each ant is to be trapped by only one antlion, the roulette wheel operator selects the antlion according to its fitness. The higher the fitness of an antlion, the higher the chance to catch ants.
- **Sliding ants towards antlion:** When the antlion realizes that there is an ant in the conic hole, it starts to shoot sands outwards the centre of the trap. Mathematically, the radius of the ants random walks is decreased adaptively according to the following equations.

$$c^t = \frac{c^t}{I}, \qquad d^t = \frac{d^t}{I} \tag{3}$$

where c^t is the minimum of all variables at t^{th} iteration, and d^t indicates the vector including the maximum of all variables at t^{th} iteration. Also, notice that $I = 10^w \frac{t}{T}$ where t is the current iteration and T is the maximum number of iterations.
- **Catching prey and re-building the trap:** Finally, when the ant becomes fitter than the antlion, the antlion catches the ant and eat its body. The antlion updates its position to the position of the hunted ant, to increase its chance for a new hunt. The following equation is proposed.

$$Antliont_j^t = Ant_i^t \quad if\ f(Ant_i^t) > f(Antlion_j^t) \tag{4}$$

- **Elit-Antlion:** It represents the best solution of all antlions in the search space, in all iterations. In each iteration, the best fitness of antlions is compared to the elit-antlion. The elit-antlion solution affects the process of random walks of the ants around the antlion, by using the roulette wheel random walk of the ant (R_A^t) and the random walk around the elit-antlion (R_E^t) according to the following equation.

$$Ant_i^t = \frac{R_A^t + R_E^t}{2} \tag{5}$$

2.2 Mean Shift Clustering

Mean shift is an iterative algorithm depending on density gradient estimation using a generalized kernel function. In fact, mean shift operates to smooth the image distribution and define its peaks. Finally, it catches the pixels that correspond to each peak. This represent the ROIs of the image [11]. In the beginning of the algorithm, a fixed kernel of width ($\boldsymbol{h}$), is used to smooth the image data. Then it searches for the local minimum and calculates the gradient of the density function.

3 The Proposed CT Liver Segmentation Approach

The proposed CT liver segmentation approach consists of some main phases. It includes preprocessing, clustering using ALO algorithm, statistical image and morpjological operations to get the segmented image, and finally mean shift phase to get ROIs. Algorithm 1 describes the steps of the proposed approach. The following subsections will describe the proposed algorithm steps in details.

Algorithm 1. The proposed liver segmentation approach

1: Prepare a statistical image, collected from all the manual segmented images.
2: Clean the image annotations using morphological operations.
3: Use ALO to get the image clusters.
4: Multiply the binary statistical image by the resulting clustered image.
5: Pick up the required clusters to get the initial segmented liver.
6: Enhance the segmented image by morphological operations to remove small regions.
7: Use Structural similarity index to calculate the accuracy of the automated segmented image compared to the manual segmented image.
8: use Mean shift technique to divide the image into a number of regions of interest.

3.1 Preprocessing Phase

This phase consists of two steps, preparing the statistical image and cleaning the image The first step in preprocessing phase is to prepare the statistical image which will be used later to get the potential area that liver might be in.

Preparing the statistical image: The liver is segmented manually in all dataset images, and approved by specialists in National Liver Institution in menofia. The manual segmented images of liver are converted into binary images. The summation of all binary images in one matrix, represents the all possible occurrences of liver in any abdominal MRI image.

Cleaning phase: Image cleaning depends on removing the patient information from the MRI image, using some morphological operations.

3.2 Ant Lion Optimization Phase (ALO)

ALO algorithm is used to cluster the intensity values of the MRI Image, helping to segment the liver. Parameter setting is the key in ALO. The parameters includes the number of the predefined clusters, the iterations number and the search agents number. K-means is used as a fitness function to help finding out the new solutions through the iterations. At the end, the best solutions will represent the centroids of the required clusters. Using ALO starts by setting the value of different parameters of clusters, search space, and iterations. Using k-means as fitness function, ALO is used to determine the image clusters. The clusters values are sorted and labelled. Finally, the distance between every pixel's intensity value and clusters values, is calculated to determine the appropriate cluster for the pixel. It should be mentioned that ALO is not affected by image noise, because noise is spread as small points. These small holes are filled in the binary image.

3.3 Statistical Image and Image Enhancement Phase

Since, the range of intensity values of liver is similar to other organs. The usage of statistical image removes a great part of other organs' tissues. The binary statistical image, which represents the liver statistical occurrence, is multiplied by the image, resulting from ALO. Then, the required clusters are picked up by manually and multiplied by the original image to get the initial segmented liver. Finally, the morphological operations are used to fill the holes of the binary image and remove the small objects outside the liver.

3.4 ROI Segmentation Phase

After segmenting the liver from the abdominal MRI image, it should be prepared for later classification process. It must be divided into small homogeneous regions surrounded by closed contours. Mean shift clustering technique segments the liver area into a number of ROIs. It starts by smoothing the image. The super-pixel clustering method gets the initial labels of the segmented image. Mean value is calculated for each region. Mean is shifted using the gradient to find the new local minimum. Finally, the vector of all local minima is calculated, and all appropriate pixels are added to the region of the nearest minimal.

4 Experimental Results and Discussion

The experiments of the proposed approach of segmentation, will be covered in this section. A set of 70 MRI images were used in the experiments to test the proposed approach. Figure 1 shows ALO image and the resulting image from applying the removal of right part close to the liver, and the multiplication of all possible statistical occurrence on the abdominal image. This process excludes a great part of the un-required organs from the image, especially the organs of stomach and spleen. It also, shows the resulting image from picking up the required clusters that represent the liver. The chosen clusters are multiplied by the original image. The user does not have to choose the cluster of the small regions of lesions inside the liver. The small fragments representing the un-chosen clusters of lesion might be holes inside the liver. When the liver is extracted, these holes can be filled easily. The last process of segmentation enhances the picked up clustered image using morphological operations. It erodes and removes the small objects in the image. Finally, Mean Shift clustering technique segments the image of the whole liver. It extracts the different homogeneous ROIs, to be the preparation for a future step of feature extraction and lesion classification. Figure 2 shows the result of the enhanced image and Segmented ROIs.

Finally, Mean Shift clustering technique segments the image of the whole liver. It extracts the different homogeneous ROIs, to be the preparation for a future step of feature extraction and lesion classification. Figure 3 shows the result of using Mean Shift to extract liver ROIs.

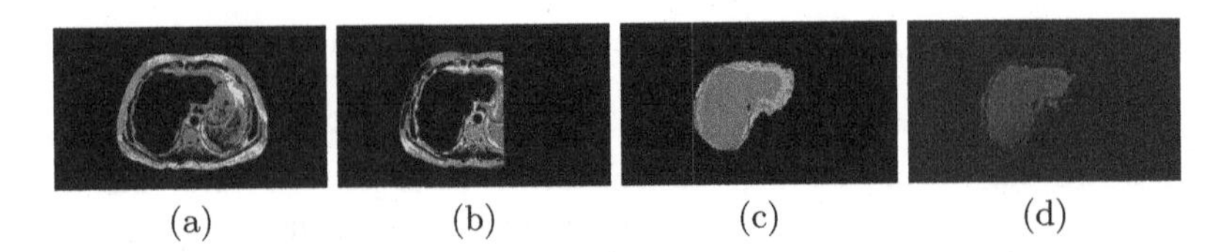

Fig. 1. Picking up the required clusters: (a) ALO image (b) Removing the right part of the liver (c) Statistical occurrence image, (d) The picked up clusters multiplied by the original image

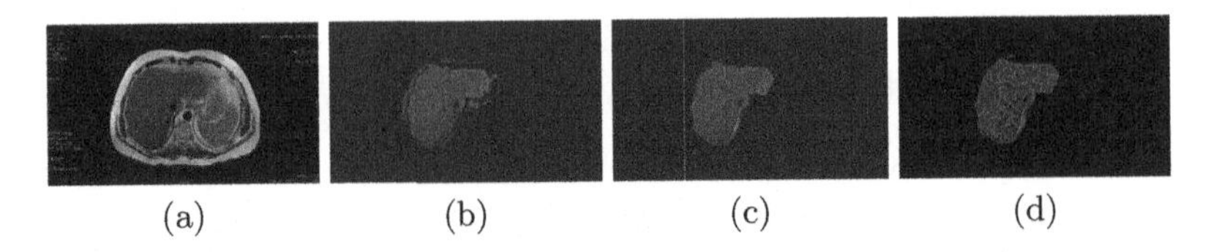

Fig. 2. Enhancing segmented image: (a) Original image (b) Picked up clusters image, (c) Enhanced image by morphological operations, (d) Mean shift segmented ROIs.

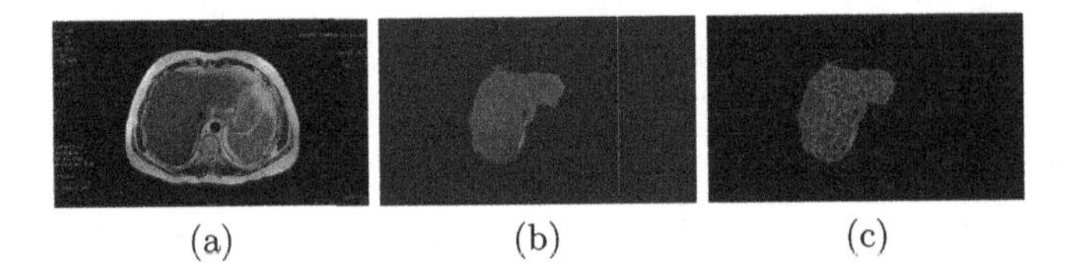

Fig. 3. Mean shift image: (a) Original image (b) Segmented image, (c) Mean shift image

Initially, The proposed algorithm is tested using 5 randomly selected MRI images from the used dataset. Each value for each parameter is tested independently on the five images. Then we get the average results of parameter's value. The best parameter values are 10 search agents, 10 iterations and 7 clusters.

Evaluation is performed using SSIM, which is defined using Eq. 6.

$$SSIM(x,y) = \frac{(2\mu_x\mu_y + c_1)(2\sigma_{xy} + c^2)}{(\mu_x^2\mu_y^2 + c_1)(\sigma_x^2\sigma_y^2 + c_2)} \quad (6)$$

where $SSIM$ is the structural similarity index measure, x is the automatic segmented image, y is the manual segmented image, μ_x is the average of x, μ_y is the average of y, σ_x is the variance of x, σ_y is the variance of y, σ_{xy} is the covariance of x and y. Besides, $c_1 = (k_1L)^2$, and $c_2 = (k_2L)^2$, where L is the dynamic range of the pixels values, $K_1 = 0.01$ and $K_2 = 0.03$ by default.

The experiments of using the proposed approach showed a good efficiency. It proved a better performance compared with other traditional and bio-inspired segmentation methods. Table 1 shows the results of the proposed approach. It shows that the performance of the proposed approach is 94.49 %. In Table 2 shows the result of the proposed approach is compared to other traditional and

Table 1. Results of proposed approach

Img no	ALO	Img no	ALO	Img no	ALO	Img no	ALO	Img no	ALO
1	0.9768	15	0.9597	29	0.9504	43	0.9563	57	0.9346
2	0.9688	16	0.917	30	0.9426	44	0.9421	58	0.9458
3	0.9796	17	0.9604	31	0.9688	45	0.9219	59	0.9174
4	0.9562	18	0.9531	32	0.9689	46	0.9632	60	0.9258
5	0.9633	19	0.9331	33	0.9385	47	0.92	61	0.9476
6	0.9646	20	0.9477	34	0.9271	48	0.9827	62	0.9668
7	0.9212	21	0.9417	35	0.947	49	0.957	63	0.9311
8	0.9607	22	0.9068	36	0.9344	50	0.885	64	0.936
9	0.9569	23	0.9484	37	0.91	51	0.951	65	0.9588
10	0.9719	24	0.9485	38	0.9381	52	0.9552	66	0.9067
11	0.93	25	0.9445	39	0.9356	53	0.9566	67	0.9253
12	0.953	26	0.9528	40	0.961	54	0.956	68	0.9167
13	0.96	27	0.9366	41	0.9767	55	0.945	69	0.9061
14	0.9428	28	0.925	42	0.9518	56	0.9474	70	0.9524
Result	0.9449								

Table 2. Comparison between the proposed approach and other approaches

Ser	Approach	Result
1	Region growing (RG) [12]	84.82
2	Wolf local thresholding + RG [4]	91.17
3	Morphological operations +RG [6]	91.20
4	Level set [12]	92.10
5	K-means +RG [7]	92.38
6	Artificial Bee Colony (ABC) [8]	93.73
7	Grey Wolf (GW) [5]	94.08
8	Proposed approach (ALO)	94.49

bio-inspired approaches. The proposed approach, compared to these approaches, achieved the best result.

5 Conclusion and Future Work

The proposed approach depends on the Ant Lion Optimizer algorithm in segmenting the liver from MRI images. It results in a clustered image, which is multiplied by the statistical image to remove a part of the abdomen that includes other organs. The required clusters representing the liver is picked up manually to get an initial segmented liver. Then the image is enhanced using some morphological operations to remove the small objects outside the liver boundary. Mean shift clustering algorithm is used finally to get the ROIs. The accuracy of the segmented image is tested using SSIM. The segmentation of liver using the proposed approach has an average accuracy rate of 94.49 % using SSIM. The future work will concentrate on a better performance using other bio-inspired algorithms.

References

1. Cuevas E. Sencin, F., Zaldivar D., Prez-Cisneros M., Sossa H.: Applied intelligence (2012)
2. Liang, Y., Yin, Y.: A new multilevel thresholding approach based on the ant colony system and the EM algorithm. Int. J. Innov. Comput. Inf. Control **9**, 1 (2013)
3. Mirjalili, S.: The ant lion optimizer. Adv. Eng. Softw. **83**, 80–98 (2015)
4. Mostafa, A., AbdElfattah, M., Fouad, A., Hassanien, A., Hefny, H.: Wolf local thresholding approach for liver image segmentation in CT images. In: Abraham, A., Wegrzyn-Wolska, K., Hassanien, A.E., Snasel, V., Alimi, A.M. (eds.) 2015 International Afro-European Conference for Industrial Advancement AECIA, vol. 427, pp. 641–651. Springer, Switzerland (2015)
5. Fouad, A.A., Mostafa, A., Ismail, S.G., Abd, E.M., Hassanien, A.: Nature inspired optimization algorithms for CT liver segmentation. In: Dey, N., Bhateja, V., Hassanien, A.E. (eds.) Medical Imaging in Clinical Applications: Algorithmic and Computer-Based Approaches, vol. 651, pp. 431–460. Springer, Switzerland (2016)
6. Mostafa, A., Abd, E.M., Fouad, A., Hassanien, A., Hefny, H.: Enhanced region growing segmentation for CT liver images. In: Gaber, T., Hassanien, A.E., El-Bendary, N., Dey, N. (eds.) The 1st International Conference on Advanced Intelligent System and Informatics, Beni Suef, Egypt, vol. 407, pp. 115–127. Springer, Switzerland (2016)
7. Mostafa, A., Abd, E.M., Fouad, A., Hassanien, A., Kim, T.: Region growing segmentation with iterative K-means for CT liver images. In: International Conference on Advanced Information Technology and Sensor Application (AITS), China (2015)
8. Mostafa, A., Fouad, A., Abd, E.M., Hassanien, A., Hefny, H., Zhue, S.Y., Schaeferf, G.: CT liver segmentation using artificial bee colony optimization. In: 19th International Conference on Knowledge Based and Intelligent Information and Engineering Systems, Procedia Computer Science, vol. 60, pp. 1622–1630 (2015)
9. Sankari, L.: Image segmentation using glowworm swarm optimization for finding initial seed. Int. J. Sci. Res. (IJSR) **3** (2014)
10. Sivaramakrishnan, A., Karnan, M.: Medical image segmentation using firefly algorithm and enhanced bee colony optimization. In: International Conference on Information and Image Processing (ICIIP), pp. 316–321 (2014)
11. Szeliski, R.: Computer vision: algorithms and applications (2010)
12. Zidan, A., Ghali, N.I., Hassanien, A., Hefny, H.: Level set-based CT liver computer aided diagnosis system. Int. J. Imaging Robot. **9** (2013)
13. Sayed, G.I., Ali, M.A., Gaber, T., Hassanien, A.E., Snasel, V.: A hybrid segmentation approach based on neutrosophic sets and modified watershed: a case of abdominal CT Liver parenchyma. In: 2015 11th International Computer Engineering Conference (ICENCO), December 2015, pp. 144–149. IEEE (2015)

Handwritten Arabic Manuscript Image Binarization Using Sine Cosine Optimization Algorithm

Mohamed Abd Elfattah[1,4(✉)], Sherihan Abuelenin[1], Aboul Ella Hassanien[2,4], and Jeng-Shyang Pan[3]

[1] Faculty of Computers and Information, Mansoura University, Mansoura, Egypt
mohabdelfatah8@gmail.com

[2] Faculty of Computers and Information, Cairo University, Giza, Egypt

[3] Fujian Provincial Key Laboratory of Big Data Mining and Applications, Fujian University of Technology, Fuzhou, China

[4] Scientific Research Group in Egypt (SRGE), Cairo, Egypt
http://www.egyptscience.net

Abstract. Historic manuscript image binarization is considered an important step due to the different kinds of degradation effects on optical character recognition (OCR) or word spotting systems. Previous methods failed on to find the optimal threshold for binarization. In this paper, we investigate the effects of sine cosine algorithm (SCA) on reducing the compactness K-means Clustering as the objective function. The SCA searches for the optimal clustering of the given handwritten manuscript image into compact clusters under some constraints. The proposed approach is evaluated and assessed on a set of selected handwritten Arabic manuscript images. The Experimental result shows that the proposed approach provides the highest value than the famous binarization methods such as; Otsu's and Niblack's in terms of F-measure, Pseudo- F-measure, PSNR, Geometric accuracy and the low value on DRD, NRM, MPM.

Keywords: Image binarization · K-means clustering · Sine cosine optimization algorithm

1 Introduction

Handwritten manuscript image binarization is executed in the preprocessing stage for handwritten manuscripts images analysis and the main goal from it to get the foreground text from the manuscript background. Excellent and cleared binarized image is the aim of any proposed approach because it results has the direct effect on the final stage of the different system such as word spotting or optical character recognition. Poorly handwritten degraded manuscript images segmentation is a very challenging task because of the high inter/intravariation between the manuscript foreground text and background of different handwritten manuscript images [1]. In [1] presented a novel image binarization technique

J. Pan et al. (eds.), *Genetic and Evolutionary Computing*, Advances in Intelligent Systems and Computing 536, DOI 10.1007/978-3-319-48490-7_32

using adaptive image contrast firstly, an adaptive contrast map is created for an input image. Secondly, the contrast map is then binarized and joint with Canny's edge map for identifying the text stroke edge pixels. Finally, a local threshold is used for document segmentation based on intensities estimation of detected text stroke edge pixels within a local window, this method applies to the different dataset and compared with the state of art methods, the result indicates to the success of it.

In our previous work, a novel handwritten binarization based on Neutrosophic [2,3] and Savoula's method [4] is presented in [5,6], it's applied to selected manuscripts from different resource, while in [7], Artificial bee colony (ABC) [8,9] is used for find the optimal threshold with the same manuscripts. The best widespread clustering technique is the K-means algorithm [13,23], the main disadvantage of it that it needs to define the initial value and the number of clusters. In this paper, we investigate a novel optimization algorithm called sine cosine algorithm based on K-means clustering as a fitness function for find the high quality binarized image (black and white). The results were compared with the famous binarization algorithm named; Otsu's [10] and Niblack's [11].

The structures of paper as follow. The sine-cosine optimization algorithm is presented in Sect. 2, The proposed binarization approach is showed in Sect. 3. The experimental result is presented in Sect. 4. Conclusion and future work are clarified in Sect. 5.

2 Sine Cosine Algorithm: Review

Sine cosine algorithm [12] is a new optimization algorithm. The updating process depends on the following Eq. 1 based on the value of r_4, if it values $\geq$0.5 depends on cosine else is routed on sine. The main advantage of this algorithm is that provides 4 random numbers for the solution instead of one solution as another the optimization algorithms. The steps of this algorithm are presented in [12].

$$X_i^{t+1} = \begin{cases} X_i^t + r_1 \times sin(r_2) \times |r_3 P_i^t - X_i^t| & r_4 < 0.5 \\ X_i^t + r_1 \times cos(r_2) \times |r_3 P_i^t - X_i^t| & r_4 \geq 0.5 \end{cases} \quad (1)$$

The r_1, r_2, r_3 and r_4 presented the random number, the position of the current solution in *i-th* dimension at *t-th* iteration presented by X_i^t, P_i is the position of the destination in *i-th* dimension, where the $||$ present the absolute value, and the value of the random number r_4 in [0,1]. The high balance among the exploration phase and the exploitation phase is the main objective of SCA that discover the eventually converge and promising regions of the search space to the global optimum, for balance between exploration and exploitation, the range of cosine and sine in the previous equation is reformed using the following Eq. 2

$$r_1 = a - t\frac{a}{T}. \quad (2)$$

where the a is constant, T denotes to the maximum number of iteration and t presents the current iteration.

3 The Handwritten Manuscript Image Binarization Proposed Approach

The handwritten manuscript binarization begins by calling the main SCA optimization algorithm and put on the selected handwritten manuscript image that minimizes the K-means clustering compactness function. The SCA explorations for the clusters, each cluster denoted by the cluster centroid, mean; and later generally, the exploration space is of 2D when examining for two clusters each denoted by single gray level. Now, the gained cluster centers are used for assembling a white and black image that characterizes the handwritten manuscript, where the foreground denoted by the darkest cluster. The cluster centers are updated on each iteration until the high quality binarized image is created.

The sine-cosine algorithm is searching for find an optimal of a given fitness function of a space of n-D. Equation 3 is K-means clustering [13] where used here as fitness function for clustering. The fitness function normally in the clustering encodes the intra-cluster compactness.

$$J = \sum_{j=1}^{x} \sum_{i=1}^{k} \| x_i^j - c_j \|^2 \tag{3}$$

Where $\| x_i^j - c_j \|^2$ represents a distance measure between the cluster center and data point, and a pointer of the distance of the n data points from their own cluster centers.

4 Implementation and Discussion

In this study. The handwritten manuscript Arabic images are selected and collected from [14]. These images were collected and selected based on the level of noise which each image suffers from one or more kind of noise such as; faded ink, shadow, smears, uneven illumination. Figure 1 shows samples of the collected and tested handwritten manuscript images.

Table 1 shows the SCA parameter's setting value which provides the high quality binarized image. Seven different performance measures are used here for evaluating the proposed approach; Geometric accuracy [15], Distance Reciprocal Distortion (DRD) [17], pseudo-F measure [16,18], F-measure [18,19,23], Peak Signal to Noise Ratio (PSNR) [18] and Misclassification penalty metric (MPM) [20] and Negative Rate Metric (NRM) [20].

Table 2 shows the performance of the proposed approach on the selected handwritten manuscript, each row from the table represents the name of the image and the value of each performance measure and finally, the average over 5 images is presented in the last row. Which approve that the proposed approach is able to present the binarized image is very similar to the ground truth image.

In Table 3, the performance comparison between the proposed approach and two other methods is presented, we can see the value of F-measure, p-FM, PSNR, Geometric accuracy is higher than other methods and low in DRD, NRM, MPM

Table 1. Sine cosine parameter's setting

Ser.	Parameter	Setting
1	No. of search agent's	15
2	Maximum iteration	10
3	No. of clusters	2
4	Dimension	2
5	Range	[0 255]

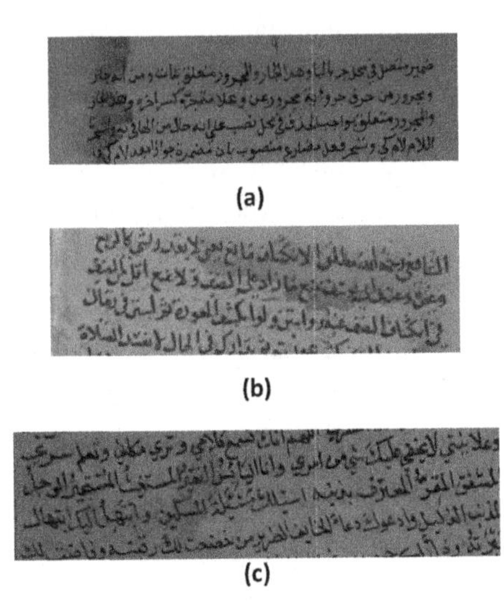

Fig. 1. Samples of handwritten images [14]

Table 2. Performance of the proposed approach

Image name.	F-Measure	p-FM	PSNR	DRD	MPM	GA	NRM
Manu 1	92.08	95.91	16.34	1.86	0.69	0.94	0.04
Manu 2	95.05	95.92	18.51	1.43	0.60	0.98	0.01
Manu 3	95.53	96.88	20.66	0.54	0.08	0.95	0.01
Manu 4	94.02	95.26	16.41	1.27	0.96	0.96	0.03
Manu 5	98.99	98.89	26.47	0.25	3.25	0.99	0.002
Average	95.13	96.57	19.67	1.07	1.11	0.96	0.01

Table 3. Comparative analysis

Method name.	F-Measure	p-FM	PSNR	DRD	MPM	GA	NRM
Otsu's [10]	90.08	91.33	15.85	1.87	2.81	0.95	0.03
Niblack's [11]	83.47	87.55	14.08	2.79	5.21	0.89	0.09
Proposed approach	95.13	96.57	19.67	1.07	1.11	0.96	0.01

Fig. 2. Results (a) Original handwritten degraded image (b) Ground truth (c) Otsu's binarized image (d) Niblack's binarized image (e) Proposed approach binarized image

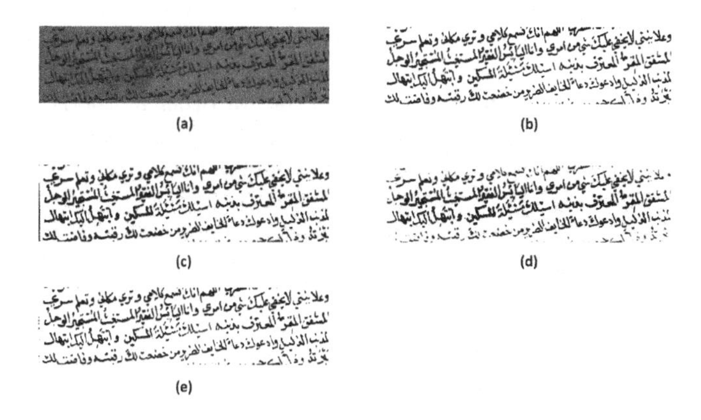

Fig. 3. Results (a) Original handwritten degraded image (b) Ground truth (c) Otsu's binarized image (d) Niblack's binarized image (e) Proposed approach binarized image

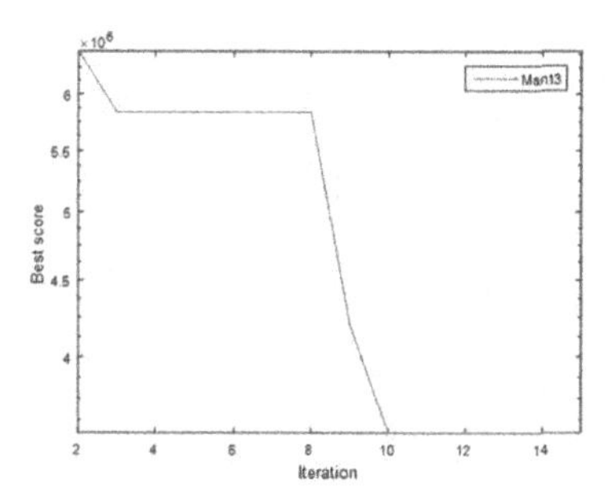

Fig. 4. Converge curve of Manu13

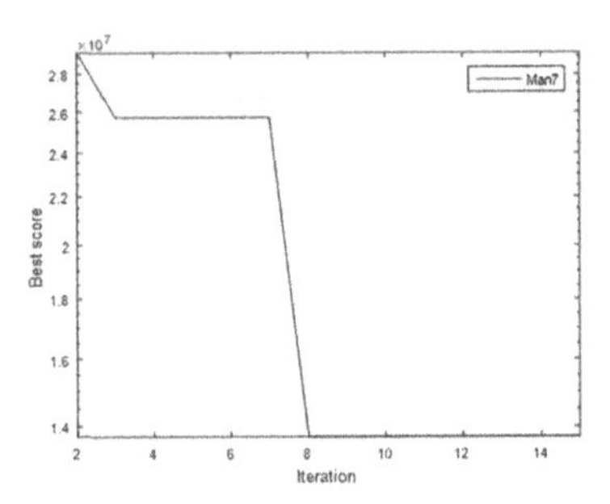

Fig. 5. Converge curve of Manu7

which indicates to the powerful of the SCA optimization for processing the different kinds of noise on handwritten manuscripts images.

In Figs. 2 and 3, the visual inspection between the original degraded handwritten manuscript image and the result of the proposed approach and the result from Otsu's [10] and Niblack's [11] are presented, the results denotes to the success of the proposed approach for dealing with different kinds of noise, The original image in Fig. 2 is different than original image in Fig. 3, however, Results indicate the superiority of the proposed method for other methods.

For more analysis and assessment of the proposed approach, the converge curve of it is presented in Figs. 4 and 5. This figure shows the powerful of the proposed approach for finding the best solution in little time. The fewer parameters and the strength of sine cosine optimization algorithm could avoid local extremum problem, and it able to reach to an excellent result of image binarization. From the previous analysis of the proposed approach result, the sine cosine optimization algorithm is very useful in this problem and able to avoid the main drawback of K-means (local minima problem). The main advantage of SCA algorithm is making a number of solution for our problem, the number is 4 solution which provides with high exploration.

Through the previous result analysis, the proposed approach is able to minimize the objective function and can provide the excellent binarized image, because it is transferred with speed motion between the exploration and exploration phase which is given to the proposed approach more strength to dealing with different fitness function in the future.

5 Conclusion and Future Work

A novel handwritten manuscript image binarization is presented using the sine cosine algorithm to minimize the possibilistic K-means clustering fitness function. The results prove the fast convergence of the sine-cosine optimize algorithm (SCA) and the robustness of the binarized image, also, the proposed approach is able to extract the weak part of the handwritten manuscript image. In future work, some fitness function will be added to the proposed approach like enhancement image. We can investigate the proposed approach in many application such as; medical application [21], also, we can make a hybrid version with any another optimization algorithm. In addition to we can investigate it improve the accuracy of prediction system as in [22] based on it for select the most important attributes.

References

1. Su, B., Lu, S., Tan, C.L.: Robust document image binarization technique for degraded document images. IEEE Trans. Image Process. **22**(4), 1408–1417 (2013)
2. Guo, Y., Cheng, H.-D.: New neutrosophic approach to image segmentation. Pattern Recogn. **42**(5), 587–595 (2009)
3. Smarandache, F.: A Unifying Field in Logics Neutrosophic Logic. Neutrosophy, Neutrosophic Set, Neutrosophic Probability, 3rd edn. American Research Press, Santa Fe (2003)
4. Sauvola, J., Pietikinen, M.: Adaptive document image binarization. Pattern Recogn. **33**(2), 225–236 (2000)
5. Amin, K.M., Abd Elfattah, M., Hassanien, A.E., Schaefer, G.: A binarization algorithm for historical arabic manuscript images using a neutrosophic approach. In: The 9th International Conference on Computer Engineering and Systems (ICCES), Egypt, pp. 266–270. IEEE (2014)
6. Hassanien, A.E., Abd Elfattah, M., Amin, K.M., Mohamed, S.: A novel hybrid binarization technique for images of historical Arabic manuscripts. Stud. Inform. Control **24**(3), 271–282 (2015). ISSN 1220–1766
7. Abd Elfattah, M., Hassanien, A.E., Mostafa, A., Ali, A.F., Amin, K.M., Mohamed, S.: Artificial bee colony optimizer for historical Arabic manuscript images binarization. In: The 11th International Conference on Computer Engineering (ICENCO), Egypt, pp. 251–255. IEEE (2015)
8. Karaboga, D.: An idea based on honey bee swarm for numerical optimization. Technical report University, Engineering Faculty, Computer Eng Department (2005)
9. Karaboga, D., Basturk, B.: A powerful and efficient algorithm for numerical function optimization: artificial bee colony (ABC) algorithm. J. Glob. Optim. **39**(3), 459–471 (2007)
10. Otsu, N.: A thresholding selection method from gray-level histogram. IEEE Trans. Syst. Man Cybern. **9**(1), 62–66 (1979)
11. Niblack, W.: An introduction to digital image processing, pp. 115–116. Prentice-Hall, Englewood Cliffs (1986)
12. Mirjalili, S.: SCA: a sine cosine algorithm for solving optimization problems. Knowl. Based Syst. **96**, 120–133 (2016). Elsevier
13. MacQueen, J.: Some methods for classification and analysis of multivariate observations. In: Proceedings of the Fifth Berkeley Symposium on Mathematical Statistics and Probability, vol. 1(14), pp. 281–297 (1967)
14. http://wqf.me/. Accesssed 12 Apr 2016, 8.00 P.M
15. Paredes, R., Kavallieratou, E., Lins, R.D.: ICFHR 2010 contest: quantitative evaluation of binarization algorithms. In: 2010 International Conference on Frontiers in Handwriting Recognition (ICFHR), pp. 733–736. IEEE (2010)
16. Ntirogiannis, K., Gatos, B., Pratikakis, I.: An objective evaluation methodology for document image binarization techniques. In: The 8th IAPR International Workshop on Document Analysis Systems (DAS 2008), Nara Prefectural New Public Hall, Nara, Japan, 17–19 September 2008, pp. 217–224 (2008)

17. Lu, H., Kot, A.C., Shi, Y.Q.: Distance-reciprocal distortion measure for binary document images. IEEE Sig. Process. Lett. **11**(2), 228–231 (2004)
18. Ntirogiannis, K., Gatos, B., Pratikakis, I.: Performance evaluation methodology for historical document image binarization. IEEE Trans. Image Process. **22**(2), 595–609 (2013)
19. Ntirogiannis, K., Gatos, B., Pratikakis, I.: ICFHR2014 competition on handwritten document image binarization (H-DIBCO 2014). In: 2014 14th International Conference on Frontiers in Handwriting Recognition (ICFHR), pp. 809–813. IEEE (2014)
20. Pratikakis, I., Gatos, B., Ntirogiannis, K.: H-DIBCO 2010-handwritten document image binarization competition. In: 2010 International Conference on Frontiers in Handwriting Recognition (ICFHR), pp. 727–732. IEEE (2010)
21. Mostafa, A., Fouad, A., Abd Elfattah, M., Hassanien, A.E., Hefny, H., Zhu, S.Y., Schaefer, G.: CT liver segmentation using artificial bee colony optimisation. Procedia Comput. Sci. **60**, 1622–1630 (2015)
22. Abd Elfattah, M., Waly, M.I., Elsoud, M.A.A., Hassanien, A.E., Tolba, M.F., Platos, J., Schaefer, G.: An improved prediction approach for progression of ocular hypertension to primary open angle glaucoma. In: Kömer, P., Abraham, A., Snášel, V. (eds.) Proceedings of the Fifth International Conference on Innovations in Bio-Inspired Computing and Applications IBICA 2014. AISC, vol. 303, pp. 405–412. Springer, Heidelberg (2014). doi:10.1007/978-3-319-08156-4_40
23. Gaber, T., Ismail, G., Anter, A., Soliman, M., Ali, M., Semary, N., Hassanien, A.E., Snasel, V.: Thermogram breast cancer prediction approach based on Neutrosophic sets and fuzzy c-means algorithm. In: 2015 37th Annual International Conference of the IEEE Engineering in Medicine and Biology Society (EMBC), pp. 4254–4257. IEEE, August 2015

An Adaptive Approach for Community Detection Based on Chicken Swarm Optimization Algorithm

Khaled Ahmed[1,3](✉), Aboul Ella Hassanien[1,3], Ehab Ezzat[1], and Pei-Wei Tsai[2]

[1] Faculty of Computers and Information, Cairo University, Giza, Egypt
khaled.elahmed@gmail.com, aboitcairo@gmail.com, Admin@drehab.net
[2] College of Information Science and Engineering, Fujian University of Technology, Fuzhou, China
pwtsai@fjut.edu.cn
[3] Scientific Research Group in Egypt (SRGE), Cairo, Egypt
http://www.egyptscience.net

Abstract. This paper presents an adaptive approach based on chicken swarm optimization algorithm (ACSO) for community detection problem in complex social networks. The proposed approach is able to define dynamically the number of communities for complex social network. The basic chicken swarm algorithm by its nature is continuous which can't fit for community detection domain so it needs to be redesigned as a discrete chicken swarm for a better exploration of the search space. Locus-based adjacency scheme is used for encoding and decoding tasks while NMI and Modularity are used as an objective function.

The proposed approach is executed over four popular cited benchmarks data sets with different size of small, medium and large scale data sets such as Zachary karate club, Bottlenose dolphin, American college football and Facebook. Experimental results are measured with quality measures such as NMI, Modularity and Ground truth. ACSO's results are compared with eight well-known community detection algorithms such as A discrete BAT, Artificial fish swarm, Infomap, Fast Greedy, label propagation, Walktrap, Multilevel and A discrete Krill herd Algorithm. ACSO has achieved high accuracy and quality results for community detection and community structure for complex social networks.

Keywords: Nature inspired intelligence · Chicken swarm optimization, Community detection, Community structure, Social networks analysis

1 Introduction

Social network is graph of nodes and edges [1] which can lead to discover many valuable insights. These nodes represent online user's profiles and these edges represent the interaction between nodes [2]. Community structure is to define complex network's topology in order to divide this complex network into a set

J. Pan et al. (eds.), *Genetic and Evolutionary Computing*, Advances in Intelligent Systems and Computing 536, DOI 10.1007/978-3-319-48490-7_33

of clusters, which have common features such as the most interacted together nodes [3,4].

Many researchers tackled community detection problem using traditional algorithms and techniques, such as graph partitioning, spectral clustering, hierarchical clustering and divisive algorithms, such as Newman-Girvan algorithm, Modularity-based, greedy techniques and simulated annealing. There are also dynamic techniques, which used for community detection problem, such as spin model and synchronization model [5,6].

On other hand some researchers tackled community detection problem using Bio-inspired swarm intelligence algorithms, which are inspired from nature [7]. In [8] presents Bat optimization algorithm, in [9] presents Artificial Fish swarm algorithm and a discrete krill herd algorithm [10].

This paper presents a new adaptive community detection approach (ACSO) based on chicken swarm optimization algorithm. Chicken swarm optimization algorithm is continuous in nature [10], so the algorithm is redesigned into a discrete chicken swarm optimization algorithm for better exploration of the search space and to be able to perform community detection task.

The rest of the paper is organized as follows: Sect. 2 providing a brief introduction of community detection problem and states the traditional chicken swarm algorithm. Section 3 discuss the proposed algorithm and the parameters redesign. Section 4 presents experiments results with comparative analysis. Conclusion and future works is presented in Sect. 5.

2 Community Detection Problem:review

2.1 Problem Statement

Community detection is the process of clustering social network's nodes based on their connectivity and interactivity with other nodes, which means nodes with high density of connected edges will be in same cluster [5].

The proposed problem is to define dynamically the number of communities within complex networks, which have many challenges such as heterogeneity and evaluation of these complex networks [11]. Measuring objective quality functions values such as Modularity, NMI and Ground truth is an important challenge [6,9,12].

2.2 Traditional Chicken Swarm Algorithm and Its Behavior

Chicken swarm optimization algorithm is a nature inspired swarm algorithm. It simulates the behavior of chicken swarm in nature. Chicken swarm moves in a hierarchy groups of one rooster with two hens and chicks searching for food so food has an effect on the herd movements. Chicken swarm movements are simulated into three basic steps: (1) Best fitness will be rooster (R), (2) Worst fitness will be chicks (C), and (3) The rest of swarms will be hens and chick's mothers (H) [13]. The behavior of the chicken swarm is illustrated in Fig. 1, where (F) represents food location.

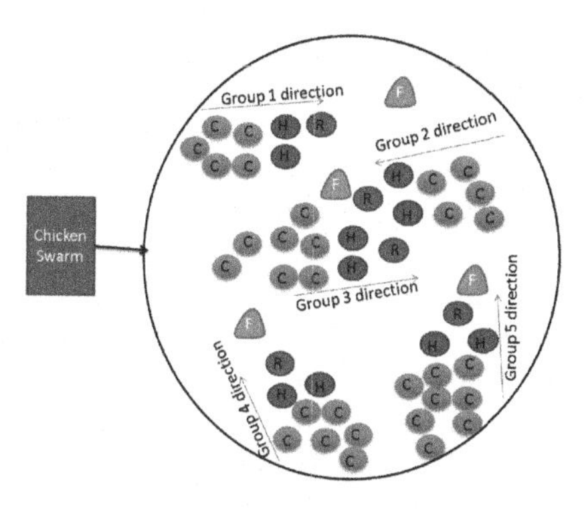

Fig. 1. Chicken Swarm Behavior.

The motion of the rooster with the best fitness value is defined as the following:

$$X_{i,j}^{t+1} = X_{i,j}^{t} * (1 + Randn(0, \sigma^2)) \tag{1}$$

where X is the related position for the rooster, $Randn$ is an add value between zero and $sigma^2$ standard deviation.

$$\sigma^2 = \begin{cases} 1, if\, f_i < f_k, \\ Exp\frac{(f_k - f_i)}{abc((f_i) + \in)} \end{cases} \tag{2}$$

where k rooster index, i is the related position and $\in$ is the smallest integer number. The motion for hens are defined as the following:

$$X_{i,j}^{t+1} = X_{i,j}^{t} + S1 * Rand * (X_{r1,j}^{t} - X_{i,j}^{t}) + S2 * Rand * (X_{r2,j}^{t} - X_{i,j}^{t}) \tag{3}$$

$$S1 = Exp\frac{(f_k - f_{r1})}{abc((f_i) + \in)} \tag{4}$$

$$S2 = Exp(f_{r2} - f_i) \tag{5}$$

Where $Rand$ is a random value between [0, 1], r_1 is the index of the rooster, r_2 the index of the rooster or hen which randomly is chosen, where r_1 not equal r_2. The foraging motion of the chicks which follow the their mothers are defined as the following:

$$X_{i,j}^{t+1} = X_{i,j}^{t} + FL * (X_{m,j}^{t} - X_{i,j}^{t}) \tag{6}$$

where FL is random value between 0 and 2 and m is the chick's mother index.

The foraging motion of the chicks is improved in [14], as the chicks can update their position by following their rooster index as well.

$$X_{i,j}^{t+1} = W * X_{i,j}^{t} + FL * (X_{m,j}^{t} - X_{i,j}^{t}) + C * (X_{r,j}^{t} - X_{i,j}^{t}) \tag{7}$$

where w is value between [0.4 to 0.9], FL is random value between 0.4 and 1, m is the mother index of the chick, C equal 0.4 and r is the rooster index.

3 The Proposed Adaptive Chicken Swarm Algorithm

3.1 Contributions of ACSO

1. Presents a new community detection algorithm based on Chicken swarm algorithm.
2. Redesigns Chicken swarm algorithm and its parameters to fit for community clustering and converts Basic Chicken swarm with its continuous features by nature into a discrete swarm algorithm by using the encoding and decoding scheme.

Community detection problem has a discrete nature, so the basic chicken swarm with its continuous features have to be redesigned. Parameters' values for our experiment are $I = 50$ (initial number of iteration), $I_{max} = 300$ (maximum number of iteration), swarm size is initialized according to the input data set size, chicks positions are initialized randomly, while food position is calculated based on fitness values [15], where $W = 0.6$,$FL = 0.7$,$C = 0.4$, rooster =0.15, hens=0.4, chicks-mother=0.3 and the chicks=0.78 of the swarm size.

Algorithm 1. ACSO algorithm steps.

Input : A network G= (V,E)
Output: Dividing and grouping network's nodes into suitable communities
Initialize chicken swarm in search space with its values and features ;
while $i < swarm.Count()$ **do**
    store the best solution at iteration i;
    for $Chicks[0] \leftarrow 0$ **to** $chicks[swarm.size]$ **do**
        store the best fitness index
        if(i == rooster) update rooster position using Eq. 2;
        else if(i == hen) update hen position using Eq. 3;
        else update the chicks position using Eq. 7;
        Updating the swarm positions;
    Get the worst fitness i_{worst} index; replace the worst fitness index i_{worst} with the best fitness index i_{best}
    i=i+1; until i less than max iterations;
return the best solution

3.2 Parameters Redesign

ACSO model searches for the best fitness index of chicken swarm to store its index then search in same iteration for the worst fitness index then ACSO model replaces the worst index with the best index which lead to more exploration in the search space and enhance the model accuracy for community detection.

The locus-based adjacency scheme is used for encoding, decoding tasks and representing results as a set of communities. In [9,10], food location is calculated based on fitness function [15], According to [14] the chick may follow a mother or a rooster in the search space which lead for higher probability to find food. Basic swarm divides the search space into 4 % for hens, 3 % for mothers, 15 % Rooster and the rest for chicks. Experimental results show that, by increasing the percentage of roosters and hens and minimizing the number of chicks this will lead to more accurate community detection results, ACSO has used 30 % for rooster, 15 % for Hens and the rest for chicks from the input swarm.

4 Experimental Results and Discussion

The experiments were implemented on a computer with Intel Core 2 GHz and 2 GB memory. Experiment aim is to achieve higher accurate results of NMI and Modularity than Ground truth value for each data sets. Experiments are executed over popular small, medium and big size data sets. Zachary Karate Club data set [16]: consists of contains 34 nodes and 78 edges. Bottlenose Dolphin data set [17]: consists of 62 bottlenose dolphins behaviors over seven years. American College football data set [18]: represents football game between American college consists of 8:12 teams. Facebook data set [19]: representation of Facebook users as nodes and connections as edges this data set contains 3959 nodes and 84243 edges. All of the previous data sets contains ground truth value which helps in measuring model accuracy [9,10].

Experiment is executed over previous data sets for 19 times and average results are recorded using ACSO and the other eight algorithms such as A discrete BAT [8], Artificial fish swarm [9], Infomap [6], Fast Greedy [20], label propagation [21], Walktrap [3], Multilevel [22], A discrete Krill herd Algorithm [10]. Table 1 summarizes the results of ACSO algorithm over the four data set. The results show that ACSO has achieved a higher promising results of NMI and Modularity than the Ground truth for each benchmark such as Zachary Karate Club, Bottlenose Dolphin and American College football while it has achieved quite promising result of NMI and Modularity over Facebook which is a large scale data set.

ACSO has achieved good results cause of it's adaptive mechanism which is illustrated in Sect. 3 which starts with an input social network data set to be encoded by adjacency schema to convert the continuous features of the swarm into a discrete to fit for community detection problem, mapping the social networks into swarm graph, applying the chicken swarm with it's updated parameters to this graph, calculate the fitness and NMI, Modularity against Ground truth, decode and represent the graph with the results.

From experimental results ACSO can be extended for more accurate results over large scale data set by using a chaotic map which will help in a better exploration of the search space and changing in chicken swarm parameters values such as swarm size and the hierarchical representation of the swarm.

Experimental results of ACSO for community detection are presented in Fig. 2 presents ACSO over Collage Football data set, Fig. 3 presents ACSO over

Table 1. The quality measures average of 19 runs, modularity and NMI results of ACSO algorithm

Measures-data sets	Zachary Karate	Bottlenose Dolphin	College Football	Facebook
Modularity	0.563	0.592	0.582	0.782
Ground truth	0.421	0.395	0.563	0.723
NMI	0.652	0.621	0.731	0.774

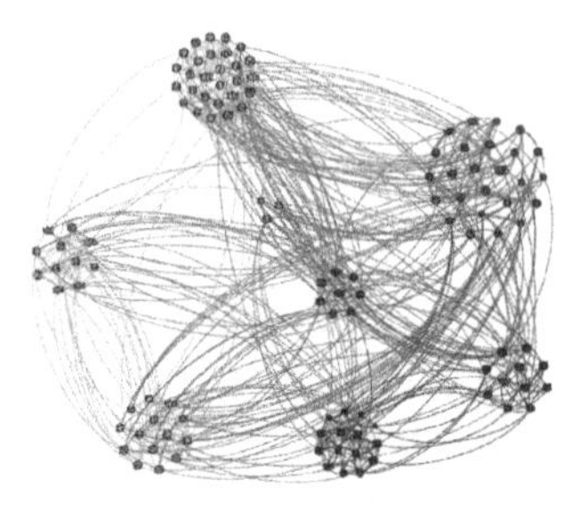

Fig. 2. ACSO for American college football data set.

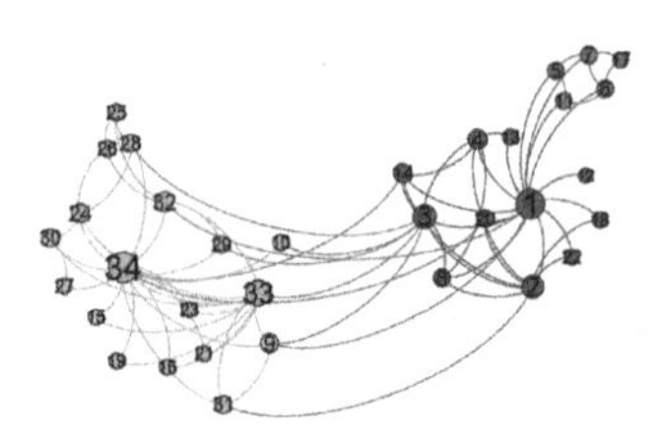

Fig. 3. ACSO for Zachary Karate Club data set.

Fig. 4. ACSO for Facebook data set.

Fig. 5. ACSO for Bottlenose Dolphin data set.

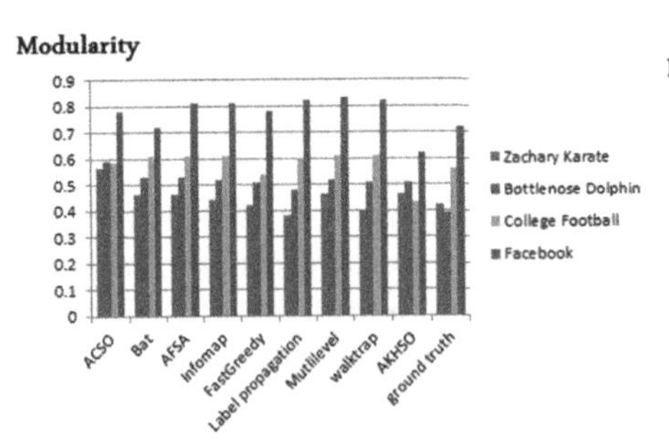

Fig. 6. Modularity results.

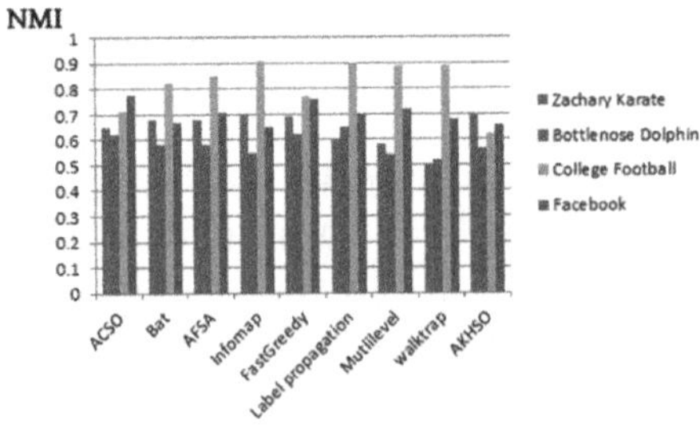

Fig. 7. NMI results.

Zachary Karate data set, Fig. 4 presents ACSO over Facebook data set and Fig. 5 presents ACSO for Bottlenose Dolphin data set. Figure 6 illustrates Modularity results, Fig. 7 illustrates NMI results for the eight algorithms and for ACSO.

5 Conclusion and Future Work

This research presents an adaptive chicken swarm algorithm for complex social network's community detection. This research presents a new community detection algorithm based on chicken swarm algorithm Our approach has achieved efficient results and quality measures of NMI and Modularity which are compared against ground truth for each benchmark data set. ACSO has achieved high results for community detection for small and medium networks data set while it has achieved medium results for big networks data set as Eqs. 4 and 5 need to be redesigned. Our future works are enhancing ACSO's performance over big data sets by using chaotic map and presenting a hybrid swarm model for community detection problem.

References

1. Narayanan, S., Venkataramanan, N., Sun, E.: Automatically generating nodes and edges in an integrated social graph, May 22 2012. US Patent 8,185,558
2. Scott, J.: Social Network Analysis. Sage, London (2012)
3. Papadopoulos, S., Kompatsiaris, Y., Vakali, A., Spyridonos, P.: Community detection in social media. J. Data Min. Knowl. Discov. **24**(3), 515–554 (2012). Springer
4. Zhao, Z., Feng, S., Wang, Q., Huang, J.Z., Williams, G.J., Fan, J.: Topic oriented community detection through social objects and link analysis in social networks. J. Knowl. Based Syst. **26**, 164–173 (2012). Elsevier
5. Malliaros, F.D., Vazirgiannis, M.: Clustering and community detection in directed networks: A survey. J. Phys. Rep. **533**(4), 95–142 (2013). Elsevier
6. Xie, J., Kelley, S., Szymanski, B.K.: Overlapping community detection in networks: The state-of-the-art and comparative study. J. ACM Comput. Surv. (CSUR) **45**(4), 43 (2013)
7. Yang, X.-S., Cui, Z., Xiao, R., Gandomi, A.H., Karamanoglu, M.: Swarm Intelligence and Bio-inspired Computation: Theory and Applications. Newnes, Oxford (2013)
8. Hassan, E.A., Hafez, A.I., Hassanien, A.E., Fahmy, A.A.: A discrete bat algorithm for the community detection problem. In: Onieva, E., Santos, I., Osaba, E., Quintián, H., Corchado, E. (eds.) HAIS 2015. LNCS (LNAI), vol. 9121, pp. 188–199. Springer, Heidelberg (2015). doi:10.1007/978-3-319-19644-2_16
9. Hassan, E.A., Hafez, A.I., Hassanien, A.E., Fahmy, A.A.: Community detection algorithm based on artificial fish swarm optimization. In: Filev, D., Jabłkowski, J., Kacprzyk, J., Krawczak, M., Popchev, I., Rutkowski, L., Sgurev, V., Sotirova, E., Szynkarczyk, P., Zadrozny, S. (eds.) Intelligent Systems'2014. AISC, vol. 323, pp. 509–521. Springer, Heidelberg (2015). doi:10.1007/978-3-319-11310-4_44
10. Ahmed, K., Hafez, A.I., Hassanien, A.E.: A discrete krill herd optimization algorithm for community detection. In: 2015 11th International Computer Engineering Conference (ICENCO), pp. 297–302. IEEE (2015)
11. Steinhaeuser, K., Chawla, N.V.: Identifying and evaluating community structure in complex networks. J. Pattern Recogn. Lett. **31**(5), 413–421 (2010). Elsevier
12. Amelio, A., Pizzuti, C.: Is normalized mutual information a fair measure for comparing community detection methods? In: Proceedings of the 2015 IEEE/ACM International Conference on Advances in Social Networks Analysis and Mining 2015, pp. 1584–1585. ACM (2015)

13. Meng, X., Liu, Y., Gao, X., Zhang, H.: A new bio-inspired algorithm: chicken swarm optimization. In: Tan, Y., Shi, Y., Coello, C.A.C. (eds.) ICSI 2014. LNCS, vol. 8794, pp. 86–94. Springer, Heidelberg (2014). doi:10.1007/978-3-319-11857-4_10
14. Dinghui, W., Kong, F., Gao, W., Shen, Y., Ji, Z.: Improved chicken swarm optimization. In: 2015 IEEE International Conference on Cyber Technology in Automation, Control, and Intelligent Systems (CYBER), pp. 681–686. IEEE (2015)
15. Gandomi, A.H., Alavi, A.H.: Krill herd: a new bio-inspired optimization algorithm. J. Commun. Nonlinear Sci. Numer. Simul. **17**(12), 4831–4845 (2012). Elsevier
16. Zachary, W.W.: An information flow model for conflict and fission in small groups. J. Anthropol. Res. **33**(4), 452–473 (1977). JSTOR
17. Lusseau, D., Schneider, K., Boisseau, O.J., Haase, P., Slooten, E., Dawson, S.M.: The bottlenose dolphin community of doubtful sound features a large proportion of long-lasting associations. J. Behav. Ecol. Sociobiol. **54**(4), 396–405 (2003). Springer
18. Girvan, M., Newman, M.E.J.: Community structure in social and biological networks. J. Natl. Acad. Sci. **99**(12), 7821–7826 (2002)
19. Leskovec, J., Krevl, A.: SNAP Datasets: Stanford large network dataset collection, June 2014. http://snap.stanford.edu/data
20. Orman, G.K., Labatut, V., Cherifi, H.: Qualitative comparison of community detection algorithms. In: Cherifi, H., Zain, J.M., El-Qawasmeh, E. (eds.) DICTAP 2011. CCIS, vol. 167, pp. 265–279. Springer, Heidelberg (2011). doi:10.1007/978-3-642-22027-2_23
21. Gregory, S.: Finding overlapping communities in networks by label propagation. J. New Phys. **12**(10), 103018 (2010). IOP Publishing
22. Noack, A., Rotta, R.: Multi-level algorithms for modularity clustering. In: Vahrenhold, J. (ed.) SEA 2009. LNCS, vol. 5526, pp. 257–268. Springer, Heidelberg (2009). doi:10.1007/978-3-642-02011-7_24

A Fully-Automated Zebra Animal Identification Approach Based on SIFT Features

Alaa Tharwat[1,6](✉), Tarek Gaber[2,6], Aboul Ella Hassanien[3,6], Gerald Schaefer[4], and Jeng-Shyang Pan[5]

[1] Faculty of Engineering, Suez Canal University, Ismailia, Egypt
engalaatharwat@hotmail.com
[2] Faculty of Computers and Informatics, Suez Canal University, Ismailia, Egypt
[3] Faculty of Computers and Information, Cairo University, Giza, Egypt
[4] Department of Computer Science, Loughborough University, Loughborough, UK
[5] Fujian Provincial Key Laboratory of Big Data Mining and Applications, Fujian University of Technology, Fuzhou, China
[6] Scientific Research Group in Egypt (SRGE), Cairo, Egypt
http://www.egyptscience.net

Abstract. Zoonoses (Zoonotic diseases) transferred from animals to human leads to death of many people every year. Controlling and tracking infected animals may save millions of human's life. One way to help achieve this is to develop an automatic animal identification/recognition systems. In this paper, a fully automated zebra animal identification approach is proposed. In this approach, the Scale Invariant Feature Transform (SIFT) feature extraction method is used to compute the features of 2D zebra images. A matching between training and testing images is calculated based on Support Vector Machine (SVM), Decision Tree (DT), and Fuzzy k-Nearest Neighbour (Fk-NN) classifiers. The experimental results show that the proposed approach is superior than other existed ones in terms of identification accuracy and the automation as our approach is fully automated while the other zebra identification systems are semi-automated or manual. The proposed approach achieved high recognition rate and the SVM classifier in this application is better than the other two classifiers.

Keywords: Animal identification · Zebra animal identification · Biometrics · SIFT

1 Introduction

One of the main objectives for governments is the tracking of animals for health purposes and localization of animals to control the infection of many Zoonotic diseases (Zoonoses) that transfer from animals to human and healthy animals. Controlling and tracking animals may decrease this infection and save millions of humans and animals. One way to help achieve this need is the automatic animal identification/recognition and traceability systems. Animal identification

J. Pan et al. (eds.), *Genetic and Evolutionary Computing*, Advances in Intelligent Systems and Computing 536, DOI 10.1007/978-3-319-48490-7_34

is important to save wild animals life, especially those that are in danger, and to protect animal species, illegal killing, and selling and buying animals [1–3].

There are three methods to identify animals. The first is the *mechanical method* such as each ear notching and tattoos. This type of identification is widely used, but it has many implications for animal welfare. Thus, many of the mechanical methods are painful, discomfort, increase the infection diseases, and requires a proper restraining of the animals [4]. The second type of animal identification methods is the *electronic method*. A typical method is RFID, which is based on attaching a device with or inside the animal. The electronic methods also painful and requires a proper restraining of the animals, hence it is against welfare [5]. The third type of animal identification methods is the *biometric*. Such methods have been used to identify animals using their physical characteristics or behavioural actions [3]. However, collecting data in biometric methods need not to restrain animals and its comfort, unique, collectable and high performance. There are several studies on zebra animal identification based on images. Foster *et al.* [5], proposed a zebra identification system based on manually selecting six predefined points on each image (Landmarks). Their database images consist of 20 zebras and they achieved a matching rate equal to $79.8 \pm 12.5\,\%$. Moreover, in [6], the user must draw manually a box around any part of the animal's body or ROI. They extracted the features from each ROI as a StripeCode. Their proposed method achieved good results against eigenface and matching multi-scale histograms. Using a database of 85 plains zebras, StripeSpotter achieved a median correct rank of 4.

In this paper, a fully automated zebra identification approach is proposed. The proposed approach makes use of the Scale Invariant Feature Transform (SIFT) technique to extract features from the images of zebra animals. Due to the high dimensionality of the extracted features, PCA is used to transform the features into a lower dimensional space. Three different classifiers are then used to match the training and testing samples, namely, SVM, Fk-NN, and DT.

The rest of the paper is organized as follows. Section 2 highlights the preliminaries of the proposed approach, which is presented in Sect. 3. The experimental results and discussion are given is Sect. 4. The conclusions of the paper are presented in Sect. 5.

2 Preliminaries

2.1 Scale Invariant Features Transform (SIFT)

Scale Invariant Feature Transform (SIFT) considered one of the robust features against many challenges such as scaling and rotation [7]. The first step in SIFT method is creating the difference of Gaussian (DoG). This step can be achieved by filtering the original image with Gaussian functions of many different scales and calculate the difference of Gaussian or the difference between two nearby scales separated by a constant multiplicative factor k as follows, $D(x, y, \sigma) = L(x, y, k\sigma) - L(x, y, \sigma)$, where $L(x, y, \sigma)$ and $L(x, y, k\sigma)$, are two images that produced from the convolution of Gaussian functions with an input image $I(x, y)$

using σ and $k\sigma$, respectively, as follows, $L(x, y, \sigma) = G(x, y, \sigma) * I(x, y)$, where the Gaussian function is denoted by $G(x, y, \sigma) = \frac{1}{2\pi\sigma^2} exp[-\frac{x^2+y^2}{\sigma^2}]$ [7]. Next, interest points in DoG pyramids, called keypoints, are detected by comparing each point with the pixels of all its 26 neighbours to compute the extrema point. However, some keypoints represent noise which needs to be eliminated by finding those that have low contrast or are poorly localized on an edge. One or more orientations is assigned to the keypoints based on local image properties. An orientation histogram is formed from the gradient orientations of the sample points within a region around the keypoint [7]. The final step in SIFT is to create descriptions for the patch that is compact, highly distinctive and to be robust to changes in illumination and camera viewpoint. The image gradient magnitudes and orientations are sampled around the keypoint location and illustrated with small arrows at each sample location [7].

2.2 Support Vector Machine (SVM) Classifier

The main idea of SVM is to maximize the decision margin between different classes. The aim of SVM is to search for an optimal hyperplane or a decision surface with the maximal margin to separate two classes and has a maximum distance to the closest points in the training set. The closest points to the optimal decision boundary are called support vectors, which require to solve the optimization problem in Eq. (1).

$$maximize \sum_{i=1}^{n} \alpha_i - \frac{1}{2} \sum_{i,j=1}^{n} \alpha_i \alpha_j y_i y_j (x_i, x_j) \text{ , s.t. } \sum_{i=1}^{n} \alpha_i y_i, 0 \leq \alpha_i \leq C \quad (1)$$

where, α_i is the weight assigned to the training sample x_i. The support vectors are the training samples x_i that have higher weights ($\alpha_i > 0$); C is a regularization parameter used to find a trade-off between the training accuracy and the model complexity [8].

2.3 Decision Tree (DT) Classifier

Decision tree classifier is represented by a tree which consists of nodes, branches, and leaves. Each node represents a test through a question or threshold on an attribute to divide and classify the data at that node based on one feature into two branches. Each leaf represents the class label. The path from the root of the tree to each leaf represents one rule [9]. The goal of the DT classifier is to test each feature of the training samples to find the best feature that divides the data into two branches. After computing which feature is robust against all other features, the training samples are divided into two branches, the left branch satisfies the threshold while the right one is not. This process is repeated recursively until all leaves are determined if the features are robust enough.

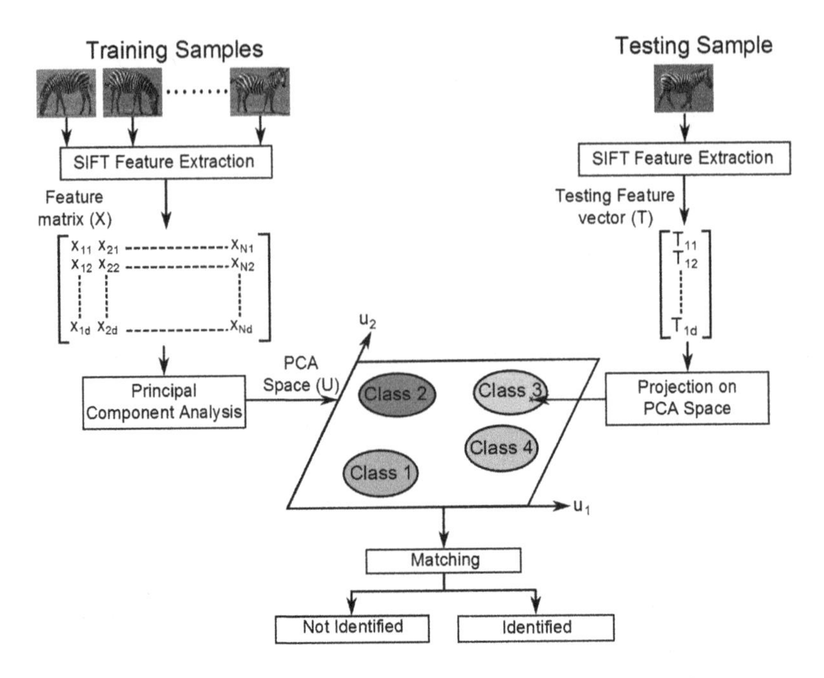

Fig. 1. A block diagram of zebra identification system based on SIFT feature extraction method.

2.4 Fuzzy k- Nearest Neighbor (Fk-NN) Classifier

The fuzzy k-NN (Fk-NN) classifier is based on assigning a membership value to the unlabelled pattern. The membership value provides the system with information to determine more accurate decision. The fuzzy membership determines the fraction or weight of an unlabelled pattern belongs to each class as follows:

$$\mu(x)_i = \frac{\sum_{j=0}^{k-1} \mu_{ij} \left(1/\left\|x - x_j\right\|^{2/(m-1))}\right)}{\sum_{j=0}^{k-1} \left(1/\left\|x - x_j\right\|^{2/(m-1))}\right)} \tag{2}$$

where $\mu(x)_i$ is the membership of the test vector x, to class ω_i, $\|x - x_j\|$ is the L-norm distance between the test vector x, and the kth nearest neighbor vector x_j, and $m > 1$ is a real number that determines how heavily the distance is weighted when calculating each neighbor's contribution to the membership value. The value μ_{ij} is the membership of the jth neighbor to the ith class [10–12].

3 Proposed Approach: Automated Zebra Identification

As illustrated in Fig. 1, the proposed system mainly consists of two phases: Training and Testing. In the training phase, features of all training zebra images are extracted using SIFT feature extraction technique. The output of this operation is represented by one feature vector (column or row) for each training image. The PCA [13–15], dimensionality reduction method, is then used to reduce the dimension of feature vectors and build a final training feature matrix ($\Gamma = UX$), where U represents the PCA space and X represents the original feature vectors.

In the testing phase, an unknown or test image will be checked against the training model. SIFT features are first extracted from the unknown image and represented by one feature vector (T). The testing feature vector is then projected on the PCA space (U), which is calculated during the training phase, to reduce the dimensionality of this vector. The output of this process is a new feature vector denoted by $y = UT$. Finally, the testing feature vector (y) is matched or classified with the training feature matrix (Γ) to identify the final decision.

4 Experimental Results

4.1 Experimental Setup

Our proposed approach have been tested on zebra images database [6] which have been collected over a period of seven days at the Ol'Pejeta Conservancy in Laikipia, Kenya. The images with different poses and positions are captured from each zebra animal. In our experiments, 200 color images with different sizes collected from 20 zebra animals (ten images for each zebra) are used. In all experiments, leave-one-out technique is used (i.e. for each experiment use $N - 1$ images for training and the remaining image for testing).

4.2 Experimental Scenarios

In this section, three scenarios will be presented. In all experiments, SIFT feature extraction method is used. Three different classifiers, SVM based on Gaussian Kernel, Decision Tree, and Fk-NN, are applied to these features.

In the first experiment, different values of PeakThr parameter are tested. PeakThr represents the amount of contrast to accept a keypoint, thus it is an effective parameter and changing it will change the number of features to be extracted. To evaluate this, three sub-experiments have been run and it was clear that the default (i.e. the optimum) value of PeakThr is 0.0. In addition, when the value of PeakThr parameter is increased, the number of features are decreased and more keypoints are eliminated. Figure 1 summarizes the identification rate when the PeakThr parameter ranged from 0 to 0.25.

As shown in Table 1, it can be noticed that the number of features, which can be extracted by SIFT, are inversely proportional to the PeakThr parameter and the maximum number of features generated when the PeakThr equal to zero, while it decreased when the PeakThr increased as shown in Table 1. Thus, according to the PeakThr parameter, the accuracy of our approach decreases when the value of PeakThr is increased. It is also proved that the best accuracy is achieved when the PeakThr value equals to zero.

In the second experiment, the patch size parameter is investigated. The problem of this parameter is not choosing the patch size, giving the suitable number of features, but testing many patch sizes to determine the best patch size giving the best accuracy. For example, increasing the patch size will consider the SIFT features as global features, thus the details of the image are not extracted well. On the other hand, decreasing the patch size parameter will not extract more features, which has a bad impact on the identification rate. To evaluate this parameter with our dataset, a number of values of it are investigated to determine its effect on the accuracy and CPU time of our approach. Figure 2 summarizes the accuracy and CPU time with different patch sizes.

From Fig. 2, showing a comparison between the accuracy and the CPU when our approach is evaluated using different patch sizes, it can be seen that (a) the identification rate rises up as the number of the sub-images grows (i.e. the patch size is decreased) until a peak appears at this peak the accuracy begins to decrease; (b) if the patch size is large, then the SIFT features can be considered as global features, which will decrease the identification rate; (c) if the patch size is too small, this may lead to less information for identification. So it is necessary to find an appropriate scale to reach an ideal recognition performance; (d) the CPU time is decreased when the patch size is decreased because it needs less computation and memory storage.

The third experiment is conducted to test the effect of the number of angles and bins which are proportional to the keypoint dimension. The feature vector at each keypoint is equal to $bins \times bins \times angles$. So, increasing the number of the angles and bins will extract features from different angles (orientations), i.e. increasing the number of the extracted features and vice versa. Consequently, the identification rate will decrease especially when the images are rotated because the features will be more sensitive to rotation. To determine, the best values for the angles and bins, a number of experiments with a different number of angles are conducted to evaluate the accuracy of the proposed approach. Table 1 presents the identification rate at different numbers of angles.

Table 1 reflects the importance of applying the PCA dimensionality reduction method in the proposed approach. As it can be seen that when using angles of two, four, and eight angles, the dimensions of extracted feature vectors are 30752, 61504, 123008 features, respectively. Thus, PCA is used to solve the high dimensionality problem. The table indicates that the accuracy of the proposed

Table 1. Identification rate (in %) of zebra animals using SVM classifier with gaussian kernel according to different angles and PeakThr values.

Classifier	Angle			PeakThr					
	2	4	8	0	0.05	0.1	0.15	0.2	0.25
SVM	96.7	98.2	100	100	100	99.3	98.2	96.7	96.7
Fk-NN	95.8	98.2	100	100	99.3	98.2	98.2	96.7	95.8
DT	95.8	97.1	100	98.2	98.2	98.2	97.1	97.1	95.8

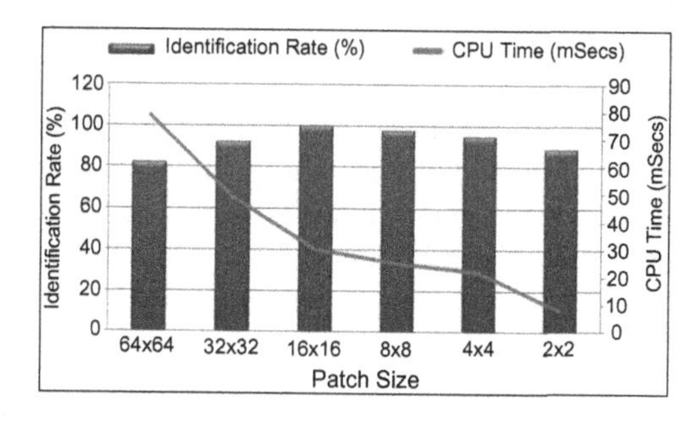

Fig. 2. Identification rate and CPU time of zebra animals identification using SVM classifier according to different patch sizes.

approach decreases when the number of angles is decreased, i.e. decreasing the angles lead to minimum orientations for each keypoint. Based on the experiments, with the two, four, and eight angles, it is found that SIFT with eight angles parameter gives the robust feature against orientation.

Another type of evaluation has been performed by comparing our approach with the most related ones [1,5]. The summary of this comparison is shown in Table 2. From this comparison, the following remarks can be noticed. Firstly, the proposed approach has achieved identification rate better than both of Foster's and Jonathan's approaches. Secondly, our proposed approach provides a fully automated zebra identification while the other two approaches only support semi-automated identification.

Table 2. Identification rate of our proposed model and state-of-the-art models

Author	Dataset	Type	Result
[1]	824 images (86 Zebras)	Semi-Automatic	**99 %**
[5]	50 images (20 Zebras)	Semi-Automatic	**79.8+12.5 %**
Proposed approach	200 images (20 Zebras)	Fully Automatic	**99.3 %–100 %**

5 Conclusion

In this paper, a fully automated approach for identifying zebra animals using digital images has been proposed. This approach makes use of SIFT technique to extract robust features, invariant to rotation and scaling, hence overcoming the problem of uncontrolled zebras. The PCA is also used to address the problem of dimensionality reduction of extracted features. The approach has been tested against SIFT parameters and SVM classifier and it has been found that (a) the best accuracy is achieved when PeakThr=0, patch size is 16×16, and number of angles equals to eight. The results have also proved that increasing the patch size will increase the CPU time and decrease the identification rate, while decreasing number of angles decreases the accuracy. Moreover, beside the

SVM, the experimental results of the SIFT with DT, and Fk-NN classifiers have shown that the SVM classifier is the best with accuracy around 99.30 % while the DT and Fuzzy k-NN have achieved accuracy of 96.4 % and 97.9 %, respectively. Thus, it could be concluded that the proposed approach has achieved an excellent identification rate against many different orientations, shifting, and challenges that comes from an uncontrolled environment.

References

1. Crall, J.P., Stewart, C.V., Berger-Wolf, T.Y., Rubenstein, D.I., Sundaresan, S.R.: Hotspotter-patterned species instance recognition. In: Proceedings of IEEE Workshop on Applications of Computer Vision (WACV), 230–237 (2013)
2. Tharwat, A., Gaber, T., Awad, Y.M., Dey, N., Hassanien, A.E.: Plants identification using feature fusion technique and bagging classifier. In: Gaber, T., Hassanien, A.E., El-Bendary, N., Dey, N. (eds.) The 1st International Conference on Advanced Intelligent System and Informatics (AISI2015), November 28-30, 2015, Beni Suef, Egypt. AISC, vol. 407, pp. 461–471. Springer, Heidelberg (2016). doi:10.1007/978-3-319-26690-9_41
3. Ahmed, S., Gaber, T., Tharwat, A., Hassanien, A.E., Snáel, V.: Muzzle-based cattle identification using speed up robust feature approach. In: 2015 Proceedings of the International Conference on Intelligent Networking and Collaborative Systems (INCOS), pp. 99–104. IEEE (2015)
4. Tharwat, A., Gaber, T., Hassanien, A.E.: Two biometric approaches for cattle identification based on features and classifiers fusion. Int. J. Image Min. **1**(4), 342–365 (2015)
5. Foster, G., Krijger, H., Bangay, S.: Zebra fingerprints: towards a computer-aided identification system for individual zebra. Afr. J. Ecol. **45**(2), 225–227 (2007)
6. Lahiri, M., Tantipathananandh, C., Warungu, R., Rubenstein, D.I., Berger-Wolf, T.Y.: Biometric animal databases from field photographs: identification of individual zebra in the wild. In: Proceedings of the 1st ACM International Conference on Multimedia Retrieval, Article no. 6. ACM (2011)
7. Tharwat, A., Mahdi, H., El Hennawy, A., Hassanien, A.E.: Face sketch recognition using local invariant features. In: 2015 7th International Conference of Soft Computing and Pattern Recognition (SoCPaR), 117–122. IEEE (2015)
8. Semary, N.A., Tharwat, A., Elhariri, E., Hassanien, A.E.: Fruit-based tomato grading system using features fusion and support vector machine. In: Filev, D., Jabłkowski, J., Kacprzyk, J., Krawczak, M., Popchev, I., Rutkowski, L., Sgurev, V., Sotirova, E., Szynkarczyk, P., Zadrozny, S. (eds.) Intelligent Systems'2014. AISC, vol. 323, pp. 401–410. Springer, Heidelberg (2015). doi:10.1007/978-3-319-11310-4_35
9. Kuncheva, L.I.: Combining Pattern Classifiers: Methods and Algorithms, 1st edn. Wiley, New York (2004)
10. Keller, J.M., Gray, M.R., Givens, J.A.: A fuzzy k-nearest neighbor algorithm. IEEE Trans. Syst. Man Cybern. **4**, 580–585 (1985)
11. Gaber, T., Tharwat, A., Hassanien, A.E., Snasel, V.: Biometric cattle identification approach based on weber's local descriptor and adaboost classifier. Comput. Electron. Agric. **122**, 55–66 (2016)
12. Tharwat, A., Ghanem, A.M., Hassanien, A.E.: Three different classifiers for facial age estimation based on k-nearest neighbor. In: 2013 9th International Computer Engineering Conference (ICENCO), pp. 55–60. IEEE (2013)

13. Tharwat, A., Ibrahim, A., Ali, H.: Personal identification using ear images based on fast and accurate principal component analysis. In: 2012 Proceedings of 8th International Conference on Informatics and Systems (INFOS), pp. 56–59. IEEE (2012)
14. Tharwat, A., Ibrahim, A., Hassanien, A.E., Schaefer, G.: Ear recognition using block-based principal component analysis and decision fusion. In: Kryszkiewicz, M., Bandyopadhyay, S., Rybinski, H., Pal, S.K. (eds.) PReMI 2015. LNCS, vol. 9124, pp. 246–254. Springer, Heidelberg (2015). doi:10.1007/978-3-319-19941-2_24
15. Gaber, T., Tharwat, A., Snasel, V., Hassanien, A.E.: Plant identification: two dimensional-based vs. one dimensional-based feature extraction methods. In: Herrero, Á., Sedano, J., Baruque, B., Quintián, H., Corchado, E. (eds.) Premi 2015. AISC, vol. 368, pp. 375–385. Springer, Heidelberg (2015). doi:10.1007/978-3-319-19719-7_33

Alzheimer's Disease Diagnosis Based on Moth Flame Optimization

Gehad Ismail Sayed[1,3(✉)], Aboul Ella Hassanien[1,3], Tamer M. Nassef[2], and Jeng-Shyang Pan[4]

[1] Faculty of Computers and Information, Cairo University, Giza, Egypt
gehad.ismail@egyptscience.net
[2] Faculty of Computer and Software Engineering, Misr University for Science and Technology, Giza, Egypt
[3] Scientific Research Group in Egypt (SRGE), Cairo, Egypt
[4] Fujian Provincial Key Laboratory of Big Data Mining and Applications, Fujian University of Technology, Fuzhou, China
http://www.egyptscience.net

Abstract. Alzheimer's disease (AD) is the most cause of dementia affecting senior's age staring from 65 and over. The standard criteria for detecting AD is tedious and time consuming. In this paper, an automatic system for AD diagnosis is proposed. A principle of moth-flame optimization is used as features selection algorithm and support vector machine classifier is adopted to distinguish three kinds of classes including Normal, AD and Cognitive Impairment. The main objective of this paper is to aid physicians in detecting AD and to compare two different anatomical views of the brain and identify the best representative one. The performance of this algorithm is evaluated and compared with grey wolf optimizer and genetic algorithm. A benchmark dataset consists of 20 patients for each class is adopted. The experimental results show the efficiency of the proposed system in terms of Recall, Precision, Accuracy and F-Score.

Keywords: Alzheimer's disease · Swarm optimization algorithms · Features selection · Grey wolf optimizer (GWO) · Moth flame optimization (MFO) · Genetic algorithm (GA)

1 Introduction

Alzheimer's disease (AD) is the most common cause of dementia which affecting seniors age staring from 65 and over. AD progression has raised a great medical research interest recently in the United States. The main symptoms of AD are tissue brain loss and nerve cell death which resulted in shrinking brain tissue and reduces larger ventricles. In earlier AD stages, usually patients are having an amnestic Mild Cognitive Impairment (MCI). The standard criteria for detecting AD from affected subjects are usually based on neuropsychological assessment and cognitive tests and often followed by a brain scan [13]. This procedure is

J. Pan et al. (eds.), *Genetic and Evolutionary Computing*, Advances in Intelligent Systems and Computing 536, DOI 10.1007/978-3-319-48490-7_35

tedious and time consuming. Also it is reported that the early detection of AD is the most effective way to treat it [16].

Several systems for AD diagnosis are proposed in literature. Their diagnosis have been achieved, using mostly the voxel intensity (VI) and ROI of MRI 3D or PET images as a features [5,15], histograms of the gradient [3], physical Characteristics such as shape and size, textural analysis features [4]. Different methods used features employed to Support Vector Machines (SVMs), artificial neural network (ANN), k- Nearest-Neighbors and Nave Bayes in the classification step [1,17,19] by used different data sets like Oasis Brain Dataset (OASIS) and AD Neuroimaging Initiative (ADNI) [12], the accuracy of all classifiers techniques were different by different data sets.

The main contribution of the paper is (1) to aid physicians in detecting AD through proving an automated system for AD detection and (2) to compare two different anatomical views of the brain and identify the best representative view of AD. In this paper, three bio-inspired algorithms including Grey Wolf Optimizer (GWO), Moth Flame Optimization (MFO) and Genetic Algorithm (GA) are presented and compared with each other.

The remainder of this paper is organized as follows. Section 2 provides the basic concepts of the MFO. While Sect. 3 presents the proposed Alzheimers disease diagnosis system. The experimental results and discussion are presented in Sect. 4. Finally, the conclusion and future work are discussed in Sect. 5.

2 Preliminaries: Moth Flame Optimization

MFO is developed by S. Mirjalili in 2015 [11]. Moths are fancy insects that are very similar to the butterflies family. In nature, there are more than 160,000 various species of this insect. Larvae and adult are the two main milestones in their lifetime. The larvae is converted to moth by cocoons. Special navigation methods in night is the most interesting fact about moths. They used a mechanism called transverse orientation for their navigation. The moths flies using a fixed angle with respect to the moon, which is a very effective mechanism for long travelling distances in a straight line.

Let the candidate solutions are moths and the problem's variables are the position of moths in the space. The P function is the main function where moths moves around the search space. Each moth update his position with respect to flame using the following equation:

$$M_i = P(M_i, F_j) \tag{1}$$

where M_i indicates the $i-th$ moth and F_j is $j-th$ flame.

$$P(M_i, F_j) = D_i.e^{bt}.cos(2\pi t) + F_j \tag{2}$$

where D_i is the distance of the $i-th$ moth for the $j-th$ flame, t is a random number in [−1,1] and b is a constant for defining the shape of the P. D is calculated using the following equation.

$$D_i = |F_j - M_i| \tag{3}$$

where M_i is the $i-th$ moth, F_j indicates the $j-th$ flame, and D_i indicates the distance of the $i-th$ moth for the $j-th$ flame.

Another concern, the moths update their position with respect to n different locations in the search space can degrade the best promising solutions exploitation. So the number of flames is adaptively decreased over the course of iterations using the following formula:

$$flame_{no} = round(N - I * \frac{N-1}{T}) \tag{4}$$

where I is the current number of iterations, T is the maximum number of iterations, N is the maximum number of flames and $round()$ used to round to the nearest integer.

3 The Proposed Alzheimer's Disease Diagnosis System

The proposed Alzheimer's disease diagnosis system is comprised of four main phases: preprocessing, features extraction, features selection and finally classification phase. These phases are described in detail in the following section along with the steps involved with the characteristics feature for each phase.

3.1 Preprocessing and Feature Extraction Phase

Noise is always undesirable. Removing these kind of noise with preserving edges plays a vital role in image processing field. Median filter is one of the simplest and most popular systems for removing noise like salt and pepper noise. In this paper, MRI image will be first resized to 150*150 in order to reduce computation time, then, median filter with window size 3*3 is applied, and finally the image background from brain region is removed by using threshold with value equal to 0.4 (trial and error), then the columns and rows having $value = 0$ are removed. Four different types of features (Statistical, Texture, Gabor and fractal features) are extracted from enhanced MRI image obtained from previous phase. The total number of extracted features are 134. These features are first six moments, median, mean, skewness and standard deviation [20], 22 haralick texture features extracted from gray level co-occurrence matrix (GLCM) including (energy, entropy, correlation,...etc.) [6], 20 features of grey run length matrix (GLRLM) [14], 40 absolute Gabor coefficient features were extracted from the Gabor wavelet with the number of the direction and the frequencies equals 8 and 4, respectively [10] and finally 20 features extracted from fractal theory [9].

3.2 Moth-Flame Based Features Selection Phase

Features selection is an important task before classification phase. A large number of extracted features are usually resulted in irrelevant, relevant and redundant features reducing the classification performance. However, feature selection algorithms can solve this problem through selecting only relevant features thus reduce the training computational time and simplify the learned classifier [8]. The principles of Moth Flame optimization are used for selecting the optimal subset features. Each features subset represents a solution in the search space and each subset has different size with different combinational of features randomly selected from index 1 to 134 (same as total number of extracted features). The solution space represents all possible features selections. The proposed features selection algorithm uses classification accuracy calculated on average from K-nearest neighbors (KNN) and 7-fold cross validation method as the fitness function. MFO iterates to discover new solutions (exploration) and exploitees this solution until reach the optimal solution. The overall proposed MFO features selection algorithm is described in Algorithm 1.

Algorithm 1. Moth Flame Optimization Features Selection Algorithm

1: initialize number of search agents n and the maximum number of iterations Max_{itr}
2: **for** $(i = 1 : n)$ **do**
3: Assign randomly a different feature combination subset to each search agent $\boldsymbol{X}_i(t)$
4: Evaluate the fitness function of each search agent $f(\boldsymbol{X}_i)$
5: **end for**
6: **for** $(t < Max_{itr})$ **do**
7: update flame no using Eq. (4)
8: Rank the agents by sorting their fitness values and assign the values of the first value (highest accuracy results)
9: Update flames positions according to the moth
10: Decrease the parameter a from -1 to -2
11: **for** $(i = 1 : n)$ **do**
12: **for** $(j = 1 : d)$ **do**
13: Calculate D using Eq. (3) with respect to moth
14: Update $M(i, j)$ using Eqs. (1) and (2) with respect to moth
15: **end for**
16: **end for**
17: Set $t = t + 1$
18: **end for**
19: Produce the best flame position F

The parameters setting for all adopted bio-inspired algorithms are Number of Search Agents (Population) = 50, range = [1,134], Number of Iterations = 10, Dimension = 134 and selection, crossover and mutation rate = 0.6 for GA. These parameters are found the best value based on trial and error method. As it can be seen, the common parameters for all adopted bio-inspired algorithms are initialized identically. Also the used fitness function is same too. Through this, it can be made a fair comparison of the performance of each bio-inspired algorithm.

3.3 Support Vector Machine-Based Classification Phase

The selected features obtained from previous phase are used to feed SVM [18] with different kernel functions; Rbf, polynomial and linear. Moreover, in order to evaluate the robustness of the proposed approach one of cross validation methods used. 7-fold is the adopted cross validation method. In k-fold, the original dataset is partitioned randomly into k equal sized sub-samples.

4 Experimental Results and Discussion

4.1 Dataset Description and Measurements

Dataset Description. A benchmark dataset consist of 20 patients for each case (Normal, Alzheimer's Disease, Cognitive Impairment) has been adopted. We use MRI dicom files 3D-T1 and the scanning was done for head and neck. It was made with 1.5 Telsa MRI systems with head surface coil. The scan parameter was adapted for every patient to fulfill the best spatial resolution for the image (TE 20, TR450, and FOV 130/1.7). The dataset are extracted from the National Alzheimers Coordinating Center (NACC) [2], in this study dataset consists of 60 subjects.

Measurements. Several measurements are used to evaluate the performance of the proposed features selector algorithms calculated from confusion matrix. These measurements are Accuracy, Precision, Recall and F-Score [7].

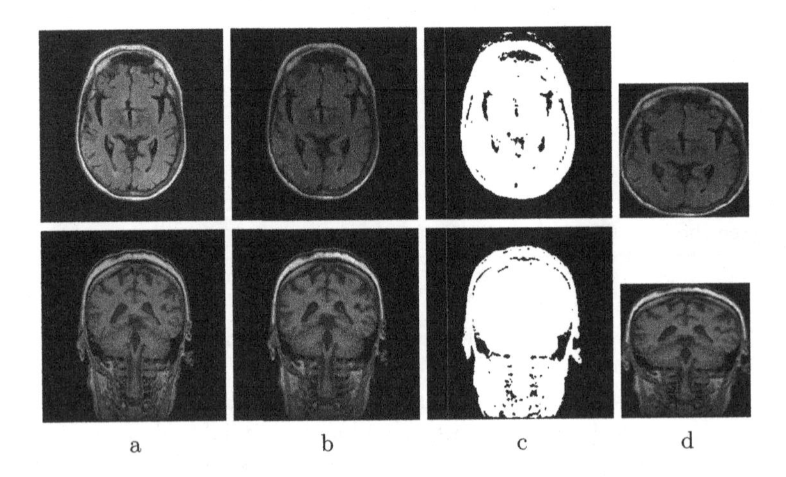

Fig. 1. Preprocessing phase results; (a) Original image, (b) Median filter result, (c) Image after converted to binary and (d) Extracted ROI

Table 1. Comparison between three features selectors algorithms and without using them for different brain views in terms of Accuracy, Precision, Recall and F-Score

			Accuracy	Precision	Recall	F-Score
All features	Axial	RBF	36.09	35.66	33.51	34.55
		Linear	60.49	60.18	60.04	60.11
		Polynomial	62.37	61.55	61.54	61.5
	Coronal cut view	RBF	39.9	39.67	39.24	36.1
		Linear	44.37	46.74	44.48	43.95
		Polynomial	39.81	40.18	39.28	38.22
	Coronal cut and axial view	RBF	35.76	25.8	30.67	28.03
		Linear	56.38	52.29	47.84	49.96
		Polynomial	60.02	54.41	52.63	53.51
GA	Axial view	RBF	36.44	nan	35.23	nan
		Linear	52.87	52.41	51.48	51.95
		Polynomial	57.28	58.45	57.67	58.06
	Coronal view	RBF	36.21	34.21	35.75	34.96
		Linear	37.07	35.35	37.02	36.17
	Coronal and axial view	Polynomial	53.85	53.99	54.01	24
		RBF	51.37	47.74	49.12	48.42
		Linear	59.66	53.66	51.28	52.44
		Polynomial	**63.55**	**57.85**	**55.21**	**56.5**
GWO	Axial view	RBF	36.75	78.55	35.11	48.53
		Linear	61.86	61.74	61.48	61.61
		Polynomial	67.05	62.21	59.44	62.19
	Coronal view	RBF	44.83	46.59	44.35	45.44
		Linear	41.38	40.47	41.05	40.76
		Polynomial	48.28	51.26	47.98	49.57
	Coronal and axial view	RBF	45.45	54.7	40.42	46.49
		Linear	63.64	58.78	56.39	57.59
		Polynomial	62.93	63.05	62.57	62.81
MFO	Axial view	RBF	40	78.57	40	53.01
		Linear	73.33	73.07	73.33	73.19
		Polynomial	**78.33**	**78.32**	**78.33**	**78.33**
	Coronal cut view	RBF	42.24	45.9	42.11	43.92
		Linear	44.35	46.05	43.91	44.95
		Polynomial	**59.48**	**62.7**	**59.49**	**61.05**
	Coronal and axial view	RBF	45.45	39.18	36.53	37.81
		Linear	58.88	52.72	50.19	51.42
		Polynomial	61.36	54.82	50.14	52.38

4.2 Results and Analysis

Three experiments are proposed and evaluated. Each experiment uses different brain view including (1) coronal view, (2) axial view and (3) coronal and axial view. Moreover three features selectors are evaluated and compared with each others too. Also three kernel functions are adopted namely gaussian radial basis (RBF), linear and polynomial kernel function. It worth to mention, the default parameters of these function are used for example; for RBF, sigma = 1 and default scaling factor and for polynomial, the order = 3. The rest of parameters are sequential minimal optimization separating hyperplane method, 20000 maximum iterations and 5000 kernel cache limit. Figure 1 shows the obtained results from preprocessing phase; (a) shows the original MRI image where first row is axial view image and second row is coronal cut view image, (b) shows the original image after applying median filter with window size 3*3, (c) shows the image after converting the image to binary using ostu' thresholding applying open morphology to remove small objects and (d) shows final extracted ROI image after removing rows columns with zero values. Table 1 compares the obtained results produced after using features selectors algorithm and before using any of them (All Features) for coronal, axial and both coronal and axial views in terms of Accuracy, Recall, Precision and F-Score. As it can be observed from the previous three figures, polynomial kernel function is best kernel for all three experiments. MFO is best features selector algorithm for axial view and coronal view. However, GA is the best features selector algorithm for both coronal and axial views. Also axial view is the best brain view to identify AD.

5 Conclusion and Future Work

In this paper, a fully automatic CAD system which uses MFO algorithms as features selectors algorithm has been presented. An application of Alzheimer's disease diagnosis from MRI images has been chosen and the scheme have been applied to see their ability and accuracy to detect AD from MRI images Support Vector Machine with its kernel functions was adopted to classify MRI images into Normal or AD or MCI. In addition to, two different anatomical views of the brain including coronal cut and axial have been compared. The experimental results show that the SVM-polynomial kernel function is best one which gives highest results in terms of accuracy, precision, recall and f-score for all proposed bio-inspired algorithms. In the future, there is a research direction of increasing the MRI slice images dataset used in order to test the reliability of the proposed CAD system. Also, more enhancements could be provided to the system by using modified versions of different swarms algorithms.

References

1. Abdullah, B.A.: Textural based SVM for MS lesion segmentation in FLAIR MRIs. Open J. Med. Imaging **1**, 26–42 (2011)
2. Beekly, D., Ramos, E., Lee, W.: The national alzheimer's coordinating center (nacc) database: the uniform data set. Alzheimer Dis. Assoc. Disord. **21**(3), 249–258 (2007)

3. Bicacro, E., Silveira, M., Marques, J.S.: Alternative feature extraction methods in 3D brain image-based diagnosis of Alzheimer's disease. In: 19th IEEE International Conference on Image Processing (ICIP), Orlando, FL, pp. 1237–1240 (2012)
4. Davatzikos, C., Fan, Y., Wu, X., Shen, D., Resnick, S.M.: Detection of prodromal Alzheimer's disease via pattern classification of magnetic resonance imaging. Neurobiol. Aging **29**(4), 514–523 (2006)
5. Fan, Y., Resnick, S.M., Wu, X., Davatzikos, C.: Structural and functional biomarkers of prodromal Alzheimer's disease: a high-dimensional pattern classification study. NeuroImage **41**(2), 277–285 (2008)
6. Galloway, M.: Texture analysis using grey level run length. Comput. Graph. Image Process. **4**, 172–179 (1975)
7. Geyer, L.H., DeWald, C.G.: Feature lists and confusion matrices. Percept. Psychophysics **14**(3), 471–482 (1973)
8. Gheyas, I.A., Smith, L.S.: Feature subset selection in large dimensionality domains. Pattern Recogn. **43**(1), 5–13 (2010)
9. Iftekharuddin, K., Zheng, J., Islam, M., Ogg, R.: Fractal-based brain tumor detection in multimodal MRI. Appl. Math. Comput. **207**, 23–41 (2009)
10. Malviya, A., Joshi, A.: Gabor wavelet system for automatic brain tumor detection. Int. J. Emerg. Technol. Adv. Eng. **4**, 826–831 (2014)
11. Mirjalili, S.M.: Moth-flame optimization algorithm: a novel nature-inspired heuristic paradigm. Knowl. Based Syst. **89**, 228–249 (2015). Elsevier
12. Nassef, T.M.: New segmentation approach to extract human mandible bones based on actual computed tomography data. Am. J. Biomed. Eng. **2**(5), 197–201 (2012)
13. Schroeter, M.L., Stein, T., Maslowski, N., Neumann, J.: Neural correlates of Alzheimer's disease and mild cognitive impairment a meta-analysis including 1351 patients. NeuroImage **47**(4), 1196–1206 (2009)
14. Sheethal, M., Kannan, B., Varghese, A., Sobha, T.: Intelligent classification algorithm of human brain MRI with efficient wavelet based feature extraction using local binary pattern. In: International Conference on Control Communication and Computing (ICCC), Thiruvananthapuram, pp. 368–375 (2013)
15. Silveira, M., Marques, J.: Boosting Alzheimer disease diagnosis using pet images. In: International Conference on Pattern Recognition, Istanbul, pp. 2556–2559 (2010)
16. Solomon, P., Murphyb, C.: Early diagnosis and treatment of Alzheimer's disease. Expert Rev. Neurother. **8**, 769–780 (2008)
17. Sun, Y., Bhanu, B., Bhanu, S.: Symmetry-integrated injury detection for brain MRI. In: 16th IEEE International Conference on Image Processing (ICIP), Cairo, pp. 661–664 (2009)
18. Tharwat, A., Gaber, T., Hassanien, A.E.: Cattle identification based on muzzle images using gabor features and SVM classifier. In: Hassanien, A.E., Tolba, M.F., Taher Azar, A. (eds.) AMLTA 2014. CCIS, vol. 488, pp. 236–247. Springer, Heidelberg (2014). doi:10.1007/978-3-319-13461-1_23
19. Wang, B., Yong, Z., Yupu, Y.: Generalized nearest neighbor rule for pattern classification. In: 7th World Congress on Intelligent Control and Automation, Chongqing, pp. 8465–8470 (2008)
20. Zulpe, N., Pawar, V.: GLCM textural features for brain tumor classification. Int. J. Comput. Sci. Issues **9**(3), 354–359 (2012)

Breast Cancer Diagnosis Approach Based on Meta-Heuristic Optimization Algorithm Inspired by the Bubble-Net Hunting Strategy of Whales

Gehad Ismail Sayed[1,3(✉)], Ashraf Darwish[2], Aboul Ella Hassanien[1], and Jeng-Shyang Pan[4]

[1] Faculty of Computers and Information, Cairo University, Giza, Egypt
gehad.ismail@egyptscience.net
[2] Faculty of Science, Helwan University, Helwan, Egypt
[3] Scientific Research Group in Egypt (SRGE), Cairo, Egypt
[4] Fujian Provincial Key Laboratory of Big Data Mining and Applications, Fujian University of Technology, Fuzhou, China
http://www.egyptscience.net

Abstract. This paper proposes a novel meta-heuristic optimization algorithm, called Whale Optimization Algorithm (WOA) to select optimal feature subset for classification purposes of Wisconsin Breast Cancer Database (WBCD). WOA is considered one of the recent bio-inspired optimization algorithms presented in 2016. A set of measurements are used to evaluate the different algorithm over WBCD from the UCI repository. These measurements are precision, accuracy, recall and f-measure. The obtained results are analyzed and compared with those from other algorithms published in breast cancer diagnosis. The experimental results show that WOA algorithm is very competitive for breast cancer diagnosis. Also it has been compared with seven well known features selection algorithms; genetic algorithm (GA), principle component analysis (PCA), mutual information (MI), statistical dependency (SD), random subset feature selection (RSFS), sequential floating forward selection (SFFS) and Sequential Forward Selection (SFS). It obtains overall 98.77 % accuracy, 99.15 % precision, 98.64 % recall and 98.9 % f-score.

Keywords: Whale optimization algorithm · Feature selection · Breast cancer diagnosis

1 Introduction

Meta-heuristic optimization algorithms are becoming popular in different applications such as medical applications in the last years. These algorithms are becoming popular due to (i) they depend on simple concepts and are easy to implement; (ii) they don't require gradient information; (iii) they can bypass local optima; (iv) they can be utilized in a wide range of applications covering different disciplines. Nature-inspired meta-heuristic optimization algorithms

J. Pan et al. (eds.), *Genetic and Evolutionary Computing*, Advances in Intelligent Systems and Computing 536, DOI 10.1007/978-3-319-48490-7_36

solve problems through mimicking physical or biological phenomena. The most popular algorithm is Particle Swarm Optimization (PSO) which originally developed by Kennedy et al. [8]. The main inspiration of PSO came from the social behavior of bird flocking [19]. Where a number of candidate solutions called particles fly around in the search space to find the optimal position called best solution. Meanwhile, these candidate solutions trace the best solution in their paths. Another popular swarm optimization algorithm is Ant Colony Optimization (ACO), first proposed by Dorigo et al. [3].

Features selection aims to identify the important features and removing irrelevant (redundant) ones from the original dataset features [4]. The features selection objectives are improving system performance, data dimensionality reduction and good data understanding for different machine learning applications. The real world applications, data representation often uses too many features with redundancy features, which means certain features can take the role of another and the unnecessary features can be removed. Moreover, the relevant (interdependence) features have an influence on the output and contain important information that will be obscure if any of them is excluded. Various heuristic algorithms mimic the behavior of physical and biological systems in the nature, also it has been proposed as strong solutions for global optimizations.

In optimization algorithms, it is necessary to have a convenient balance between exploitation and exploration. In a bee swarm algorithms, different behaviors of the bees give us the possibility to create robust balancing technique between exploration and exploitation. Artificial fish swarm (AFS) algorithm mimics the stimulant reaction by controlling the tail and fin. AFS is a robust stochastic technique based on the fish movement and its intelligence during the food finding process [11]. This paper presents a new meta-heuristic optimization algorithm (namely, Whale Optimization Algorithm (WOA)) mimicking the hunting behavior of humpback whales for features selection of clinical breast cancer datasets. Optimization results demonstrate that WOA is very competitive compared to the state-of-the-art optimization methods.

The rest of the paper is structured as follows. In Sect. 2, the inspiration of the proposed method is first discussed. Then, the mathematical model is provided. The proposed method is presented and results discussed in Sects. 3 and 4, respectively. Section 5 summarizes the main findings of this paper.

2 Preliminaries

A. Inspiration Analysis

Whales are one of the biggest mammals in the world. There are seven different main mammal'species such as Minke, killer, humpback, Sei, finback, blue and right [16]. Whales are predators, where breathing from the oceans'surface so that they never sleep. According to [14], whales and spindle cells of human have similar cells in certain areas of their brains. These spindle cells are responsible for judgment, social behaviors and emotions in the humans. However, whales have twice number of adult human'spindle cells that is why the main reason of whales'smartness [7].

The special hunting method called bubble-net feeding is considered the most interesting thing about the humpback whales [20]. Humpback whales prefer to hunt school of the small fishes or krill that close to the surface. It has been noticed that the foraging is made by creating '9'-shaped path or kind of distinctive bubbles along a circle. Such a behavior was only investigated before 2011 based on the observation from surface. On the other hand, authors in [7] investigated such a behavior by utilizing tag sensors. In this work the spiral bubble-net feeding maneuver is modeled mathematically to perform optimization.

B. Mathematical Model and Optimization Algorithm

In this section the mathematical model of spiral bubble-net feeding maneuver, search for prey and encircling prey is first provided. The WOA algorithm is then presented.

1. Encircling Prey. Humpback whales can detect the prey'location and encircle them. This behavior is mathematicaly represented by the following equations:

$$\boldsymbol{D} = |\boldsymbol{C}\boldsymbol{X}^*(t) - \boldsymbol{A}\boldsymbol{X}(t)| \tag{1}$$

$$\boldsymbol{X}(t+1) = \boldsymbol{X}^*(t) - \boldsymbol{X}\boldsymbol{D} \tag{2}$$

where A and C are coefficient vectors, t indicates the current iteration, X^* is the position vector of the optimal solution obtained so far $\boldsymbol{X}^*$ is the position vector. The position vector X^* is updated in each iteration if there exist a better solution. The vectors A and C are calculated using the follows:

$$\boldsymbol{A} = 2\boldsymbol{a} \cdot \boldsymbol{r} - \boldsymbol{a} \tag{3}$$

$$\boldsymbol{C} = 2 \cdot \boldsymbol{r} \tag{4}$$

where a is linearly decreased from 2 to 0 through iterations (in both exploitation and exploration phases) and $\boldsymbol{r}$ is a random vector in [0,1].

2. Bubble-Net Attacking Method (Exploitation Phase)

$$\boldsymbol{X}(t+1) = \begin{cases} \boldsymbol{X}^*(t) - \boldsymbol{A}\boldsymbol{D} & \text{if } p < 0.5 \\ \boldsymbol{D} \cdot e^{bt} \cdot cos(2\Pi t) + \boldsymbol{X}^*(t) & \text{if } p \geq 0.5 \end{cases} \tag{5}$$

where p is a random number between 0 and 1. In addition to the bubble-net method, the humpback whales search for prey randomly. The mathematical formulas of the search is as follows.

3. Search for Prey (Exploration Phase): The same approach based on the variation of the $\boldsymbol{A}$ vector can be utilized to search for prey (exploration). The mathematical formulas are as follows:

$$\boldsymbol{D} = \boldsymbol{C} \cdot \boldsymbol{X}_r\boldsymbol{and} - \boldsymbol{X} \tag{6}$$

$$\boldsymbol{X}(t+1) = \boldsymbol{X}_r\boldsymbol{and} - \boldsymbol{A} \cdot \boldsymbol{D} \tag{7}$$

where $\boldsymbol{X_{rand}}$ is a random position vector chosen from the current population.

2.1 The Proposed Model

The proposed diagnosis model consists of two main phases; features selection and classification. First the model starts from taking the clinical breast cancer dataset as input, then WOA bio-inspired algorithm is adopted to select salient features, then the selected features are used to feed support vector machine (SVM) [18]. Finally the obtained results are evaluated using four different measurements such as precision, accuracy, recall and f-measure.

Features Selection Stage. In this paper, WOA algorithms is used as features selection algorithms based wrapper mode. The best position is the subset which gives the highest fitness value (Classification accuracy in our case) which obtained from KNN classifier. The initial parameter setting for WOA are, population size = 50 with number of iteration = 10, dimension = 30 and boundary range from 1 to 30.

Classification. In this paper, SVM [2] is the adopted classifier and leave-one-out [6] is the adopted cross validation method where used to evaluate the robustness of the proposed model. Four different kernels function are used including linear, quadratic, polynomial and rbf. The best kernel function is determined based on the experimental results.

3 Evaluation Metrics

Four measurements calculated from confusion matrix are used to evaluate the performance of the proposed model. These measurements are recall, precision, f-measure and accuracy [5]. TP indicates true positive, FP indicates false positive, FN indicates false negative and TN indicates true negative. F-measure represents the harmonic mean of the recall and precision which measures the weighted average between recall and precision. The Best value of f-measure is 1 and the worst one is 0.

4 Experimental Results and Discussion

In this paper, the wisconsin breast cancer diagnosis dataset (WBCD) from the UCI machine learning repository is adopted to evaluate the proposed system [21]. This data has number of attributes equals to 32 with number of instances equals to 596 and two classes. The adopted dataset has no missing values, so no needs for preprocessing. The experiments were implemented in MATLAB-R2012 on a computer with Intel Core 2 GHz and 2 GB memory.

Figure 1 shows WOA convergence curve. As can be seen, as the number of iteration increases, the highest fitness value (best score) obtains. Also, WOA converge at iteration 7 with score equals 97.18 %. Figure 2 compares the obtained results of using selected features from WOA using different kernel functions.

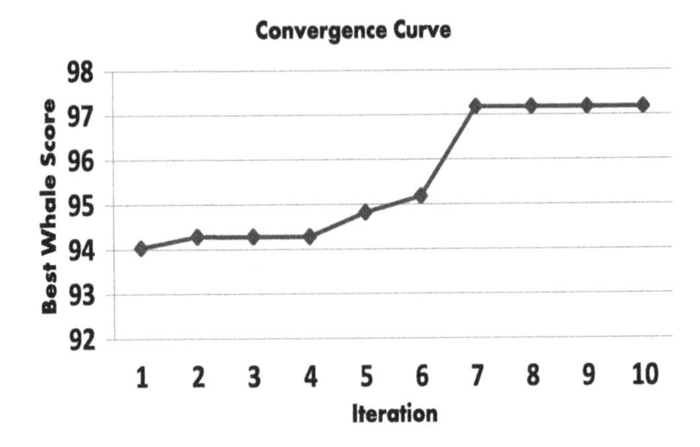

Fig. 1. WOA convergence curve

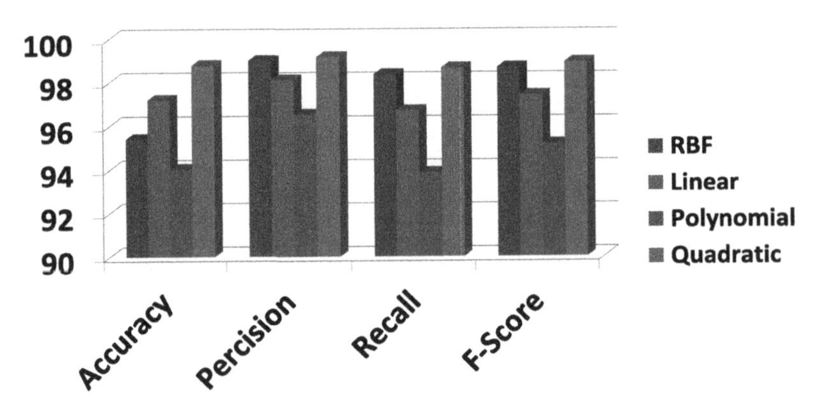

Fig. 2. The obtained results of WOA in terms of accuracy, recall, precision and f-measure using different kernel functions

As can be observed, quadratic is the optimal kernel function as it obtains the highest accuracy.

The next experiment show the superior of WOA compared with other features selection algorithms. Figure 3 compares the obtained results using different features selection algorithms including those based on filter and wrapper methods in terms of accuracy, recall, precision and f-measure. These algorithms are genetic algorithm (GA), principle component analysis (PCA), mutual information (MI), statistical dependency (SD), random subset feature selection (RSFS), sequential floating forward selection (SFFS) and Sequential Forward Selection (SFS). As can be seen, the selected features using WOA owns the highest results. Also GA is in second place and SFS is the worst one.

The last experiment proof the high performance of the proposed algorithm compared with other presented in literature. Table 1 compares the obtained results from the proposed algorithm with those presented in literature in terms

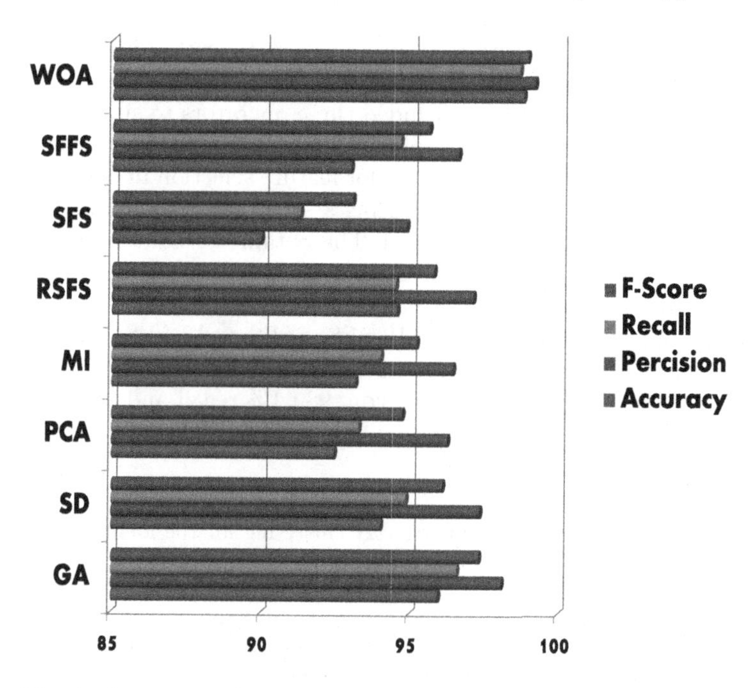

Fig. 3. Comparison between different features selection algorithms

Table 1. Comparison with existing works

Model (Reference)	Classifier	Accuracy (%)
D. Lavanya et al. [10]	CART	93.49
E. Osuna et al. [13]	C4.5	94.74
D. Lavanya et al. [9]	Ensemble decision tree	95.96
B. Ster et al. [17]	Linear discreet analysis	96.8
V. Narayan et al. [12]	Neuron-Fuzzy	95.06
J. Abonyi et al. [1]	Supervised fuzzy clustering	95.57
G. Salamaet al. [15]	SMO	97.71
G. Sayed et al.	SVM-Quadratic	98.77

of accuracy. As can be seen, our algorithms obtains highest result compared with the others.

5 Conclusion

Feature selection is one of the most effective methods to improve system performance and enhance the data representation in terms of specified criteria. The main objective of this paper is to select the best features subset. This paper presented a new swarm-based optimization algorithm inspired by humpback whales'the hunting behavior mechanism. The proposed method (named as Whale Optimization Algorithm) included three operators to simulate the encircling prey, search for prey and humpback whales'bubble-net foraging behavior. The proposed model is applied and used for feature selection in machine learning domain. The evaluation is performed using a set of evaluation criteria to assess different aspects of the proposed model. The obtained results find out that the proposed features selection algorithm performs better than those presented in literature. Also it has been compared to with seven well known features selection algorithm including SD, PCA, GA, MI, SFS, SFFS and RSFS. The experimental results shows good performance of selected features using WOA. It obtains overall 98.77 % accuracy, 99.15 % precision, 98.64 % recall and 98.9 % f-score.

References

1. Abonyi, J., Szeifert, F.: Supervised fuzzy clustering for the identification of fuzzy classifiers. Pattern Recogn. Lett. **14**(24), 2195–2207 (2003)
2. Bottou, L., Cortes, C., Denker, J., Drucker, H.: Comparison of classifier methods: a case study in handwritten digit recognition. In: Proceedings of the 12th IAPR International Conference on Pattern Recognition, pp. 77–82, Jerusalem (1994)
3. Dorigo, M., Birattari, M., Stutzle, T.: Ant colony optimization. IEEE Comput. Intell. Mag. **1**(4), 28–39 (2006)
4. Emary, E., Zawbaa, H.M., Hassanien, A.E.: A binary grey wolf optimization approaches for feature selection. Neurocomputing **172**, 371–381 (2016)
5. Fawcett, T.: An introduction to ROC analysis. Pattern Recogn. Lett. **27**, 861–874 (2006)
6. Gawely, C.: Leave-one-out cross-validation based model selection criteria for weighted LS-SYMS. In: International Joint Conference on Neural Networks (IJCNN), pp. 1661–1668, Madrid (2006)
7. Goldbogen, J., Friedlaender, A., Calambokidis, J., Mckenna, M., Simon, M., Nowacek, M.: Integrative approaches to the study of baleen whale diving behavior, feeding performance, and foraging ecology. Bio-Science **63**, 90–100 (2013)
8. Kennedy, J., Eberhart, R.: Particle swarm optimization. In: Proceedings of the 1995 IEEE International Conference on Neural Networks, pp. 1942–1948, Perth, WA (1995)
9. Lavanya, D.: Ensemble decision tree classifier for breast cancer data. Int. J. Inf. Technol. Convergence Serv. **2**(1), 17–24 (2012)
10. Lavanya, D., Rani, D.K.: Analysis of feature selection with classification: breast cancer datasets. Indian J. Comput. Sci. Eng. (IJCSE) **2**(5), 756–763 (2011)
11. Lin, L., Gen, M.: Auto-tuning strategy for evolutionary algorithms: balancing between exploration and exploitation. Soft Comput. **13**(2), 157–168. Springer (2009)
12. Narayan, V., Chunekar, H., Ambulgekar, P.: Approach of neural network to diagnose breast cancer on three different data set. In: International Conference on Advances in Recent Technologies in Communication and Computing, pp. 893–895, Kottayam, Kerala (2009)
13. Osuna, E., Freund, R., Girosi, F.: Training support vector machines: application to face detection. In: Proceedings of Computer Vision and Pattern Recognition, pp. 130–136, San Juan (1997)

14. Hof, P., Van der Gucht, E.: Structure of the cerebral cortex of the Humpback whale, Megaptera novaeangliae (Cetacea, Mysticeti, Balaenopteridae). Anat Rec. (Hoboken) **290**(1), 1–31 (2007)
15. Salama, G., Abdelhalim, M.B., Zeid, M.: Breast cancer diagnosis on three different datasets using multi-classifiers. Int. J. Comput. Inf. Technol. **1**(1), 36–43 (2012)
16. Seyedali, M., Andrew, L.: The whale optimization algorithm. Adv. Eng. Softw. **95**, 51–67. Elsevier (2016)
17. Ster, B., Dobnikar, A.: Neural networks in medical diagnosis: comparison with other methods. In: Proceedings of the International Conference on Engineering Applications of Neural Networks, pp. 427–430 (1996)
18. Tharwat, A., Gaber, T., Hassanien, A.E.: Cattle identification based on muzzle images using gabor features and SVM classifier. In: Hassanien, A.E., Tolba, M.F., Azar, A.T. (eds.) International Conference on Advanced Machine Learning Technologies and Applications, vol. 488, pp. 236–247. Springer, Switzerland (2014)
19. Wang, K.J.: A hybrid classifier combining smote with pso to estimate 5 year survivability of breast cancer patients. Appl. Soft Comput. **20**, 15–24 (2014)
20. Watkins, W., Schevill, W.: Aerial observation of feeding behavior in four baleen whales: eubalaena glacialis, balaenoptera borealis, megaptera novaean- gliae, and balaenoptera physalus. J. Mammal. **60**(1), 155–163 (1979)
21. Wolberg, W., Mangasarian, O.: UCI machine learning repository, Irvine, CA. http://archive.ics.uci.edu/ml

Author Index

J. Pan et al. (eds.), *Genetic and Evolutionary Computing*, Advances in Intelligent Systems and Computing 536, DOI 10.1007/978-3-319-48490-7

Printed in the United States
By Bookmasters